Privacy on the Line

The right of the people to be secure in their persons, houses, papers and effects against unreasonable searches and seizures shall not be violated, and no Warrants shall issue but upon probable cause, supported by Oath or affirmation, and particularly describing the place to be searched, and the persons or things to be seized.

Fourth Amendment
United States Constitution

The evil incident to invasion of privacy of the telephone is far greater than that involved in tampering with the mails. Whenever a telephone line is tapped, the privacy of persons at both ends of the line is invaded, and all conversations between them upon any subject, and although proper, confidential and privileged, may be overheard. Moreover, the tapping of one man's telephone line involves the tapping of the telephone of every other person whom he may call or who may call him. As a means of espionage, writs of assistance and general warrants are but puny instruments of tyranny and oppression when compared with wire-tapping.

Justice Louis Brandeis
dissenting opinion in
Olmstead v. United States
(277 US 438, 1928, pp. 475–476)

Senator Herman Talmadge: Do you remember when we were in law school, we studied a famous principle of law that came from England and also is well known in this country, that no matter how humble a man's cottage is, that even the King of England cannot enter without his consent.

Witness John Ehrlichman: I am afraid that has been considerably eroded over the years, has it not?

Senator Talmadge: Down in my country we still think of it as a pretty legitimate piece of law.

United States Senate
Select Committee on Presidential
Campaign Activities, Hearings,
Phase 1: Watergate Investigation,
Ninety-Third Congress, First
Session, 1973, p. 2601

Privacy on the Line
The Politics of Wiretapping and Encryption

Updated and Expanded Edition

Whitfield Diffie
Susan Landau

The MIT Press
Cambridge, Massachusetts
London, England

Composed in LATEX 2$_\varepsilon$ by the authors.
Set in Sabon by Loyola Graphics of San Bruno, California.
Printed and bound in the United States of America.

Library of Congress Cataloging-in-Publication Data

Diffie, Whitfield.
    Privacy on the line : the politics of wiretapping and encryption / Whitfield
Diffie, Susan Landau. — Updated and expanded ed.
        p.   cm.
    Includes bibliographical references and index.
    ISBN 978-0-262-04240-6 (hardcover : alk. paper)

        1. Electronic intelligence—United States. 2. Wiretapping—United States.
3. Data encryption (Computer science)—Law and legislation—United States. 4.
Electronic surveillance—United States—Political aspects. 5. Telecommunication—
Political aspects—United States. 6. Privacy, Right of—United States. I. Landau,
Susan Eva. II. Title. III. Title: Politics of wiretapping and encryption.

UB256.U6D54 2007
342.7308'58—dc22   2006035514

This book is dedicated to our spouses, Mary Fischer and Neil Immerman.

Contents

Preface to the Updated
and Expanded Edition

It would be difficult to find a more fundamental theme in the contemporary world than the migration of human activity from physical, face-to-face contact into the virtual world of electronic (and digital) telecommunications. Globalization would not be possible without the high-quality, reliable, and inexpensive telephone service that has been made possible by optical fibers and computerized central offices. In the industrialized world and beyond, governments, businesses, universities, and other institutions have made the World Wide Web a centerpiece of their communications with the public.

One of the critical issues raised by this transformation is what effect it will have on privacy and security. The digitization of the world has made the effortless privacy of interpersonal conversations a thing of the past and enabled spying on a global scale never before seen. The decisions we make as we lay the foundations of the new world will have an impact on the structure of human society that transcends that of any previous technological development. If, in designing our new world, we do not take privacy and security into account in a way that reflects the primacy of the individual, our technology will enforce a social order in which the individual is subordinate to the institutions whose interests were put foremost in the design.

The first edition of *Privacy on the Line* was written at a time in which the issue seemed simple. The primary technology for protecting telecommunications privacy was cryptography, and the right to use cryptography for the protection of personal and business privacy seemed in jeopardy. The battle had two fronts, and we set out to explore them both.

The more visible front was chronologically second but stood first in most people's minds. The US government's plan for *key escrow* sought to use its standard-setting power—backed by its substantial purchasing power—to make cryptographic systems with built-in government master keys ubiquitous. Had the plan succeeded, it might plausibly have been extended to outlaw systems that did not have this provision.

The less visible but economically more significant front was export control. Exporting of cryptographic products had been tightly controlled for decades but, until the sudden need for cryptography in commercial uses that followed the opening up of the Internet this had, by and large, only the intended effect of inhibiting the exporting of cryptographic equipment intended for military customers. As low-cost integrated circuits brought high-grade cryptography within the reach of many commercial products, its use expanded steadily. Businesses oriented toward making consumer products now found themselves forced by the export laws to bear the unrewarding expense of producing separate products for export and for domestic consumption.

The first edition was written in the midst of this political struggle over whether individuals and commercial enterprises had a right to protect their communications with cryptography or whether governments had the right to limit its use to prevent possible interference with their law-enforcement and intelligence activities. The preface to that edition gives a flavor of the situation as it stood at that time.

A book written in the midst of events will always become outdated, sometimes quite quickly. Just the short interval between the appearance of the original edition and the first paperbound edition saw a striking sequence of events.

- The existing encryption standard was decisively shown to be inadequate. In an event noteworthy for its neatly orchestrated publicity, the Electronic Frontier Foundation revealed that it had built the often-designed *DES Cracker*—a specialized computer capable of producing DES keys from cipher text in (at worst) just over a week.

- The secret Skipjack algorithm that underlay the key-escrow plan was declassified, apparently in order to allow the Department of

Defense to save money by using software encryption to secure email in the *Military Message System*.

- The National Institute of Standards and Technology's plans for an Advanced Encryption Standard (AES) to replace DES by a cipher with blocks twice as long and a key nearly five times as large made dramatic progress, with fifteen designs accepted for first-round evaluation and presented at a public conference as well as published on the Web.

More dramatic events were to follow shortly. In September 2000, the American export control rules were revised to place less emphasis on the strength of cryptography and more on the end users and the degree of customization provided. Selling off-the-shelf hardware and software to commercial users throughout the industrialized world became relatively easy, while selling customized equipment, particularly to governments, continued to be burdened with a lengthy approval process. The scheme was clever because foreign military organizations—the major target of export control—had well-established cryptographic traditions and usually wanted to employ their own cryptographic algorithms rather than those in common use in the commercial world.

An important ingredient in the demise of export control was the unexpected exposure of a multi-national (though primarily US-controlled) signals intelligence network called Echelon that appeared to be organized for the interception of commercial rather than military traffic. Never mind that the world's military were making ever increasing use of commercial channels; it looked to the Europeans as though they were being spied on. Their response was a new emphasis on secure communications, and one important step was decreased regulation.

By comparison with export regulation, key escrow merely faded from view without being officially withdrawn or renounced. When the National Institute of Standards and Technology began the process of replacing the quarter-century-old Data Encryption Standard with a new system, it placed a high level of security at the top of its requirements. The resulting *Advanced Encryption Standard* was adopted in late 2001 and has since been approved for national-security applications as well as civilian ones.

Even as these events were under way, it was clear to observers that the underlying issues had not been resolved and that other, non-cryptographic, aspects of communications privacy were evolving in a different direction. Although regulatory jockeying and lawsuits delayed its full implementation, the Communications Assistance for Law Enforcement Act was, for the first time, forcing the major telecommunications companies to build wiretapping into the infrastructure of the American communications system. In a disquieting parallel development, the FBI had begun demanding the right to implement wiretap orders by installing its own hardware on the premises of Internet Service Providers rather than presenting the order to the ISPs and allowing them to comply using their own technology. Critics feared that the new technique would lift a layer of scrutiny from the wiretap process. If the ISP were not doing the monitoring, they would not know what was being monitored, and would be unable to challenge overbroad interception.

Cryptography, free from oppressive regulations, was going nowhere fast. Although SSL (the Secure Socket Layer protocol used to protect Internet commerce) is perhaps the most widely deployed cryptographic mechanism of all time, the application of cryptography to protecting Internet communications—and electronic communications overall—is spotty. Some Web transactions and most VPN connections are encrypted, but only a small fraction of email, voice, or video communications, or even Web browsing, is protected.

There are many proximate causes of the changed aspect of communications privacy. In the late 1990s, the world, particularly the United States, was in the midst of a massive economic boom. The collapse of the Soviet Union had given America the sense that it had no real enemies, and, despite vicious civil wars in Africa and Eastern Europe, the world seemed more peaceful than it had been in decades.

The September 2001 attack on the United States ended that sense of peace and initiated an era of widespread fear, fear that inclined the population toward accepting greater encroachment on their liberties and supporting more ambitious intelligence programs. At the time of this writing, the activities of the intelligence community (what they are and what they should be) have become a subject of debate in the courts, the Congress, and the press.

The debate has moved beyond the attempt to suppress access to strong cryptography. In the United States such access is now supported, in principle, by government policy. In Britain, a state right to access encrypted information was included in the Regulation of Investigatory Powers Act. The law authorizes expanded surveillance, and one clause requires individuals to divulge cryptographic keys on demand.

The political battle in the United States now focuses on decline of the once-rigid wall separating foreign intelligence from domestic law enforcement. There is acceptance of the increasing use of facilities originally built for spying on other countries to spy on targets inside the United States. Along with the shift in policy comes a steady push to extend the built-in wiretapping approach of the Communications Assistance for Law Enforcement Act from the conventional telephone system to the Internet.

In an effort to provide supporting material for the conduct of the new debate, we have brought out this updated and expanded edition, adding two new chapters and changing the existing ones in varying degrees to reflect new developments.

Preface to the First Edition

In the spring of 1993, the White House announced an unprecedented plan for both promoting and controlling the use of secret codes to keep communications private. The plan, formally called *key escrow* but popularly known as "Clipper" after its star component, the Clipper chip, was to adopt a new federal standard for encryption, a standard that would ensure that the government could always read encrypted messages if it chose.

The Clipper proposal was met by a storm of protest. It was criticized by some as an outrageous violation of civil liberties, by some because the standard could only be implemented in hardware, and by still others on a wide variety of grounds. Despite the opposition, Clipper seemed, in a sense, to have won. After a mandatory public comment period, which produced two letters in favor and 300 against, the standard was adopted. In a more fundamental sense, however, the Clipper program seemed to have lost. Aside from the 9000 telephone security devices that the FBI purchased in an attempt to seed the market, very little Clipper-based equipment has been built.

The Clipper debate proved to be the opening engagement in an ongoing battle about the right to use encryption. Having tried to use its buying power and standards-making authority to impose key escrow, the government turned to the only other non-legislative tool available: export control.

The United States has approximately 5% of the world's population. In light of this, it is not surprising that, although the country's share of the world economy is way out of proportion to its population, most major

US corporations sell more than half their products in other countries. This makes the larger part of their markets subject to export-control laws.

It is also true that a key competitive strategy in modern business is to eliminate unnecessary versions of products. Duplication can be particularly costly in high-technology products such as computer software. If US corporations are unable to export the same versions of their products that they sell at home, the effect is a significant increase in costs. The government's subsequent attempts to achieve key escrow have turned on this fact.

In January 1997, the administration began to permit the export of some unescrowed encryption products for 2 years to companies that submit detailed plans for developing escrowed products within that time.

Why is all this important? Why should anyone who is not in the cryptography business be concerned about regulation of the export of cryptographic equipment? The answer lies in the rush to put society online.

For most of human history, most communication between individuals was conducted face to face. For a few thousand years some has been conducted in writing, but this is in many respects a poor substitute. Letters took weeks, months, or even years to travel long distances. The fact that a letter might be opened en route and thus was less private than a whisper was just one of many limitations.

For a little more than 100 years, some human communication has been carried by electronic media, particularly the telephone. This has brought to remote communication an immediacy that approximates face-to-face contact. The quality of telecommunication continues to improve, and the portion of relationships in which telecommunication is the primary mode of communication continues to increase. We are moving the fabric of our society into electronic channels as quickly as we can.

When telecommunication was merely an adjunct to physical communication, it was possible to hedge about privacy. When two people meet frequently as well as talking regularly by telephone, they can reserve indiscreet remarks for their face-to-face meetings. But as telecommunication becomes more the rule than the exception, this becomes less feasible. In a future society (which may not be far off) in which most communication is telecommunication and many close relationships are between people who never meet in person, it becomes impossible. If people are to enjoy the same effortless privacy in the future that they enjoyed in the past, the means to protect that privacy must be built into their communication systems.

Were the discussion to stop here, the conclusion would be self-evident: we should design all our communication systems to guarantee confidentiality. Personal privacy, however, is not everyone's paramount concern. There are powerful elements of society—police and military organizations—that make use of intercepted communications in what they consider the protection of public safety. These groups view the ready availability of strong cryptography as threatening their ability to perform their functions. Moreover, these once-distinct government activities are drawing closer together in response to the perceived threat of international terrorism. Not surprisingly, this emerging coalition sees individual access to cryptography more as a curse than a blessing.

We see no simple resolution of this conflict. The debate so far has been largely an argument among partisans, all anxious to bias the evidence in their own favor. This is also a field with an extraordinary number of secrets. Neither the police and the spies, who oppose widespread cryptography, nor the big corporations, which support it, are the most open and forthcoming of society's institutions.

In this book, we attempt to lift enough veils to permit the reader to develop an informed opinion on the subject. We examine the social function of privacy: how it underlies other aspects of a free and democratic society and what happens when it is lost. We explore how intelligence and law-enforcement organizations intercept communications, what use they make of them, and what problems cryptography might create. We also describe how cryptography works and how it can be used to protect the secrets of both individuals and organizations.

If we have succeeded, the reader will come away from our book with a new understanding of an issue that, despite the publicity it has received in the past few years, has seemed mysterious and confusing.

Acknowledgements

We would like to thank Marc Rotenberg, David Sobel, and David Banisar of the Electronic Privacy Information Center for their much-seeing eye on developments in Washington, for uncovering mountains of useful information through their Freedom of Information Act suits, and their willingness to answer the innumerable questions of a pair of legal novices; Lynn McNulty for explaining what really happened at the National Institutes of Standards and Technology; Elizabeth Rindskopf, former general counsel at the National Security Agency, for her straightforward attempts to explain the views of that inscrutable organization; David Burnham, for sharing an investigative reporter's sensibility on how to discuss these matters; Andrew Grosso for giving us a taste of the prosecutor's viewpoint and a lot of his expertise; and the government-documents librarians at the University of Massachusetts in Amherst, Bill Thompson, Len Adams, and Terrie Billel, for their ability to turn wild-goose chases into careful hunts that yielded ripe, rich fowl.

Privacy on the Line

1

Introduction

In the early nineteenth century it took six weeks for the British government to send a message from London to its representative in Delhi. In the late nineteenth century, the telegraph cut this time to days, then to hours. Today, at the dawn of the twenty-first century, the time has been cut to a fraction of a second and the service is available not just to the government but to most of the citizens. In a century and a half, we have gone from a world in which people separated by distance could communicate only through the slow process of sending letters to one in which they can communicate quickly, directly, and interactively—almost as though they were standing face to face. In the near future we may take the next step and move into a world in which computer-mediated interaction may offer such advantages over meeting face to face that it will supplant an even larger part of face-to-face interaction.

The result is that we now conduct more and more of our communications, whether personal, business, or civic, via electronic channels. The availability of telecommunication has transformed government, giving administrators real-time access to their employees and representatives in remote parts of the world. It has transformed commerce, facilitating worldwide enterprises and beginning the internationalization that became the byword of business a decade ago. It has transformed warfare, giving generals the ability to operate from the safety of rear areas and admirals the capacity to control fleets scattered across oceans. It has transformed personal relationships, allowing friends and family to converse with an immediacy that belies the fact they are thousands of miles apart.

These developments in technology have also had a profound impact on privacy. To attempt to function in modern society without employing telecommunication is to be eccentric. Most people use the telephone (including cellphones) daily, and many make constant use of electronic mail and the World Wide Web. These communications are by their essential nature interceptable. A typical telephone call travels over many miles of wire, of which only a few feet are under the control of the people talking. For most of its journey the signal is in the hands of one or more telephone companies, who will give it a reasonable degree of protection, but who can readily listen to it or record it and will from time to time do so. Many a call travels by radio for some part of its journey. The radio link may be at an end, in the form of a cordless, or cellular telephone, or it may be in the middle, in the form of a microwave link or a satellite hop. In either case, the call's vulnerability to interception is increased, and many people, using many kinds of radio equipment, will have the ability to listen in.

The vulnerability of long-distance communication is nothing new; remote communication has always been subject to interception. Couriers have been waylaid, seals have been broken, and letters have been read. But before the electronic era conversing in complete privacy required neither special equipment nor advanced planning. Walking a short distance away from other people and looking around to be sure that no one was hiding nearby was sufficient. Before tape recorders, parabolic microphones, and laser interferometers, it was not possible to intercept a conversation held out of sight and earshot of other people. No matter how much George III might have wanted to learn the contents of Hancock's private conversations with Adams, he had no hope of doing so unless he could induce one or the other to defect to the Crown.

Achieving comparable assurance of privacy in today's world—a world in which many of the most personal and sensitive conversations are carried on by people thousands of miles apart—requires both advanced planning and complex equipment. Most important, privacy in long-distance communication is not something the conversants can achieve on their own. A secure telephone is a complicated device combining a voice digitizer, cryptography, and a modem. Building one is as much beyond the abilities of most potential users as building a television set is beyond the abilities of most viewers. In general, secure communication facilities are complex and require numerous people, many of whom must be trusted, for their construction and maintenance.

The vulnerability of telephone calls is the vulnerability of something that did not exist before the late 1800s. Unfortunately, holding a conversation face to face is not the guarantee of privacy it once was. The same electronic technologies that have made telecommunication possible have also given us a wide range of listening devices that make finding a private place to talk difficult indeed. Technology has changed the rules for the old game as well as for the new.

Telecommunication and to a lesser extent face-to-face communication suffer from another vulnerability that did not exist when the United States was founded: the possibility that one party to a conversation is recording it without the consent of the others. Before the development of sound recording, even one of the parties to a conversation had limited ability to reveal what had been said. Notes, an outline, or even a transcript would typically be only one person's word against another's. Audio and video recordings have changed the standards of evidence and opened the way for the repetition—sometimes to a very broad audience —of remarks that the utterer did not expect to be repeated.

The result is that privacy of conversation is no longer, as it was 200 years ago, a fact of life. It is now something over which society has a large and ever-increasing measure of control—a privilege that governments can grant or deny rather than a rule of nature over which they have no influence.

Society's response to these developments has been both to exploit them for various ends and to regulate them. It has tried to replace the fact of inviolably private communications with a "right to communicate privately." In the process, however, society has stopped short of creating an absolute right comparable to the reality of a former day. Society has placed controls on the use of technology to violate privacy by either the government or the citizens, but has also allowed it under many circumstances. Police employ wiretapping in criminal investigations, and intelligence agencies intercept foreign, and occasionally domestic, communications on a grand scale. Both regard their activities as a natural prerogative of the state, necessary for an orderly society. Many who are not spies or police have a different perception of electronic surveillance. They see wiretapping not as a tool for law and order but as an instrument of the police state.

The ill ease that many people (including a number who were members of Congress at the time the federal wiretapping law was passed) feel when contemplating police use of wiretaps is rooted in awareness of the abuses to which wiretapping can be put. Unlike a search, the fact of whose occurrence is usually obvious, a wiretap is intrusive precisely because its invisibility to its victim undermines accountability. Totalitarian regimes have given us abundant evidence that the use of wiretaps and even the fear of their use can stifle free speech. Nor is the political use of electronic surveillance a particularly remote problem—the Watergate scandal is only the most recent example in contemporary American history of its use by the party in power in its attempts to stay in power.[1]

The fundamental similarity between the government's power to intercept communications and its ability to search physical premises has long been recognized. The Fourth Amendment to the US Constitution takes this ability for granted and places controls on the government's power of search. Similar controls have subsequently been placed by law on the use of wiretaps. There is, however, no suggestion in the Fourth Amendment of a guarantee that government searchers will find what they seek. Just as people have always been free to protect the things they consider private by hiding them or storing them with friends, they have been free to protect their conversations from being overheard.

Today, a new development in communication technology promises —or threatens, depending on your point of view—to restore some of the privacy lost to earlier technical advances. This development is electronic cryptography, a collection of practical and inexpensive techniques for encoding communications so that they can be understood only by their intended recipients. Modern cryptography also serves to provide anonymity to certain transactions.

Technology rarely exists in a vacuum, however. The rise of cryptography has been accompanied, and often driven, by a host of other phenomena.

Ease of communication, electronic as well as physical, has ushered in an era of international markets and multinational corporations. Today's business is characterized by an unprecedented freedom of movement for both people and goods. More than one-fourth of the gross national product of the United States, for example, comes from either foreign trade or return on foreign investment (Dam and Lin 1996, p. 28). When foreign sales rival or exceed domestic ones, corporations open new divisions in proximity to markets, materials, or labor.

Security of electronic communication is as essential in this environment as security of transportation and storage have been to businesses throughout history. The communication system must ensure that orders for goods and services are genuine, guarantee that payments are credited to the proper accounts, and protect the privacy of business plans and personal information. These needs are all the more pressing today because, as governments have come to view the economic battlefield as an extension of the military one, industry has become a direct target of foreign espionage (Dam and Lin 1996, p. 33; Schweizer 1993, pp. 15–20; Williams 1992).

The rising importance of intellectual property has expanded the role of electronic communications in business. The communication systems with which we have been familiar all our lives—the telephone and the mail on one hand, ships, trains, trucks, and airplanes on the other— serve quite different sorts of business needs. The business function of the former has lain primarily in negotiation of commercial transactions, that of the latter in delivery of goods and services.[2] Today these distinctions are blurring. A larger and larger fraction of our commerce is commerce in information, so delivery of goods and services by electronic media is becoming more and more common. To support this delivery, the media themselves are becoming more unified. These phenomena are commonly referred to as the development of a "Global Information Infrastructure."

Both the negotiation and the delivery aspects of commercial communications have long required security. In the pre-electronic world, the validity of letters was established by seals, letterheads, and signatures; that of negotiators was established by personal recognition or letters of reference. Goods were typically protected by less subtle mechanisms. In past centuries, merchant ships carried cannon, and port cities were fortified. Today, warehouses are locked, airports are guarded, and roads are patrolled.

The growth of an information economy merges the channels used for business negotiation with those used to deliver goods and services. Much of what is now bought and sold is information, such as computer programs and knowledge about consumers' buying habits. The security of information has become an end in itself rather than just a means for ensuring the security of people and property.

In parallel with the growth of a commerce in information, there is a development that makes security harder to achieve: the rising demand for mobility in communication. Traveling executives sit down at workstations they have never seen before and expect the same environment that is on the desks in their offices. They carry cellular telephones and communicate constantly by radio. They haul out laptop computers and connect to the Internet from locations around the globe. With each such action they expose their information to threats of eavesdropping and falsification barely known until the 1990s. It is the lack of security for these increasingly common activities that we encounter when we hear that most cellular telephone calls in major metropolitan areas are overheard or even recorded by eavesdroppers with scanners, that a new virus is destroying data on the disks of personal computers, or that industrial spies have broken into a database half a world away.

The growing awareness of security, particularly in regard to Internet communications, has given rise to an explosion in the market for cryptography and in the development of products to satisfy that market. Software examples include Lotus Notes, the Netscape browser, and the seamless encryption interface in the popular Skype VoIP service. Hardware encryption is used in satellite TV decoders, in automatic teller machines, in point-of-sale terminals, and in smart cards. One researcher estimates that the commercial market for cryptography—still in its infancy—has already outstripped the military market.[3]

Cryptography's good fortune has not been to everybody's liking. Its detractors see its potential use by criminals, terrorists, and unfriendly foreign countries as outweighing its benefits to commerce and privacy. Two groups in particular have emerged in opposition to the easy availability of strong cryptography: the national-security community and the law-enforcement community.

The Allies' ability to understand German and Japanese communications, even when they were encoded with the enemies' best cryptographic systems, is widely seen as having been crucial to the course of World War II. Since that time, the practice of communications intelligence has grown steadily. Today it accounts for one of the largest slices of the US intelligence budget.[4]

The availability of wiretaps—legal or otherwise—for more than a lifetime has given us generations of police who cannot imagine a world without them. Confronted with even the suggestion of losing this tool, they respond in the same way one would expect of a modern doctor faced with the prospect of returning to a world without MRIs, CT scans, blood panels, and the numerous other diagnostic tests that characterize modern medicine.

The US government's initial response was a series of programs designed to maintain its eavesdropping capabilities. The centerpiece of those efforts, initially called *key escrow* and later *key recovery*, is a scheme that provides the users of cryptographic equipment with protection against most intruders but guarantees that the government is always in possession of a set of "spare keys" with which it can read the communications if it wishes. The effect is very much like that of the little keyhole in the back of the combination locks used on the lockers of schoolchildren. The children open the locks with the combinations, which is supposed to keep the other children out, but the teachers can always look in the lockers by using the key.

The first of these "spare keys" was the Clipper program, which made the term Clipper virtually synonymous with key escrow. The program was made public on Friday, April 16, 1993, on the front page of the *New York Times* and in press releases from the White House and other organizations. The proposal was to adopt a new federal standard for protecting communications. It called for the use of a cryptographic system embodying a "back door" that would allow the government to decrypt messages for law-enforcement and national-security purposes. Subsequently adopted over virtually unanimous opposition, the "Escrowed Encryption Standard" did not prove popular; most of the equipment implementing it was bought by the government in an unsuccessful attempt to seed the market.

Business objected to the Clipper scheme on every possible ground. First of all, its workings were secret. This meant that the algorithm had to be implemented in tamper-resistant hardware, which was unappealing not only to the software industry but also to hardware manufacturers. Because of the secrecy and the tamper resistance, the Clipper chip's functions could not readily be integrated into other chips. And the scheme entailed the cost of adding a chip to each product—typically several times the cost of the chip itself.

Perhaps most important was the fact that Clipper's back door was accessible to the US government and only to the US government. This made it unlikely that Clipper products would appeal to foreign customers and undercut one of its major selling points. The Clipper chip, unlike most cryptographic equipment, was supposed to be exportable.

The White House saw the objections, which came from almost every quarter, as falling into two classes: those concerned with privacy and civil liberties and those concerned with business. In subsequent proposals, it attempted to address the business objections while flatly rejecting the civil-liberties position and maintaining the view that the government has the right not only to intercept citizens' communications but also to ensure that it will be able to understand the intercepted material. In all these proposals the executive branch attempted to use export controls—the only significant controls it had over cryptography under US law—to pressure industry to accommodate its desires.

The explosion in cryptography and the US government's attempts to control it gave rise to a debate between those who hail the new technology's contribution to privacy, business, and security and those who fear both its interference with the work of police and its adverse effect on the collection of intelligence. Positions have often been extreme. The advocates of unfettered cryptography maintain that a free society depends on privacy to protect freedom of association, artistic creativity, and political discussion. The advocates of control hold that there will be no freedom at all unless we can protect ourselves from criminals, terrorists, and foreign threats. Many have tried to present themselves as seeking to maintain or restore the status quo. For the police, the status quo is the continued ability to wiretap. For civil libertarians, it is the ready availability of conversational privacy that prevailed at the time of the country's founding. The fact that if cryptography has the potential to interfere with police investigations it also has the potential to prevent crimes and thus make society more secure was often overlooked.

At the turn of the century, the argument seemed to have been won by the civil-liberties and business interests. Export controls were relaxed and revised, moving their focus away from the strength of security systems and toward a regime that preferred allowed commercial sales while restricting government ones. The new regime encouraged uniform *retail* offerings while discouraging customized products that could accommodate the needs of organizations that already had an installed base of cryptographic equipment.

The argument was won, in no small part, because the national-security establishment decided that the widespread use of strong encryption, difficult though it make certain aspects of intelligence, was, in the end, ultimately in the nation's interest.

The attempt to push key escrow was quietly dropped. Skipjack, the secret cryptographic algorithm underlying the Escrowed Encryption Standard, was declassified. More significantly, the aging Data Encryption Standard was replaced not with Skipjack but with a new algorithm of seemingly unbounded security.

Sober minds knew that the victory could not be so complete as it appeared. Police and intelligence agencies had begun to realize that their eavesdropping problem was not so much one of overcoming the protection of communications as of acquiring the data in the first place. The exploding diversity of communications technologies as well as the explosion in the volume of communications had the interceptors running to keep up. The interceptors' response was to offload the difficulty onto the communications carriers by applying a law adopted in the early 1990s to areas beyond those originally intended. These moves have reinvigorated—and fundamentally changed—the privacy-versus-intelligence argument, moving it, at least for the moment, away from cryptography and toward the expansion of interception technology. Should current law-enforcement efforts be successful, however, the issue of restrictions on the use of cryptography is sure to recur.

Had telecommunication merely given us a new option, the fact that the new medium lacked privacy would be at most regrettable—similar, perhaps, to the fact that telecommunication cannot provide physical contact, either friendly or hostile.[5] The problem arises from the fact that telecommunication has transformed society. It has made possible long-distance relationships between people who rarely or never meet in person. Without secure telecommunication, these people are effectively denied the possibility of private conversation.

The issues are not cut and dried, and no amount of calling a tail a leg will make telecommunication equivalent to face-to-face communication. Any attempt to force such an equivalence and establish an absolute right of private conversation is doomed to failure. The interceptability of communications is as much a fact of life in the electronic era as the inviolability of private conversation was in the pre-electronic. On the other hand, if we deny the fact that telecommunication, whatever its new properties, is rooted in face-to-face conversation and shares much of its social function, we will doom ourselves to a world in which truly private conversation is a rarity—a perquisite belonging exclusively to the well-traveled rich.

Ultimately, to make good policy we must consider the sort of world in which we want to live and what effects our actions will, indeed can, have in bringing about such a world. Such consideration depends on awareness of many factors, including the technology of cryptography and electronic surveillance, the aims and practices of intelligence and law enforcement, and the history of society's attempts to deal with similar problems over more than a century.

2

Cryptography

The Basics

What does it mean to say that communication is secure? In most circumstances, it means that the communication is free from eavesdropping—that the information exchanged is kept private or confidential and does not, in the course of communication, become known to anyone other than the sender and the receiver. The techniques required to achieve this vary substantially, depending on whether the medium of communication is sound, writing, pictures, or some other form of data.

Security can be obtained in a variety of ways. The most common form of secure communication is the private conversation. Although it has become more difficult, in the age of electronics, to be sure of conversational privacy, it is still easier to have privacy in a face-to-face conversation than in any other sort. For a telephone conversation to be private, the speakers must at least have privacy at their respective ends of the line.

The security of conventional handwritten letters is a bit different. It is harder to remain unobserved while reading over the shoulder of someone writing a letter than it is to remain unobserved while listening to someone talk. Unless two people are communicating by passing notes back and forth while sitting in the same room (something people rarely do unless they suspect they are being spied upon), the easiest way to discover what they have written is to intercept the message as it travels from one to the other. In the case of messages written on paper, the primary means of protection is using a trusted means of transport, whether this is a private courier or a state-run mail system. Within this trusted transport medium, the message is further protected by an envelope, whose function is not so much to prevent entry as to ensure that entry will not go undetected. As we shall see, in the case of electronic messages there is no satisfactory analog to the envelope.

Physical protection is also used to guard electronic messages. A message traveling through copper wires is less vulnerable to interception than one carried by radio. A message traveling through optical fiber is less vulnerable still. Even a message that is sent by radio may be protected by an appropriate choice of frequencies and routes.[1]

There is, however, another possibility. A message may be put at risk of falling into the hands of opponents but may be disguised in such a way that even if unintended parties are able to intercept the message they will not be able to understand it. This is the domain of cryptography.

Less well known than the problem of keeping messages private is the problem of guaranteeing that messages are genuine—of being sure that they really come from the people from whom they appear to have come and that no one has altered them along the way. These properties of communication, called *authenticity* and *integrity*, are arguably more important than privacy. Nonetheless, we will devote more attention to privacy than to authenticity, for several reasons. Although privacy is of limited use in a conversation with someone you do not know,[2] it is generally more difficult to falsify communications than merely to intercept them. Sending a message exposes the sender to discovery in a way that receiving a message does not, because the message invariably exists as evidence of the fact that it was sent. This makes violations of authenticity difficult to achieve under circumstances in which violations of privacy are easy. The foremost reason we focus on privacy, however, is that the right to use cryptography for authentication is not in question; the right to use it for privacy is.

Encrypting a message is often described, by analogy with written messages, as placing it in an envelope, but the analogy is not entirely adequate. A well-encrypted message is far harder to open than one surrounded by paper; however, if the encryption is broken, the break leaves no traces. It is this difference between the functioning of cryptography and the functioning of physical protection mechanisms that gave rise to the policy issues so prominent in discussions of cryptography in the 1990s.

We shall describe cryptography for the moment only by what it does: a transformation of a message that makes the message incomprehensible to anyone who is not in possession of secret information that is needed to restore the message to its normal *plaintext* or *cleartext* form. The secret information is called the *key*, and its function is very similar to the function of a door key in a lock: it unlocks the message so that the recipient can read it.

The analogy with locks and keys is particularly apt in another respect. The lock and the key are distinct components of a system that controls the use of doors, cabinets, cars, and other things. The lock is a moderately complex mechanical device with numerous moving parts—about two dozen in the case of a normal door lock. The key is a single piece of metal. There are, on the other hand, far fewer types of locks than cuts of keys. Most doors use one of a dozen popular brands of locks, each of which can be keyed to accept one of a million different possible keys. A lock is typically far more expensive than its key, and more expensive to replace, particularly with a lock of a different kind. Perhaps the most important distinction between locks and keys is that locks are not, in principle, secret. Locks are easily recognizable even if they do not display their brand names, and there is no reason to be concerned that people know what type of lock you use on your front door. The cut of the key, on the other hand, is a secret, and any locksmith or burglar who knows it can make a duplicate that will open the door.

In exactly the same way, cryptographic systems are divided into the so-called *general system* (or just *system*) and the *specific key* (or *key*). It has been a principle of cryptography for more than a century (Kerckhoffs 1883) that the general system should not be regarded as secret.[3] The keys, in contrast, must be kept secret, because anyone who is in possession of them will be able to read encrypted messages, just as anyone who is in possession of a door key can open the door.

There is a distinction that is particularly prominent in the older literature between codes and ciphers. *Codes* in this terminology are transformations that replace the words and phrases in a human language with alphabetic or numeric *code groups*. *Ciphers* are transformations that operate on smaller components of a message, such as bits, bytes, characters, and groups of characters. The distinction is not always entirely clear.[4] Most of the systems we discuss are cipher systems, but codes appear from time to time in historical discussions.

Cryptography in the Small

In using cryptography to achieve secure communication, scale is everything. Two people who meet occasionally and usually communicate by postcard can make use of their infrequent contacts to exchange the secret keys that they will later use to encrypt what they write on their cards. This basic case, in which a small number of correspondents exchange messages of small size, is worth examining in some detail.

Suppose that, as you are about to embark on a journey, a reporter friend asks you for help. Within the next few months there is going to be a demonstration in the city you will be visiting. Your friend has great respect for your powers of investigation and is sure you can learn the time and place of the demonstration. There is one problem. The police are trying to learn just the same information so that they can stop the demonstration. If you call your friend and mention what you have learned, the police will surely overhear since they are tapping all the telephones. What should you do?

Clearly you must encrypt your conversation—encode it in such a way that no one who receives it (except your friend) will be able to understand it. Your friend has told you, however, that the police have all the government's resources available to them. You must, therefore, encrypt your conversation well to keep the police from reading it.

Cryptography on this scale is both theoretically and practically easy. Suppose, for simplicity's sake, that the city you are visiting has a very regular structure with numbered avenues running north-south and numbered streets running east-west. Any address in the city can therefore be given as a pair of numbers, the first representing the avenue and the second representing the street. The time, of course, can be represented by the year, month, day, hour, and minute. (The demonstration is expected to be brief but effective.) Your message will therefore have the form year month day hour minute avenue street,

where each of these elements is a two-digit number. Perhaps

19 99 12 30 15 25 01 44

means that the demonstration will take place at 3:25 P.M. on December 30, 1999, on the corner of 1st Avenue and 44th Street. Note that every digit in this message is significant. For example, were the demonstration to be only a few days later, four of the digits would change and the value of the year would be 2000 rather than 1999. The digits in some positions, however, are limited in their values. For example, the first digit of the day of the month can only be 0, 1, 2, or 3, and the first digit of the minute can never be more than 5.

In order to encrypt a message of this sort, you and your friend need only agree on a key that will transform the string of 14 digits into some other string of 14 digits. This key must be selected before your departure, and you must carry it with you and keep it secret.

In order to carry out the transformation you will add the digits of the message one at a time to the digits of the key, without carrying.

Suppose that the key is

64 25 83 09 76 23 55 72

and you add the message

19 99 12 30 15 25 01 44.

You will get

73 14 95 39 81 48 56 16

as the cryptogram. Note that in some cases the addition produced a number greater than 10. In the third place, for example, the sum is 11. In such cases the 1 in the tens place is simply thrown away, rather than "carried" into the next place as in ordinary arithmetic.

Once you have learned the time and place of the demonstration, you convey your message by calling and reading the encrypted version over the phone.[5]

To decrypt the received string of numbers, your reporter friend will take the cryptogram

73 14 95 39 81 48 56 16

and subtract (without "borrowing") exactly the same numbers you added:

64 25 83 09 76 23 55 72.

This will yield the original message:

19 99 12 30 15 25 01 44.

Assume that the police have intercepted your phone call. No matter what computers or codebreaking skills they may possess, they have no hope of recovering the underlying message; there simply is not enough information in what they have received. The fact that the key was chosen entirely at random means that for any possible message there is a key that would produce any observed cryptogram. For example, if the time and place of the demonstration had instead been January 11, 2000 at 5th Avenue and 23rd Street

20 00 01 11 10 45 05 23

and the key had been:

53 14 94 28 71 03 51 93

the cryptogram would have come out exactly the same.

A cryptosystem of this kind is called a *one-time* system because it is perfectly secure if used only once. If it is used even twice, the results are likely to be disastrous.

Suppose that, in your travels, you learned about not one demonstration but two. Since you have brought only one key along on your trip, you use it for both messages. Unfortunately for the second demonstration, the police cryptanalysts figure out what is happening. When the first demonstration occurs and, despite the secrecy of its planning, gets mysteriously good coverage in the foreign press, they consider the demonstration in light of your two telephone calls. By combining your first message with the date of the first demonstration, they extract what they presume to be a key. When they decrypt your second message with the same key, they get another place and time[6]; with that information, they can prevent the second demonstration before it begins.

If you follow sound procedures, this sort of error will never occur. You might perhaps carry only one copy of the key, written on the paper of a cigarette. Once the message has been sent, you can light the cigarette in the lee of the telephone booth and stroll off down the street, feeling like a real spy.

Cryptography is, of course, not limited to the occasional exchange of short messages between friends. It may require millions of bits of information to be encrypted in order to protect a first-run movie or a semiconductor mask file.[7] It may require the exchange of messages among hundreds or thousands of people to protect the communications of a large corporation. Often it requires both. As cryptography grows in each of these directions, it rapidly becomes more complex.

One-Time Systems on a Larger Scale

The scenario that has just been described is entirely practical. In fact, the same procedure can be used and has been used on a much larger scale. It was a mainstay of Soviet diplomatic communications from the late 1920s until at least the early 1950s. Since their messages did not always have the convenient numerical character of those presented in our example, the Soviets first had to convert them into numerical form. This was done with *one-part codes*[8] similar to the following:

abovementioned	0000
academician	0001
acknowledge receipt	0002
arrange meeting	0003
avoid contact	0004
. . .	. . .

These four-digit code groups were then added to four-digit key groups in the same digit-by-digit fashion employed in our example above.

Using one-time systems in a large network creates a number of serious problems. First, although it is trivial to produce a few dozen or a few hundred random digits by throwing 20-sided dice (figure 2.1) from a fantasy games shop, it is quite another thing to manufacture millions upon

Figure 2.1
Twenty-sided dice. (Photograph by Eric Neilsen.)

millions, type them up onto sheets, and produce precisely two copies of each sheet.

Before we go further, a word about the terminology of modern cryptography is in order. In many papers, the participants in encrypted communication are personified, whether they are people, pieces of equipment, or computer processes. If there are two parties involved, they are called Alice and Bob. If more are required, they are drawn from a cast of supporting characters that includes the couple's friends Carol and Ted, along with Eve, the eavesdropper.

In a network with more than two correspondents, there is difficulty in coordinating the keys used. If three people share a body of one-time key and Alice uses some in sending a message to Bob, she must inform Carol of what she has done. If Carol does not know this, she will at some time use the same key to send a message of her own and thereby create an insecurity. The feasibility of such coordination among three people is clear; for 1000, it is not.

The way the Soviets dealt with this problem was by having all communications go through Moscow. Every embassy could communicate securely with Moscow using keys that it shared only with Moscow. If the Soviet embassy in the United States needed to communicate securely with the Soviet embassy in Mexico, it was required to send its message to Moscow and have it relayed to Mexico City. Such an arrangement makes a network less flexible and requires twice as much keying material to be expended in sending each message.

Despite the centralized approach, the Soviets got into trouble. For reasons that are still unclear, a serious mistake was made in the early months of 1942. Rather than making exactly two copies of the key sheets, they made four. These excess keys then entered the inventory and remained in use for several years. Western intelligence noted and exploited the multiple use of the keys, with disastrous results for Soviet security. Under the code name Venona, cryptanalytic study of the reused "one-time" keys went on for decades. The system was used for the most sensitive Soviet information, and the Americans and the British studied it in hopes of identifying Soviet "moles" thought to be operating at the highest levels of their intelligence establishments.[9]

One-time systems are not the only form of highly secure cryptography, and they are by no means the dominant form today. In order to avoid having to ship the titanic amounts of keying material that are required in one-time systems, most enciphering today is done by *cipher machines*: mechanical or electrical or computer devices that encode messages. One-time systems have the advantage of simplicity but the disadvantage of failing completely if used to encrypt an amount of text exceeding the size of the key. The functioning of cipher machines is more complex than that of one-time systems. This complexity is the price of a system that can protect quantities of traffic far greater than the size of the key.

A Brief History of Cryptographic Systems

Despite the vast progress of cryptography during the twentieth century, there is a remarkable continuity with systems that have been known since the Renaissance.

Cryptography is always a matter of substituting one thing for another. The earliest cryptographic systems substituted one letter of the alphabet for another in an unchanging fashion, a technique called a *simple* or

```
b   o   o   k   k   e   e   p   e   r

g  |a a|  |o o|  |b b|  t   b   w
z  |e e|  |w w|  |i i|  a   i   k
o  |z z|  |e e|  |j j|  s   j   y
n   s   s   i   i   r   r   o   r   q
```

Figure 2.2
The characteristic letter pattern of the word 'bookkeeper'.

monoliteral substitution. Simple substitutions are easy to perform, even when the computational resources are limited to pencil and paper. There are also plenty of them: some 2^{90} for a 26-character alphabet. In other words, a simple substitution cipher has a 90-bit key—far larger than anyone could have needed before the computer age. Despite these virtues, simple substitution ciphers are quite easy to break. This is because they leave many characteristics of the message, such as letter frequency and letter patterns, unchanged.

The best-known approach to solving simple substitution ciphers is to compute the frequencies of the various letters. In English, the letters of the alphabet have widely varying rates of occurrence. The letters E and T, for example, occur quite frequently, accounting for 13% and 9% of typical text, whereas J and Z account for only 2% and 1% of such text. These characteristics frequencies permit a cryptanalyst to recognize the identities of the letters despite the substitution. Analysts also make use of the preservation of letter patterns. Figure 2.2 shows that the exceptional structure of the word 'bookkeeper' remains visible when it is encrypted under a variety of cipher alphabets. Notice that in each case the resulting cryptogram shows the letter pattern 1223344536—that is the cryptogram contains three consecutive pairs of repeated letters in the middle. Admittedly, this is a word chosen for its exceptional pattern of repeated letters (it is the only English word with three pairs of repeated letters in a row); however, it is only an extreme case of a very common phenomenon that occurs in such words as 'pepper', 'papa', and 'noon'. Repetition patterns allow a skilled cryptanalyst to read words of this sort directly from the ciphertext.

The cure for the shortcomings of monoalphabetic substitution—

discovered some 500 years ago and still in use today—is to change the cipher alphabet from one letter to the next. *Polyalphabetic encryption*, as this is called, is the creation of three Renaissance scholars, Alberti, Belaso, and Trithemius (see Kahn 1967), but is commonly known by the name of another, Blaise de Vigenère—an error too deeply embedded in cryptographic terminology to admit of historical correction at this date.

The simplest form of polyalphabetic cipher employs a sequence of alphabets. The first alphabet is used to encrypt the first letter of the message, the second alphabet to encrypt the second letter, and so on. Once the supply of alphabets has been exhausted, the encipherer starts over again with the first. The cipher alphabets may either be unrelated, as in figure 2.3,[10] or may be generated by simple transformations from a single alphabet. The more distinct alphabets are used, the more secure the system, but the more it suffers from the problems of a one-time system—the amount of keying material becomes excessive.

The general form of the Vigenère system employs a number of independent cipher alphabets, as in figure 2.3, and employs some subset of them sequentially in a pattern that may or may not repeat and may or may not use all of the alphabets. Note that each alphabet is labeled at the left with a letter of the alphabet. This allows a *key word* or *key phrase* to represent a sequence of alphabets, as in figure 2.3.

Polyalphabetic ciphers call on three basic processes to achieve security: they employ a set of unrelated cipher alphabets, they derive a number of secondary alphabets from each of these primary alphabets, and they vary the use of the secondary alphabets in a more or less complex pattern. These three processes can be traded off against each other. The more complex the pattern in which the alphabets are used, the smaller the number of distinct alphabets that are needed. Of the innumerable variations of polyalphabetic ciphers that are possible, we will examine a small number of examples illustrative of the development of cryptography from the Renaissance to the twentieth century.

The simplest form of Vigenère cipher uses *direct standard alphabets* —that is to say, the ordinary alphabet (standard) in its usual order (direct).[11]

The difficulty of solving a Vigenère system is entirely dependent upon the relationship between the length of the key and the length of the

Vigenère table

	A B C D E F G H I J K L M N O P Q R S T U V W X Y Z
A	Q F A L H I M Z E T Y N B O U D X P C S K G R J V W
B	Y L K O R U C J X P A S V H B D Q G M I T E Z J W N
C	J O A M S I T Y R D N H X E W P F V Z B L G K Q U C
⋮	· · ·

Plain: d   o   d   e   c   a   h   e   d   r   o   n

Key: B↓ A↓ F↓ F↓ L↓ E↓ D↓ B↓ A↓ F↓ F↓ L↓

Cipher: O   U   A   S   M   O   P   R   L   U   Q   Y

"d" is carried to "O" under the "B" alphabet;
"o" is carried to "U" under the "A" alphabet;
"d" is carried to "A" under the "F" alphabet; etc.

Figure 2.3
Action of a Vigenère system.

message to be encrypted. After a certain point (called the *period*) in the encryption, a new sequence of message characters will be encrypted with the same sequence of key characters. The number of times this occurs is referred to as the *depth* of the message with respect to the key. The greater the depth, the easier the message is to break.

Despite the fact that Vigenère systems with short periods are not secure, they are often used. One example is the CAVE cipher still used to protect digital cellular telephone calls in the United States (Electronic Industries Association 1992).[12]

A little examination will show that the short "time and place" message with which we began was encrypted in a Vigenère system using numbers instead of letters. A central theme in cryptography has been to produce systems with long periods without having to transport the large amounts of keying material required by the one-time systems. One way of solving the problem of short periods is to encrypt the message more than once— a *multiple Vigenère system* (figure 2.4).

Although encrypting two, three, or more times makes for a vast improvement in the security of messages, it was not actually feasible when enciphering was done by hand. In practice, errors by the code clerks—

Plain:	HENRY IS HUNGRY ...
Key-1:	PAPAD UM PAPADU ...
Cipher-1:	WECRB CE WUCGUS ...
Key-2:	HIPSH IP SHIPSH ...
Cipher-2:	DMRJI KT OBKVMZ ...

Figure 2.4
Double Vigenère system.

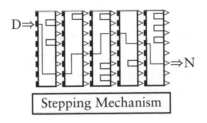

Stepping Mechanism

Figure 2.5
Rotor machine.

possibly in both the enciphering and the deciphering—made many messages unreadable. The wide use of such complex encryption techniques only appeared in the twentieth century, with the development of electromechanical enciphering equipment. The first of these was the *rotor machine*, a device that was to dominate cryptography for half a century.

The central component of a rotor machine is the *rotor*, a disk about the size of a hockey puck that serves to implement a cipher alphabet. On each face of the disk there are a number of electrical contacts corresponding to the letters of the alphabet. Each contact on the front face is wired to exactly one contact on the rear face. As an electrical signal passes through the rotor, the signal is carried to a new alphabetic position, just as a letter looked up in a cipher alphabet changes to another letter.

Rotor machines have had from three to more than ten rotors. Every rotor through which a "letter" passes represents an additional layer of encryption. Thus, the simplest rotor machines correspond to triple Vigenère systems, and the more elaborate ones may be several times as complex.

With the appearance of digital electronics after World War II, the dominance of rotor machines gave way to that of *shift registers*. A shift

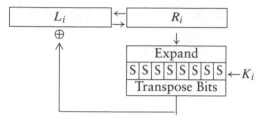

Figure 2.6
DES as a shift register.

register is an electronic device made up of a number of *cells* or *stages*, each of which holds a single 0 or 1 of information. As the shift register operates, the data shift one or more places along the register at each tick of the clock. In addition to moving left or right, some of the bits are modified by being combined with other bits. In the more modern *nonlinear* shift registers, some of the bits are looked up in tables[13] and then used to change other bits of the register, under control of a key. This process is repeated over and over until every bit has changed in a way that is a complex function of every other bit and of every bit of the key. In other words, if a single bit of either the input or the key is modified, approximately half of the output bits will change.

Typical of modern non-linear shift-register systems is the US Data Encryption Standard (USDoC 1977). DES, as it is generally known, is a *block cipher* or *electronic code book*. It takes 64 bits (8 bytes) of information as input, and, under the control of a slightly smaller (56-bit) key, produces 64 bits of output. DES is rarely operated in the basic mode in which it is described in the standard. It is intended as a primitive for building more complex modes of operation suitable for use in encrypting common data formats (USDoC 1980).

DES is a shift-register system in a modern style. In each of its 16 rounds it performs a complex operation on half of the 64 bits in its register and uses the result to alter the other half of its bits. Each round is controlled by a distinct 48-bit subset of the key.

DES expands the right-hand side of its register from 32 to 48 bits and exclusive-ors these with 48 bits selected from the key. It then contracts the 48 bits back to 32 by making eight substitutions, using six-bit-to-four-bit tables. These tables, called *S-boxes* (for *selection boxes*), are the heart of the algorithm; along with the key size, they have been sources of controversy (Bayh 1978). Finally the bits are rearranged (*transposed*) before the result is exclusive-ored with the left side of the register.

A proposed replacement for DES was Skipjack, an NSA-designed algorithm with a 64-bit blocklength and an 80-bit key.[14] Skipjack operates on its text 16 bits at a time and employs a number to techniques not present in DES. It has two types of rounds, doing eight of one then eight of the other. It has only one S-box but one equal in size to the sum of DES's eight, which it varies from occurrence to occurrence by adding in a counter. It has a simple key schedule compared with DES's complex one. Overall, Skipjack is a cleaner and more attractive algorithm than DES that might have been successful as a DES replacement had it not been introduced as part of a plan for *key escrow*. Key escrow is a mechanism for guaranteeing that some third party—in Skipjack's case the US government—is always able to read the encrypted traffic. To this end, the Skipjack algorithm was kept secret for six years to prevent unauthorized *unescrowed* implementations. When the escrow program had clearly failed, Skipjack was declassified to allow its use in software for email protection on military networks. It was too late, however, for the algorithm to gain acceptance as a replacement standard. A contest to select such a replacement was already underway; the system it produced will be discussed shortly.

Just as rotor machines look at first to be quite different from Vigenère encipherment, shift registers look superficially quite different from rotor machines. There is, however, a deep similarity. All these systems combine a process of looking things up in tables with one of "adding" them together using some sort of arithmetic operation. In shift registers, the message and the key are more thoroughly mixed together, rather as though the positions of rotors in a rotor machine were affected by the letters passing through them.

Strengths of Cryptosystems

Before we go further into cryptosystems and their use, a word about the strengths of cryptosystems is in order. A cryptosystem is considered secure when an opponent cannot break it under reasonable circumstances, in a reasonable amount of time, at a reasonable cost. The term "reasonable" is perforce vague. Despite being the most important problem in cryptography, the evaluation of cryptographic systems is the least understood. An adequate mathematical theory of cryptography has eluded cryptographers for centuries.[15]

The issue of what constitutes *reasonable* conditions for an attack on a cryptosystem is better understood than others. An opponent must, of course, be in possession of ciphertext to have any hope of discovering the underlying plaintext. If this is all that is available, we say that the cryptanalyst mounts a *ciphertext-only attack*.[16] Typically, however, the opponent knows some information about the plaintext before starting to work on the problem. A message from Alice to Bob, for example, is likely to begin "Dear Bob" and to be signed "Love, Alice." Although knowledge of such *probable words* or *cribs* is difficult to prevent, the situation may go much farther than this. Many messages, such as product announcements and press releases, are secret until a certain date and then become public. Under these circumstances an opponent has the corresponding plain text and cipher text of one message and can make use of this in attacking other messages sent using the same key. If this is the case, we say that the opponent is in a position to mount a *known-plaintext* attack.

At the time of World War II, the belligerents, not trusting their cryptosystems to resist known-plaintext attacks, imposed such *signaling rules* as "No message transmitted in cipher may ever be sent in clear." In fact a message sent in cipher could never be declassified without being *paraphrased* to reduce its cryptanalytic utility. This problem appears to have been solved for US government cryptosystems by the 1960s, permitting formerly encrypted messages to be declassified on the basis of content alone.

The sort of signaling rules formerly used by the military are entirely infeasible in a commercial environment. Fortunately, trust in modern electronic cryptosystems is sufficient that the availability of plaintext to an opponent makes no difference. In fact, we presume that the opponent can send an arbitrary number of text messages and will receive our co-operation in enciphering them or deciphering them before beginning to attack the actual message of interest. This is called the *chosen-plaintext* assumption.[17]

Workfactors

Time is the essential element in measuring computation. The question is "How much will it cost me to get the answer when I need it?" It is a rule of thumb that computing power doubles every 18 months[18]; thus a personal computer purchased for $1000 today will have twice the computing power of one purchased less than 2 years ago.[19] These improvements in speed have profound implications for cryptographic systems.

The number of operations required to break a cryptographic system is called its *workfactor*. The form or complexity of the operations is not precisely stated. They might be encryptions, as they are when the analytic process is one of searching through the keys, or they might be something entirely different. In a spirit of oversimplification, we will assume that operations are always entirely parallelizable. If two processors can do a million operations in 5 seconds, then ten processors can do the same number of operations in 1 second. For our purposes, this assumption is conservative in the sense that if it is false the problem merely becomes somewhat harder.

If a system has a workfactor of 2^{30}, it can be broken by a billion operations. These may not be elementary computer instructions; they may be complex operations requiring hundreds of instructions each. Even so, typical desktop computers today can do a billion instructions a second. If such a cryptosystem requires several hundred instructions per encryption, it can be searched in minutes. In short, breaking a system with a workfactor of 2^{30} is trivial.

A workfactor of 2^{60} means that a million processors, each doing a million operations a second, can solve the problem in a million seconds (between 11 and 12 days). It is clear that a system with a workfactor of

2^{60} can be broken today if the analytic operations are such that processors capable of executing them are worth building or already available. If the operations are encryptions, the processors might be built from available encryption processors.

On this path, systems with workfactors of 2^{90} are the first that seem beyond reach for the foreseeable future. A billion processors in parallel can certainly be imagined. A billion operations a second, even operations as complex as DES encryptions, had already been achieved around 1990 (Eberle 1992). A billion seconds, however, is 30 years—long enough to count as secure for most applications.[20]

Workfactors of 2^{120} seem beyond reach for the indefinite future. A trillionth of a second is less than one gate delay in the fastest experimental technologies; a trillion processors operating in parallel is beyond reach; a trillion seconds is 30,000 years.

The only technological development on the horizon that would be capable of bringing such computations within reach is *quantum computing*. Quantum computing makes use of *superposition* of physical states to calculate all of a set of possibilities simultaneously. To date, its application to any real problem remains a fiction. Quantum computing has the potential to destroy all of the public-key cryptosystems currently in use and to cut the effective key lengths of conventional cryptographic systems in half. Quantum computing, however, is unlikely to be an immediate threat. Breaking the most secure key management systems in use today would require factoring 2000-bit numbers; quantum computing made news when it factored the number 15 (Vandersypen et al. 2001).

Estimating the cost of searching through keys and validating the estimates by actually doing it was a sport in the cryptographic community for some time. In the fall of 1995 a group of cryptographers met and prepared an estimate of search costs, concluding that 40-bit keys (the largest that could be readily exported at the time) could easily be searched and that keys at least 70 to 90 bits long were needed to provide security for commercial applications (Blaze et al. 1996). The previous August, students at the École Polytechnique in Paris had searched out a 40-bit key. The following January, students at MIT repeated the feat using an $83,000 graphics computer. This amounted to a cost of $584 per key. At its annual conference in January 1997, RSA Data Security offered prizes for searching keyspaces of various sizes. The 40-bit prize was claimed before the conference ended and the 48-bit prize was claimed a week later. The 56-bit DES challenge lasted for only 5 months.

The US Data Encryption Standard used a 56-bit key and thus falls within the range we have described as clearly possible. The standard has been used extensively throughout the commercial world—particularly by banks, which commonly engage in billion-dollar electronic funds transfers. In such applications, the inadequacy of any algorithm with a 56-bit key is apparent. Because the National Institute of Standards and Technology (NIST) was slow to issue a replacement standard, *triple-DES* —a block cipher employing DES three times in a row with three different keys[21]—arose as a de facto standard and was formally adopted first by the Banking Security Standards Committee (ANSI X9F) of the American National Standards Institute (ANSI 1998) and the National Institute of Standards and Technology (FIPS 46-3).

Eventually, DES was replaced by a new cipher using much longer keys. This system, which is called the Advanced Encryption Standard or AES and will be discussed shortly, has largely ended the game of searching keyspaces.

Lifetimes of Cryptosystems
In designing a cryptographic system, there are two important lifetime issues to consider: how long the system will be in use and how long the messages it encrypts will remain secret.

Cryptographic systems and cryptographic equipment often have very long lifetimes. The Sigaba system, introduced before World War II, was in use until the early 1960s. The KL-7, a later rotor machine, served from the 1950s to the 1980s. DES was a US standard for some 25 years. It is still in widespread use, and it may be for decades.[22] Other systems that are neither formal standards nor under the tight control of organizations such as the American military probably have longer lifetimes still.[23]

Secrets can also have very long lifetimes. The Venona messages were studied for nearly 40 years in hopes that they would reveal the identities of spies who had been young men in the 1930s and who might have been the senior intelligence officers of the 1970s. The principles of the Sigaba system were discovered in the mid 1930s and were not made public until 1996. Much of the "H-bomb secret" has been kept since its discovery in 1950, and the trade secrets of many industrial processes are much older. In the United States, census data, income tax returns, medical records, and other personal information are supposed to be kept secret for a lifetime.

If we had set out to develop a piece of cryptographic equipment in the late 1990s, we might have expected it to be in widespread use today. We might also reasonably plan for the system to stay in use for 25 years or more. No individual piece of equipment is likely to last that long; however, if the product is successful, the standards it implements will. If the equipment is intended for the protection of a broad range of business communications, some of the messages it encrypts may be intended to remain secret for decades. The cryptosystem embodied in our equipment might thus encrypt its last message in 2030 or later, and that message might be expected to remain secret for 25 years more. The system must therefore withstand attack by a cryptanalyst in the late twenty-first century, whose mathematical and computing resources we have no way of predicting. The prospect is daunting.

The Advanced Encryption Standard

The daunting prospect was taken on by a process begun in 1997 and concluded in 2001, when the Data Encryption Standard was replaced by the Advanced Encryption Standard. The new system is an improvement over the old in every respect. It doubled the length of the block from 64 to 128 bits, increased the size of the key to between 128 and 256 bits, and substituted a mathematically based design for one dominated by engineering concerns.

No mathematical theory behind the tables at the center of DES was included in the standard or in any accompanying material. More important, no significant cryptanalytic techniques that might be applied to DES were publicly known when it appeared. This reduced any attempt to prove that DES was secure to vague generalities. If we could develop an algorithm based on a mathematical theory of the cryptanalysis of block ciphers, we could have proofs that the algorithm would resist certain types of attacks. If the attacks were sufficiently general in scope, resistance to the attacks might reasonably be described as security.

The starting point for modern cryptography was put forth by Claude Shannon, the founder of information theory, in 1949. He proposed combining *confusion* (intrinsically complex mathematical operations that perforce operate on small quantities of data) with *diffusion* (operations that spread the effects of confusion across larger data elements). In DES, the confusion was provided by a set of lookup tables and the diffusion by permutations of bits.

Beginning in the late 1980s, cryptographers began applying algebraic techniques to improve both components. These theories were based on the theory of finite fields. Slightly later two fundamental cryptanalytic techniques were developed. *Differential cryptanalysis* analyzes the pattern of changes in output resulting from changes in input; *linear cryptanalysis* makes use of a deep mathematical fact—that no transformation can be purely non-linear—to derive expressions approximating the key bits. This made it possible to give proofs that systems built using algebraic structures would resist differential and linear cryptanalysis.

Like much of cryptographic work, the research took two steps forward, and then a step back. It produced the algorithm SHARK, which had good *diffusion* (spreading the attacker's attention over large numbers of bits), but a plaintext/ciphertext attack on a simplified version of SHARK showed other problems. This led to development of the algorithm SQUARE, which in turn succumbed to attacks on its byte-oriented structure. Further improvements let to the algorithm Rijndael, which became the Advanced Encryption Standard.[24] Rijndael[25] operates approximately as follows. The input is a 128-bit block organized as a 4×4 matrix of 8-bit bytes. In each round of the Rijndael algorithm, there are four steps:

- Add in the key.

- Look the bytes up in a table.

- Rotate the rows.

- Apply a linear transformation to each column. (Landau 2004, p. 108)

Rijndael's confusion step is the table lookup; its diffusion steps are the row and column operations. The combination, done ten to fourteen times (depending on the size of the key), provides the algorithm's security; the mathematical formulation of the algorithm provides a base from which to analyze that security.

Key Management

Key management—the production, shipment, inventorying, auditing, and destruction of cryptographic keys—is an indispensable component of secure communication systems. Cipher machines make a spectacular reduction in the amount of keying material that users must ship around. This diminishes the problem of key distribution; however, it does not eliminate it, since the difficulty of distributing keys is typically more a function of how many people are involved than of whether each one has to get a large codebook or a short message.

The production of keys is the most sensitive operation in cryptography. Cryptographic keys must be kept secret and must be impossible for an opponent to predict. If they do not achieve these objectives, the results will be disastrous, regardless of how cleverly designed the cryptographic systems in which they are used. The failure of Soviet key production that led to the breaking of the Venona intercepts is one example of this. A far more recent example is the penetration of the Secure Socket Layer (SSL) protocol, which is used to secure Internet transactions by encrypting such sensitive information as credit card numbers. In the summer of 1995, a group of graduate students at the University of California at Berkeley discovered that the SSL protocol generated session keys by consulting the clock on the client machine (typically a personal computer). The time on the client machine's clock could be inferred quite accurately from other aspects of the protocol, and thus the key could be discovered easily.

In conventional cryptographic practice, reliability in key production is achieved by centralization. If all communications go to and from a central site, as in the Soviet diplomatic network, it is natural to manufacture keys at the center and ship them to the far-flung sites. The same procedure is typically followed even in cases where messages can flow directly from one field organization to another.

In US parlance, the organization that manufactures keys is called the *central facility*. The production process goes to great lengths to guarantee that the keys produced are completely random and are kept completely secret. From the central facility, keys are shipped to installations around the world through a special administrative structure called the COMSEC *materials control system*, which uses such elaborate security procedures as *two-person control*.[26] Keying material is stored by COMSEC *custodians* and constantly tracked by an elaborate system of receipts.

At one time, keys took the form of codebooks. Originally, these were produced by writing the words and phrases of the plain text on a set of index cards and shuffling the cards by hand. In the 1930s, handwritten or typed index cards were replaced by punched cards, and the shuffling was accomplished by using card sorters to sort cards at random. Later cryptographic systems used wired rotors, key lists, plugboards, and paper tape. The most recent have keys packaged in entirely electronic form.

Because keys are secret, they must have names to provide the users with a way of "talking" about which ones to use. A message sometimes contains the name of the key that is to be used in decrypting it. At other times the key is a function of the message's origin or subject matter. Thus an embassy might have one system for diplomatic messages, another for messages dealing with trade, and yet a third for the messages of the military attaché. A ship at sea might have one set of keys for communicating with shore-based facilities and another for communicating with each of the fleets with which it was likely to come in contact.

The distribution of keys follows the structure of the organization that employs them. Keys, however, usually change more often than organizational structures, and it would be confusing to ask a code clerk to remember an entirely new key name when the key was used in exactly the same way. The solution is to keep the name constant but to label each new key as to its *edition* (a term that dates from the era of codebooks but is still used today in naming purely electronic keys).[27]

The management of keys is a constant tug of war between the need for flexibility in communication and the need to maintain security by limiting keys. Typical of military communications is the US Navy's Fleet Broadcast System, which transmits constantly to American ships all over the world. The keys used by the Fleet Broadcast System are changed every day, but on any given day they are the same for every ship. This arrangement favors flexibility over security and has had its costs. In the late 1980s, a Navy chief petty officer named Jerry Whitworth sold keys from the Alameda Naval Air Station to Soviet intelligence officers. Because of the widespread distribution of those keys, the USSR was potentially able to decipher many US Navy communications. At the other extreme, keys may be issued for a communication network with only two members. The result is more secure but less flexible. If keys are distributed physically, the endpoints of the circuit must anticipate the need to communicate far enough in advance to order their keys—a process that may take days or weeks. In many circumstances (for example a secure phone system), this is an unacceptable burden and some less cumbersome solution must be found.

Dynamic Key Distribution

Suppose, in the traditional terminology of the modern cryptographic literature, that Bob wants to communicate with Alice. Bob may not know Alice; he may simply know that she is a lawyer, an investigator, or a doctor whose expertise and confidential collaboration he requires. If the proper arrangements have been made in advance, Bob and Alice can be "introduced" to each other in real time.

The mechanism that makes the introduction is called a *key management facility* (KMF).[28] It is a network resource, similar in function to a directory server, that shares a key with every member of the network.[29] In the description that follows, we will assume that Bob is using a cryptographically secure telephone to call Alice, and to simplify the description we will blur the distinction between the people and the instruments.

If Bob and Alice belong to the same community and rely on the same KMF, the process goes as follows:

- Bob calls the KMF and informs it that he wants to communicate with Alice. This call is encrypted in a key that Bob shares with the KMF.

- If the KMF finds that it "knows" Alice (i.e., that it shares a key with her), it manufactures a new key for the exclusive use of Bob and Alice and sends it to each party. In the particular case of telephones, which can have only one call in progress at a time, this is done by sending Bob a two-part message. Both halves of the message contain the same key, but the first half of the message is encrypted so that Bob can read it and the second half so that Alice can read it (Needham 1978; Rosenblum 1980).

- With the key now in hand, Bob calls Alice. The portion of the key that was encrypted so that only Alice can read it functions as a letter of introduction; Bob's phone sends it to Alice's phone at the beginning of the call.

- Bob's telephone and Alice's telephone now *hold* a key in common and can use it to make their secure phone call.

The *long-term* keys that Bob and Alice share with the KMF may reasonably be called *subscriber keys*; the ones they use for particular messages or conversations are called *session keys* or *traffic keys*. Because of the extra phone call required for Bob and Alice to acquire a common key, systems of this sort typically *cache* keys for some period of time. This allows the users to make repeated calls without the overhead of the KMF call.

This approach to keying secure phone calls is far from ideal. For one thing, if communication between any two members of a community requires the services of the KMF, it will be a busy facility indeed.[30] A far more serious problem, peculiar to security, is that Bob and Alice must place too much trust in the KMF.

Strengths and Weaknesses of Cryptography

Cryptography is the only technique capable of providing security to messages transmitted over channels entirely out of the control of either the sender or the receiver. To put this another way: the cost and applicability of cryptography depend very little on either the length or the "shape" of the path the encrypted message must follow.

Cryptography is not always the most appropriate security technique. In designing a system for securing the communications within a building, a campus, or a military base, it is appropriate to protect the signals physically by running them through shielded conduits or optical fibers. On the other hand, if the trunks of a network span great distances, cryptography is usually the most economical security mechanism and in some cases the only possible one. Physical protection techniques that are perfectly suitable for a network spanning distances of a few hundred yards will be prohibitively expensive if used to connect cities as far apart as Paris, Los Angeles, Hong Kong, and Sydney. If cryptography is used to provide security, however, the cost will be independent of the distance. Mobile and satellite communications represent an extreme case in which cryptography appears to be the only possible means of protection.

From a managerial viewpoint, security obtained through the use of physical means of protection must be bought at the cost of constant auditing of the signal path. If land lines are being leased, the customer must be constantly vigilant to the danger that the service provider will accidentally misroute the calls through a microwave or satellite channel. Ground routing is often more expensive than satellite routing and might thus be reserved for more sensitive messages. In the past, a similar phenomenon took the form of a choice between a faster route (telephone or telegraph) and a more secure one (physical shipment). If, however, a message has been properly encrypted, the communication network is free to send it any distance, via any channel, without fear of compromise.

One special case of this is particularly important: the freedom to pass an encrypted message through the hands of a potential opponent or competitor. Telecommunications suppliers, for example, often find themselves bidding against the local telephone system to provide equipment or services. In the process of bidding on these contracts, they often have no choice but to employ the services of the communications suppliers with whom they are competing.

Modern security protocols make special use of the ability to pass a message through the hands of a potential enemy. A party to communications is often asked to present cryptographically protected *credentials*. If the credentials can be deciphered correctly, the challenger who receives, deciphers, and judges them need not worry about having received them from a previously untrusted party. This is analogous to the procedure commonly employed to control international travel. When the border guards judge the traveler on the basis of a passport received from the traveler's own hands, they are placing their trust in the tamper resistance of the passport. If the passport appears intact and the picture resembles the traveler, they will not generally feel the need to conduct any further investigation into the traveler's identity.

In the world of paper documents, this mechanism is not considered adequate for all purposes. A visitor attending a secret briefing at a military installation, for example, must typically be preceded by a letter of authorization. The corresponding cryptographic process, however, is considered reliable enough to be used for the most sensitive applications.

Cryptography can best be thought of as a mechanism for extending the confidentiality and authenticity of one piece of information (the key) to another (the message). The protection of the key, which is typically small and can be handled very carefully, is extended to the message, which can then be handled much less carefully—routed through the least expensive communication channel, for example. As a consequence of this extension of security from the key to the message, the compromise of the key will likewise be extended to the message. The consequences of this compromise, however, differ markedly, depending on whether the compromised key is being used to protect privacy or authenticity.

In regard to authentication, compromise of a cryptographic key is similar to the compromise of other sorts of identifying information, such as passwords or credit cards. Suppose that on Wednesday morning the security officer of a bank learns that an authentication key in use since Monday has been stolen, but also knows that no messages have been received and authenticated with the key since the day before, when the key was known to be safe. The key will be changed immediately and no actual compromise will occur. The thieves cannot make use of a key acquired on Tuesday night to go back and initiate a wire transfer on Monday or Tuesday. If the key is used to protect the privacy of information, however, things are quite different. Even if the compromise of a key is discovered immediately, all messages ever sent in that key must be regarded as compromised. This is because no one can be sure that they were not intercepted and recorded by the same parties who later acquired the key. As soon as the compromise is discovered, the key will of course be changed, but this does far less to repair the damage than the change of an authentication key. The thieves can still read traffic they intercepted while the compromised key was in use. This gives breaches of cryptographic security the power to reach back in time and compromise the secrecy of messages sent earlier.[31] Changing the key will only prevent future messages from being read.

This vulnerability becomes especially critical in systems, like Alice and Bob's secure telephone in the previous section, that transmit some cryptographic keys encrypted under others. Although session keys in these systems typically last for only the duration of one phone call, session, or transaction, the subscriber keys may stay the same for weeks or months. If one of these is compromised at any time during its life, all the messages ever sent using session keys that were themselves sent enciphered under the subscriber key will likewise be compromised. Worse yet, if the key distribution center is compromised, the messages of every subscriber will be compromised.

Cryptography as we have described it so far has a very centralized character. If keys are distributed physically from a central facility, the center has the power to decide which elements of the network can communicate with which others. If the keys are distributed electronically and in real time, as described for secure phones, the key distribution center acquires control over communication on a virtually call-by-call basis. Implicit in this centralization is the phenomenon that is now called key escrow. The users of a cryptocommunication system may trust it to protect them against external opponents, but they can never be confident that they are protected against the system managers. There is always the possibility that the central facility will be employed to supply additional sets of keys for eavesdropping on the users.

The unique vulnerabilities of conventional cryptography and its centralization are intimately connected. In order to eliminate the former, we must also eliminate the latter.

Public-Key Cryptography

All the cryptographic systems discussed thus far—one-time systems, Vigenère systems, rotor machines, and shift registers—are *symmetric* cryptosystems: the ability to encrypt messages is inseparably linked to the ability to decrypt messages. The course of cryptography for the past 20 years has been dominated by *asymmetric* or *public-key* cryptosystems.

In a public-key cryptosystem, every message is operated on by two keys, one used to encipher the message and other to decipher it (Diffie and Hellman 1976). The keys are inverses in that anything encrypted by one can be decrypted by the other. However, given access to one of these keys, it is computationally infeasible to discover the other one. This makes possible the practice that gives public-key cryptography its name: one of the two keys can be made public without endangering the security of the other. There are two possibilities:

- If the secret key is used for deciphering, anyone with access to the public key will be able to encipher a message so that only the person with the corresponding secret key can decrypt it.

- If the secret key is used for enciphering, its holder can produce a message that anyone with access to the public key can read but only the holder of the secret key could have produced.

The latter property is what characterizes a signed message, and a public-key cryptosystem used in this way is said to provide a *digital signature.*[32]

Using Public-Key Cryptography

In a network using public-key cryptography, the secret key that each subscriber shares with the key management facility in a conventional network is replaced by a pair containing a public key and a private key. The function of the KMF is now merely to hand out public keys. Since these keys are not secret, they are not subject to compromise.

One problem remains, however: a subscriber who receives another subscriber's public key from the KMF must have a way of verifying its authenticity. This problem is solved by providing the KMF with a private signing key and providing every subscriber of the network with a copy of the corresponding public key. This enables the KMF to sign the keys it distributes, and it enables any subscriber to verify the KMF's signature.

Public-key cryptography vastly diminishes the vulnerability of the KMF. If the KMF is compromised, that compromise is the compromise of an authentication key (the KMF's signing key, the only piece of secret information the KMF knows) rather that the compromise of any key used to protect secrecy. If the signing key of the KMF is found to be compromised, that key can be changed. The network's subscribers will have to be informed of the KMF's new public key and warned not to accept any key distribution messages signed with the old one. Such reinitialization of the network may be costly, but it does not expose past network traffic to disclosure.

This use of public-key cryptosystems also has the practical benefit of reducing the load on the key distribution center. Instead of requiring the subscribers to call the KMF on the occasion of any conversation, they can be provided with a form of credentials called *certificates*. A certificate is a signed and dated message from the KMF to a subscriber containing that subscriber's public key, together with such identifying information as name and address. When two subscribers begin a call, they exchange these credentials. Each one verifies the KMF's signature on the received certificate and extracts the enclosed public key.

Although public-key cryptography as described above does much to diminish the vulnerability of the network as a whole, it does nothing to reduce the vulnerability of individual subscriber's private keys. If a subscriber's private key is compromised, it is possible that any message ever sent in the corresponding public key will be read.

For non-interactive communications, such as electronic mail, there does not appear to be any way around this problem. The person who looks up a public key in a directory and encrypts a message with it is sending the message to the holder of the corresponding private key. Possession of the correct private key is the only thing that distinguishes the receiver, and anyone who has it will be able to read the letter.

For interactive communication, however, the problem can be solved by means of another form of public-key cryptography: *Diffie-Hellman key exchange*,[33] which allows communicating parties to produce a shared secret piece of information despite the fact that all messages they exchange can be intercepted (Diffie and Hellman 1976). Alice produces a secret piece of information that she never reveals to anyone, and derives a corresponding public piece from it. She sends the public piece to Bob and receives from him a piece of public information formed in the same way. By combining their own secret information with the other's public information, Alice and Bob each arrive at a common shared piece of information that they can use as a cryptographic key. The process is analogous to a perfect strategy for bridge bidding in which the North-

South partners (each of whom knows his own hand) agree on a secret but the East-West partners (who do not know either North or South's hand) remain completely ignorant of what North and South have concluded even though they have heard their opponents' every bid and response.

Alice and Bob use the key they have negotiated to encrypt all subsequent messages, but being engaged in encrypted communication is not a sufficient condition for secure communication. Not only does neither Alice nor Bob yet have secure knowledge of the other's identity; they cannot even be sure that an intruder who has performed a key exchange with each of them is not sitting in the middle, translating between keys and recording the plain text of their conversation.

The second step, therefore, is to exchange certificates: Alice sends Bob hers, and he sends her his. Verifying the KMF's signature on the received certificates allows the receiver to be sure that the certificate is authentic but does not guarantee the authenticity of the person who sent the certificate. It remains, therefore, for Alice (for example) to verify that the person with whom she is in contact is actually the legitimate holder of Bob's certificate—the person who knows the secret key corresponding to the public one it contains. This final step is done by a process called *challenge and response* in which the challenger sends a test message and judges the legitimacy of the responder by verifying the signature on the response. In this case, it is best to use as the challenge the piece of public information from the exponential key exchange, since so doing gives assurance that the encryption is actually being performed by the same entity that engaged in the authentication process.

Packet communication, on which the Internet is based, has made complex cryptographic protocols vastly more feasible. In a call-based telephone system, the only way of providing a common resource is to make it available to be called—as an 800 number, for example. This suffers from two problems. Typical use of a common resource in cryptography is brief: call, send and receive a few hundred to a few thousand bits (usually a key or certificate), and hang up. Such a call is inherently expensive because it uses the same circuit setup machinery as a call of much greater length.[34] Second, unless the phones have conference calling capability, they can only make one call at a time, so the protocol must be organized into a sequence of non-overlapping connections for each party.[35]

Packet switching functions much like the mail but at the speed of the telephone. A packet is a small collection of data: an address, a return address, a size field, the contents, and some flags. Typical packets vary from tens to thousands of bytes in length; tiny in networks operating at thousands to millions of bytes a second. Energetic correspondents might send each other hundreds or even thousands of pages in the course of a year; similarly, devices on the Internet are free to send arbitrarily large amounts of data, a little bit at a time in the "small" packets. The analogy to a postal correspondence is a connection. Like a postal correspondence, the connection uses the resources of the communication system only when a packet (letter) is in transit. At other times, it consumes nothing but a line in an address book.

The protocols that implement communication on the Internet are organized into layers. The lower layers deal with characteristics of the physical communication system and have the suggestive names *physical* and *link*. The topmost layer is called *application* and is inhabited, as its name suggests, by the computer programs doing the communicating. In the middle are two critical layers, whose names are almost as instructive, called *network* and *transport*.

The network layer is all important; it is what defines the network. For two nodes to be on the same network they must have exactly the same concept of addressing. It is the Internet Protocol (IP) that specifies how Internet addressing works.[36] The Internet Protocol provides unreliable delivery packet delivery; it does its best to get each packet to its destination in a timely fashion. If the time runs out or if a packet is lost to equipment failure, the packet is lost. The packet will generally be retransmitted but this is the responsibility of protocols in the layer above.

The transport layer is the layer that accommodates itself to the varying characteristics of Internet traffic. The most frequently used transport-layer protocols is called the Transmission Control Protocol and provides reliable communication—through error detection codes and retransmission—on top of the unreliable service provided by the layer below.

Packet-switched systems are free from both of circuit switching's problems. One-time transmission of small numbers of bits is inexpensive and a device connected to a packet-switched system can readily be in contact with many devices at the same time. This has given rise to a variety of packet-network security protocols, including SSL and the Internet Protocol Security Protocol (IPSec).

Because the network layer is shared across the network, it is the obvious place to install cryptographic protection. The operation of the network layer is not entirely compatible with cryptography, however. Each packet carried by IP is independent of every other packet, even between the source and destination. Encryption of traffic between two points typically uses the same key for substantial periods of time. A satisfactory compromise between these factors took some time to achieve. IPsec standardization began to take shape in the late 1990s.[37]

A more natural place to put security is in the transport layer because this layer already has the facilities for associating packets and keeping track of information flowing in streams. The best-known and most widely used security mechanism on the Internet is SSL. This facility is embodied in all browsers and implements https, the secure version of the hypertext transport protocol that delivers web pages.

Communication Security

The indispensable application of cryptography is the protection of communication. How this is accomplished depends on the form of the communication network and on whether the protection is being applied by the network's owners, by the subscribers, or by communication providers who supply the network with particular communication resources (such as satellite channels).

There are three basic ways in which encryption can be applied to network communication:

- *Net keying.* Every element in the network uses the same key, which is changed at regular intervals (often daily). This is the key-management technique employed by the US Navy's Fleet Broadcast System and many other government networks.

- *Link keying.* Messages are encrypted as they go from switching centers onto communication channels and decrypted as they go back into switching centers. A message is thus encrypted and decrypted several times as it goes from sender to receiver. This technique permits addressing information as well as message contents to be encrypted and makes the traffic particularly inscrutable to anyone outside the network.

- *End-to-end keying.* Each message is encrypted as it leaves the sender and cannot be decrypted by anyone other than the receiver. This is the typical behavior of secure telephones and secure email.

The results achieved by applying cryptography to secure communication networks depend dramatically on where it is applied. The most effective way for two individuals or two organizations to communicate securely over a network they do not control is to encrypt their communications using end-to-end keying. Relying on measures applied by network management may protect them from most opponents but will always leave them vulnerable to a foe with the resources to seek out a weak point along their communication path and exploit it. On the other hand, network keying and link keying allow the network to provide the users with security services they cannot provide for themselves.

Transmission Security and Covert Communication

Through a technique known as *traffic analysis*—the study of the patterns of communication—an opponent can learn a great deal about the activities of an organization without being able to understand any individual message. The counter to traffic analysis, *transmission security* or *communications cover*—which always amounts to sending dummy traffic to conceal the pattern of real traffic (Kent 1977)—is difficult and expensive to implement on an end-to-end basis and is best left to the network infrastructure.

The most extreme form of transmission security is to conceal the existence of communication altogether. This is often done by using *frequency-hopping radios*, whose frequencies change many times a second in an unpredictable pattern. When concealment of the existence of communication is the primary objective, we speak of *covert communication*. This is the dominant concern in the communications of criminals and spies.

Supporting Technologies

In order to be effective in network security, cryptography must not only be employed for the right purposes, it must also be implemented correctly. Several *supporting technologies* play important roles in this respect.

Reliability is critical in secure communication. Failures of either cryptographic equipment or its human operators, called *busts*, are typically the most lucrative sources of cryptanalytic successes (Welchman 1982). Automation has made a major contribution to the ease of use of cryptoequipment, making human errors less likely, but rising data rates have made reliability of the equipment vastly more significant. Performing a *security failure analysis* to trace the effects of all likely failures and including self-check and failure-monitoring circuitry in essential in designing cryptographic equipment.

Secure computing practices are integral to the construction of reliable communication-security (COMSEC) equipment whose functioning can be trusted under adverse conditions. A particularly difficult aspect of the logical analysis is the detection of malicious hidden functions, and no good solution to this problem is known for equipment acquired from untrusted suppliers.

Electromagnetic shielding is essential to prevent cryptographic channels from being bypassed by radiative or conductive emissions or by accidental modulation of a ciphertext signal with a plaintext signal. The military term for this particular form of protection is Tempest.[38]

The most interesting and difficult Tempest problem is the contamination of ciphertext by plaintext. In the United States, this problem first seems to have been observed in the late 1940s or the early 1950s in an online version of a one-time-tape system called SIGTOT (Martin 1980, pp. 74–75). What they observed was probably a form of amplitude modulation in which combining a 0 in the key with a 1 in the plaintext produces a waveform that is distinguishable from the waveform produced by combining a 1 in the key with a 0 in the plaintext, even though the results are supposed to be identical. If this occurs, opponents with the right equipment can read both the plaintext and the keystream from the ciphertext signal. The effect as described is unlikely to occur in modern digital cryptosystems, but amplitude modulation is only the simplest form of plaintext contamination. Frequency and timing modulation present subtler pitfalls for the designer.

Worried as people tend to be about surprises in mathematics undermining cryptosystems, in recent years implementation has proven the richest source of out-of-the-box thinking leading to surprising compromises of cryptographic equipment. No attack exemplifies this better than Paul Kocher's *differential power analysis*. Kocher showed that by measuring the the varying power demands of a microprocessor or a dedicated cryptographic chip. This problem can be overcome in "large" pieces of cryptographic equipment that contain batteries and power supplies and present a constant demand for recharging power to the outside world. Smart cards, however, have negligible power storage and depend on a constant supply of external power and countering attacks of this kind in this environment is extremely difficult. Kocher's work has subsequently been generalized and many cases have been found in which a measurable external symptom has been found to correlate with a cryptographically significant event in algorithm execution. Often it is just a question of whether the algorithm is processing a key bit that is 0 or a key bit that is 1. In such cases, the key can be read out directly, with disastrous results for security.

Tamper resistance guarantees that COMSEC equipment will not be altered to defeat its functioning without requiring the expensive practice of guarding it constantly or locking it in safes when not in use. Tamper resistance dramatically reduces the level of trust that must be placed in personnel permitted to operate COMSEC equipment and has become a mainstay of US military cryptography.

The Operation of Secure Communications

A secure communication system must not only be designed and built correctly; it must be operated correctly over its entire lifetime. The threat posed by opponents must be assessed, not once when a system is developed, but continually. The vulnerability must be judged against advancing technology. This should include continuing cryptanalytic study of the cryptosystems, other sorts of penetration studies, and continuing reevaluation of who the opponents are and what their resources are.

Equipment that worked securely when it was new may not continue to work securely. Ongoing assessment of the functioning of installed equipment must be accompanied by a careful program of maintenance.

A vital element of a secure communication posture that is operational rather than architectural is *communication security monitoring*. This is the practice of intercepting friendly communications to monitor the effectiveness of COMSEC practices. COMSEC monitoring is a difficult and sensitive task that is rarely undertaken by non-governmental organizations.

Cryptographic Needs of Business

Many large American firms have manufacturing plants around the world and need to communicate product-design information, marketing plans, bidding data, costs and prices of parts and services, strategic plans, and orders to them. Much of this information, which is often communicated electronically, must be kept secret from competitors.[39]

In the current banking system, the transfer of currency is the transfer of electronic bits, and it could not be undertaken without adequate security. Cryptography is simply the latest manifestation of the security upon which the banking industry has always relied. Internationalization of banking complicates the security problem while exacerbating the need for security.

Except to the extent that loss of confidentiality threatens security by exposing access controls, authentication is more important in banking than data confidentiality.[40] In part, this is because exposure of one person's data typically harms only that individual, whereas failures of authentication can result in loss of assets. In part, it is because there is already substantial government monitoring of financial transactions.[41]

Electronic funds transfer is already several decades old, but bankers now face the complex security issues posed by opening the transfer mechanisms to a wider audience. Indeed, despite stringent security procedures, Citicorp has already been the victim of such a scam, losing $400,000 in a theft that occurred over several months and involved three continents.[42] After this electronic theft became public, six of Citicorp's competitors went after its largest accounts, claiming they could provide better security than Citicorp (Carley 1995; Hansell 1995).

Although banking is a highly regulated industry, that does not mean that the government and banks see eye-to-eye on cryptography. In the 1980s, the National Security Agency proposed replacing the Data Encryption Standard with a classified algorithm. Bankers, having invested heavily in designing protocols based on DES, protested, and the NSA plan was dropped. Bankers also objected to the Clipper Chip.

In the oil industry seismic data and other geographic information can be worth a small fortune. Even before they came to rely on electronic communications, oil companies were targets of electronic eavesdropping.[43] Oil companies often operate in politically unstable regions, and criminal actions against employees, including kidnappings, are not unknown. Thus, in order to minimize knowledge about their whereabouts, employees' communications must be secured (Dam and Lin 1996, p. 464).

Other businesses face different requirements. In contrast to banking, the health-care industry emphasizes confidentiality. Medicine's heavily integrated systems of insurance records, hospital files, and doctors' records require a system that preserves patient confidentiality while allowing access by a large set of users (NRC 1997). The Health Insurance Portability and Accountability Act (HIPAA) has become one of the major drivers of privacy technology in the commercial world, and protection of personal data leans heavily on encryption.

Knowledge-based industries seek to protect their heavy investments in intangible goods, and cryptography will be central to that protection.

The entertainment industry has sought to protect digitized content through a combination of legislation[44] and technology. It has embraced cryptography as the way to make first-run movies and other expensive products available online while retaining close control that prevents the viewers from making copies or making any use of the products other than those the providers approve. The essence of this technology, called *Digital Rights Management* (DRM), is to keep information in encrypted form everywhere except inside tamper-resistant devices. In Hollywood's vision, a movie would travel encrypted, perhaps all they way from initial production at the studio until it reached the processor in the screen of your TV set.

A major initiative called *Trusted Platform Technology*, which seeks to install tamper-resistant security hardware into computers, particularly personal computers, laptops, and smaller devices, has been under way for nearly a decade. This technology will be of inestimable value in securing national critical infrastructure but threatens to diminish the degree of control users can exercise over their own machines.

Although digital digital signatures can provide a guarantee that movies, recordings, and other works of art have not been modified, a digital signature can be removed without trace. In addition to direct attempts to control the use of digital materials, content providers are also using cryptography to embed *watermarks* that permit individual copies to be identified and tracked. A watermark is a sort of tamper-resistant digital signature that makes the origin of digital information difficult to conceal.

Economic espionage is not solely the province of competing companies; it is also widely practiced by governments. One of the top priorities of the French intelligence service is industrial spying. Pierre Marion, a former director of the Direction Générale de la Sécurité Extérieure, the French Intelligence Service, told NBC News that he had initiated an espionage program against US businessmen to keep France economically competitive.[45] Another French spy, Henri de Marenches, the director of the French secret service from 1970 to 1981, observed that economic spying was very profitable: "It enables the Intelligence Services to discover a process used in another country, which might have taken years and possibly millions of francs to invent or perfect." (Schweizer 1993, p. 13) The British wiretap law includes protecting the economic well-being of the country as one of the reasons to permit wiretapping.[46] The Japanese invest heavily in industrial espionage, sending hundreds of businessmen abroad to discover in detail what the competition is doing, hiring many foreign consultants to further the contacts, and electronically eavesdropping on foreign businessmen in Japan (ibid., pp. 74–82 and 84). In China, the office of a multinational company experienced a theft in which unencrypted computer files were copied (ibid., p. 471).

In many countries, telecommunications are run by the national government. This makes state electronic eavesdropping particularly simple. The governments of Japan and France are notorious for eavesdropping on the communications of US businessmen (Schweizer 1993, pp. 16 and 84). It is alleged that the Bundesnachrichtendienst, the German intelligence service, regularly wiretaps transatlantic business communications (ibid., p. 17).

Similar charges have been raised against the US government. There was suspicion in Britain for some time that the NSA establishment at Menwith Hill was being used as much to spy on the British as on the Eastern Europeans (Bamford 1982, p. 332). These suspicions became more general in 2000 with the exposure of a US intelligence network called *ECHELON* that was largely devoted to monitoring commercial communication channels around the world. (Campbell 1999) From one viewpoint this was a natural outgrowth of the growing use of commercial rather than purpose-built communication systems by military organizations around the world, but this did little to soothe the worries of those who felt they were being spied on.

In 1990 FBI Director William Sessions said that the FBI would devote greater investigative energies to the intelligence activities of "friendly" nations (Schweizer 1993, p. 4). Of course, greater protection of the goods to begin with would decrease the need to investigate thefts after they had occurred—a point that is surely not lost on the FBI.

In another sort of case, an American businessman working for a multi-national firm reported that his laptop computer was taken by customs officials of an unnamed country and returned to him three days later; as he attempted to negotiate his business deals, it became clear that his sensitive files had been copied during those three days (Dam and Lin 1996, p. 471). With the widespread use of laptops and other portable devices (AP 2006) by large numbers of workers not necessarily versed in information security, the loss of such devices and the data they contain has become an almost daily news item. The potential compromise of the sensitive data they contain has led to a federal government requirement that all data on laptops used by federal agencies be encrypted unless the data are determined to be nonsensitive by a designee of the Deputy Secretary of the agency (Johnson 2006).[47]

It has been claimed that a system like public key, in which the private key is stored with a single user, will not provide the data-recovery features that corporations require. In fact, the danger posed to corporations by this lack of data recovery for communications is minimal. With the exception of the financial industry, few businesses currently record telephone communications. Those that do can continue to do so even if the telecommunications are encrypted. Because the ends of the conversation will be unencrypted, and the recording can be done at that point, the choice of encryption—whether public key or some form of escrow— will not affect data recovery. Encrypted communications will provide corporations with at least as much security as they have now; they will not be losing information, for *they do not have such information in the first place.*

Because they rely on "written" records rather than recordings of conversations, businesses are in the same position as law-enforcement agencies. Conversations are transient whereas records endure, and wiretaps are used in far fewer criminal cases than seized records (Dam and Lin 1996, p. 84).

In the mid 1990s, talk of a global information infrastructure, an information superhighway, and Internet commerce was everywhere. The sense was that we were moving our culture into digital channels of communication. A decade later, the Internet has indeed transformed the life of the industrialized world. Email is replacing mail; websites are replacing catalogs and advertising; weblogs and search engines are replacing newspapers. Governments, corporations, and terrorist organizations make contact with their constituencies over the World Wide Web.

In the first edition of this book, we said of the move to the digital world:

> If this is to be a success, we will to have to find digital replacements for a lot of everyday physical practices. In the area of security, many of the new practices will be cryptographic.

In some cases, the correspondences are obvious. In the physical world, we close doors or stroll off somewhere by ourselves or whisper in order to have privacy. In the digital world, we encrypt. In other cases, the correspondence is not so obvious. In the physical world, we place written signatures on contracts, letters and checks. In the digital world, we add digital signatures and rejoice in the degree of similarity. Some of the correspondences are still unresolved and are controversial. A copying machine may reproduce a readable copy of a bestseller, but for most purposes the copy is a poor imitation of the original. In the digital world, if you can get your hands on a document you can copy it exactly. A system that prevents the unauthorized copying of a digital novel, however, is capable of preventing the reader from making many legitimate uses of it. Many of these issue remain unresolved.

So where do we stand? On one hand, the Secure Socket Layer protocol that underlies secure browsing (https) is perhaps the most widely deployed cryptographic system in the world. On the other, the news abounds with stories of spam, breakins, phishing, and identity theft. It is reasonable to conclude that not only will a lot of everyday security practices have to find digital replacements but a lot of new ones will have to be developed.

Digital equivalents have been found for a surprising range of human interactions, including:

- The delivery of a registered letter. (You get the letter if and only if the deliverer gets a signature on the receipt.)

- The signing of a contract. (The contract is valid only if both signatures are on both copies and neither party can get a final copy while denying one to the other.)

- Sharing of authority. (The president of the company can sign a million-dollar check, but it takes two vice presidents.)

- Power of attorney. (My lawyers have access to the contents of my safe deposit box because they are working for me.)

- Issuing of credentials. (A passport certifies that the secretary of state vouches for the bearer.)

Why Has Cryptography Taken So Long to Become a Business Success?

If cryptography is so valuable, why has it taken so long to become a business success? One answer lies in the truly remarkable qualities of communications intelligence. Unlike installing an infiltrator, breaking and entering, or going around asking questions, interception of communications is very hard to detect. People may suspect that their communications are being spied on, but they are rarely able to prove it or to convince themselves with certainty that that is happening. Even an intelligence agency typically has difficulty discovering that it is being spied on and how. During the 1960s and the 1970s, the British and American intelligence establishments were convinced that they had been penetrated by the Soviets and spent an excessive amount of their time trying to discover who the turncoats were (Wright 1987). Despite their efforts, the investigations were inconclusive.

For an organization that is not an intelligence agency, such *counterintelligence* is much harder. Merely discovering that information has been leaking is very difficult. Discovering how or to whom is far more difficult. It may even require an intelligence agency to uncover the fact of the eavesdropping; this has happened several times for American companies.[48] Selling secure communications has often been likened to selling insurance, in that the customer must pay up front for protection against something that may occur in the future. The fire, auto, and medical insurance salesmen have an advantage, however. Everyone has seen houses burn down, cars crash, and friends get seriously ill. Almost no one has seen information taken by eavesdroppers and used to bankrupt an otherwise profitable business. Indeed, admitting to break-ins has its own costs: competitors may use the seeming lack of security to woo customers.

The Problem of Standards

Cryptography also suffers from a serious standards problem. In a sense, security equipment exists to amplify minor incompatibilities into absolute non-interoperability. If a radio is not tuned to quite the right frequency, reception will suffer but communication may still be possible. If a cryptographic key is off by even one bit, communication is impossible. By themselves standards problems probably would not account for cryptography's slow start, but in combination with other factors they have played a major role. In particular, the lack of standards has contributed directly to the lack of "critical mass." A single secure telephone, just like a single telephone, is a useless piece of hardware, and even a pair has only limited applicability. Only when there is a proliferation of such devices does their value increase to the point that purchasing a security device for one's telephone is as common as purchasing a lock when buying a bicycle.

Similarly, the lack of a supporting infrastructure has slowed the adoption of secure communications. Without world-wide keying infrastructure and key-management facilities, the provision of keys is a remarkably cumbersome process, and encrypted communications are tend to be used only by thosec committed to security.

The Adverse Effect of Government Opposition

In the United States, cryptography was long hurt by the ambivalent attitude of the federal government. Despite the controversy that surrounded the adoption of the Data Encryption Standard in 1977, this standard and the others that accompanied it have made a major contribution to the deployment of cryptography both in the US and abroad. This latter aspect made the government wonder whether it had done the right thing, and in the 1980s it became an antagonist rather than a proponent of cryptographic standards.

The delays caused by the 1990s crypto wars are part of the reason for the poor state of laptop security. Now the government finds itself in the peculiar situation of insisting on ubiquitous security—and cryptography —on consumer laptops, at least those purchased by the federal government.

The Effect of the Internet

Since 1995, the market for cryptography has exploded. What has changed? Most conspicuously, the Internet.

The Internet has made global electronic commerce a day-to-day reality for a large number of people.[49] And commerce, on a large scale, can prosper only when people can deal confidently with people they have never met and have no reason to trust. The problem is made worse by the Internet's internationality. No uniform system of law or policing can patrol it. The merchants, like the cannon-carrying merchant ships of two centuries ago, must provide security themselves. The more secure people can be in their transactions, the larger those transactions will be and the more profitable the Internet will be as a business medium.

At present, the security of Internet commerce must be judged as fair to middling. The tension between strong identification and anonymity has yet to be resolved and a uniform public-key infrastructure has yet to be built. Nonetheless, the Internet now accounts for a significant fraction of most businesses' business. It has created new businesses—eBay's worldwide auctioning of low-value items, Google—and transformed others. Some used book stores have been driven out of business while others have thriven by putting their wares online.

The last decade saw substantial societal change in the usage of mobile communications technologies: cell phones, PDAs, etc. Communication patterns changed, not just among the young with cellphones and SMS, but also among the traveling public, and, at least in certain societies, an expectation has arisen that communication devices are always on and people are always available.

As bandwidths rise and computing technology advances, telecommunications increasingly have become preferred modes of communication rather than merely increasingly satisfactory substitutes for pre-electronic modes.

There has been a simultaneous explosion of online communications using the Internet. The traditional communication tools—if anything less than a century old can actually be called traditional—are email, which in less than a decade has become ubiquitous,[50] instant messaging, and voice calls using Internet technology (VoIP, or Voice over IP). There are some quite non-traditional communication methods as well, such as music-jamming sessions over the Internet and the very popular *massively multiplayer on-line role playing games (MMORPGs)*.

Although MMORPGs are called games, they are a recreation with as much in common with hanging out in the mall as with the videogames you might find there. Some MMORPGs are more focused on the prize (or on reaching "levels") than others, but MMORPGs differ from console games exactly because they enable interaction with other players. (Until the Internet was opened to commercial use, MMORPGs were limited to online services such as AOL and CompuServe, and all players in a particular game had to subscribe to that service. Once that changed, MMORPGs began to appear on the public Internet.) MMORPGs provide an online environment in which players, through their avatars (icons representing the user), slay dragons, compete for prizes, communicate, and buy, trade, and sell fictional objects of value. The fictional objects, moreover, are no longer entirely fictional. Often they are bought and sold on eBay and in other online venues.

MMORPGs are communities and yet another venue for electronic communications, one far from traditional, but widely used, especially in Asia. In the United States, roughly 19 percent of the gaming population are online gamers, but in the Asia/Pacific region, consisting of Hong Kong, Korea, Malaysia, the People's Republic of China, Singapore, and Taiwan, computer games (which means online games, as console games are virtually non-existent in the region) are extremely popular. There are two reasons driving the interest in Asia: in China, Internet cafes are open around the clock, and Korea is experiencing rapid broadband adoption (IDC 2006).

What Is Cryptography's Commercial Future?

In light of the sudden growth of cryptography, it is natural to ask how big a business it will become. In order to find the answer, we might look at an existing business.

Locksmithing is not a large fraction of the building trades. If you own a $500,000 house you are unlikely to have $500 worth of locks on it. Even if, like Oliver North, you have a fancy electronic security system, its total value probably doesn't exceed a few percent of the house's value. On the other hand, no one would argue that locks are not essential to the functioning of society. The modern suburb, in which most members of most families are gone for most of the day, would not function without them.

A harbinger in the cryptographic world is Skype, a Voice over IP communication system that encrypts all of its transmissions over the Internet automatically. Security isn't the point of Skype, as it is with Zfone[51]; it is just part of the package. Skype was built by a group of Latvians, now in their mid thirties, who grew up under Soviet rule and who take it for granted that someone is trying to listen in on their phone calls. One of the most popular Voice over IP systems, Skype, encrypts the voice traffic in every call without any explicit action by the user.

Cryptography seems to be most successful where it is following a similar course, not a prominent part of any product but ever present and essential. The Secure Socket Layer protocol in Web browsers is a candidate for being the most widely deployed cryptographic protocol of all time. It could be overtaken by the automatic encryption in Skype. Skype is also an example of the internationalization of cryptotechnology: the product originated in Latvia but is now owned by eBay. It may also be the focus of the US government's latest policy efforts to capture communications traffic, an issue we discuss in chapter 11.

3

Cryptography and Public Policy

The Legacy of World War I

Histories of cryptography usually begin by observing that cryptography is of ancient lineage, having been used by Caesar and recommended as one of the feminine arts in the Kama Sutra. As we saw in the last chapter, there is a remarkable degree of continuity in the basics of cryptographic technology over the past several centuries, and new ideas often turn out to have old roots.[1] Be this as it may, the field we know today is a creature of the twentieth century.

World War I was the first war to be fought in the era of radio. In the early years of the century, the military (particularly navies) saw the potential of the new medium of communication and adopted it wholesale. Before the advent of radio, Britain's First Sea Lord sent an admiral off with a fleet and expected to hear news in weeks or months. With radio, ships at sea were brought under much closer control by shore-based headquarters. Radio also had the advantage of being able to carry human voice, which undersea cable systems of the time could not do. Radio, however, had a serious disadvantage: anyone could listen in. Sometimes an unintended listener even got better reception than the intended recipient.

The Mechanization of Cryptography

The solution to the ubiquity of radio reception was, of course, to use cryptography. Messages were enciphered before being transmitted, and were deciphered upon reception. Radio was much more vulnerable to interception than any communication channel that had been in use before, with the result that a much larger fraction of the messages required encryption. The methods in use, codes and some hand cipher systems, had grown up over centuries of written communication in which they were usually reserved for the most sensitive traffic. In consequence, there was a crushing burden on the cipher clerks, and often a glut of traffic waiting to be deciphered.

In the years immediately after the war, inventors turned their attention to the problem of mechanizing encryption and embarked on the path that has led us to the automatic cryptography of today. As described in the previous chapter, the technique was a form of multiple-Vigenère cipher in which letters were looked up in tables by electric currents passing through wires rather than by cipher clerks.

Rotor machines arose simultaneously on both sides of the Atlantic. The inventions of Edward Hebern of Oakland, California emerged as the backbone of US cryptography in World War II. Those of Arthur Scherbius and Arvid Gerhard Damm played a similar role in Europe and gave rise to the most widely used of all rotor principles: that of the German Enigma machine (Kahn 1967, pp. 420–422).

At the time of World War I, cryptography was more an esoteric than a secret field. It was not widely understood, and codebreakers kept their intrusions into opponents' codes secret. Yet the whole culture of military secrecy, rampant today, was in its infancy, and military secrets did not belong to governments in the way they do now. Works on cryptography reasonably representative of the state of the art were published and enjoyed a status similar to that of other technical treatises. Indeed, during the war, William Frederick Friedman, the intellectual founder of the organization that eventually grew to become the National Security Agency, published groundbreaking new cryptographic discoveries as technical reports of the private Riverbank Laboratories.

The inventors of the rotor machine all took out patents, intending to sell their machines commercially. In the United States and in Europe the granting of a patent is a very public process and the publication of patents is regulated by treaty. Both earlier and later laws made it possible for the US government to declare a patent application secret and delay the granting indefinitely, but in 1919 this was not the case.

The issuing of the patents turned out to be one of the last public things about rotor machines. The commercial market for cryptographic equipment proved not to be as good as the inventors expected, and Hebern

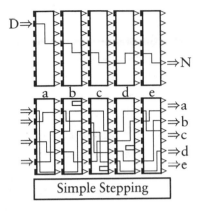

Figure 3.1
Simplified Sigaba.

tried to sell his invention to the government. As designed, Hebern's machine was not secure, and it was broken by Friedman in 1921 (Deavours and Kruh 1985, pp. 46–47). Although attacks on the machine Friedman broke can be found in the contemporary public literature, Friedman's 1925 paper on the subject is still secret.

The government bought only sample quantities of Hebern's device. The weaknesses Friedman discovered were correctable, but the Army did not share its design changes with Hebern. It developed and employed his invention without compensating him, under cover of cryptography's growing secrecy. Only in the 1950s did Hebern's heirs succeed, through a lawsuit, in recovering some of what Hebern was owed (Kahn 1967, p. 420). In the United States, development of cryptographic systems became a secret enterprise of small groups in the Army and the Navy.

Hebern's simple three-wheel machine and regular pattern of wheel movement led to Sigaba, the most secure rotor machine of World War II. In the mid 1930s Friedman designed a rotor machine, designated M-134a, whose five wheels moved under the control of a paper tape. In retrospect this can be seen to be merely an overly complicated way of achieving a one-time cryptosystem, but at the time the additional complexity must have seemed an advantage. Unfortunately, the M-134a had the same problem that goes with any one-time system: it needed lots of keying material. A young cryptanalyst named Frank Rowlett was given the task of producing matching pairs of key tapes—a task made tedious by the fact that any error in the tapes would produce not only an error in the plain text but also a disastrous loss of synchronization between the sending and receiving machines. Rowlett conceived of using one rotor machine to manufacture the pattern of rotor motions for the other, an idea that was to remain officially secret for 60 years.

The Growth of US Signals Intelligence

After World War I, the United States maintained and developed its capacity for signals intelligence (SIGINT), something unprecedented in peacetime. Equally important, the responsibility for signals intelligence was merged with the responsibility for developing codes to protect US military communications.[2] Both intelligence and security prospered under this arrangement. By making good use of signals intelligence during arms negotiations with Japan in 1921, the US pushed the Japanese all the way to their planned best offer (Kahn 1967, pp. 357–359; Yardley 1931).

The role secrecy plays in intelligence is quite different from the role it plays in security. Cryptography relies on the secrecy of cryptographic keys, just as safes rely on the secrecy of their combinations. It may also profit from the secrecy of cryptographic systems. It could hardly be easier to break a cryptosystem whose workings are not known than to break one whose workings are known. On the other hand, such secrecy has costs that usually outweigh its benefits. The workings of ordinary pin-tumbler door locks and those of the combination locks that guard bank vaults are familiar to locksmiths and burglars alike, yet both are widespread and successful security devices. It is essentially impossible, on the other hand, to practice intelligence without secrecy. If you reveal the fact that you are spying on people or, worse yet, precisely how you are doing it, it is almost certain that, if those people care at all about security, they will change their communication practices in order to make your task more difficult.

A celebrated example of this occurred in Britain in 1927. The Security Service (MI5) raided the London office of the All-Russian Cooperative Society, whose trade activities were a key element of Soviet spying on British industry. Parliament imprudently demanded to know what had led MI5 to such an action and publicly extracted the fact that Britain had been reading Soviet messages for years. In the words of R. V. Jones,

Britain's head of scientific intelligence during World War II, that incident showed the Soviets the weakness of "their diplomatic codes, which were changed a fortnight later and were not since readable, at least up to the time that I left MI6 in 1946" (Corson et al. 1985, pp. 286 and 440).

The integration of communications intelligence (a field in which it is advantageous to keep even the basic techniques secret) and communication security (a field whose need for secrecy is far narrower) is the apparent cause of the general increase in the secrecy of cryptography that occurred from the 1920s through the 1960s.

World War II

World War II was a triumph for America's codemakers and codebreakers. On the defensive side, American high-grade cryptosystems, Sigaba in particular, survived the war, apparently unread by our opponents. On the offensive side, communications intelligence contributed decisively to victory in both the Atlantic and the Pacific. The United States routinely read high-level Japanese traffic in both military and diplomatic systems, and collaborated extensively with Great Britain in reading German communications.

Cryptography and signals intelligence during World War II have been written about extensively, but a little-known project undertaken during the same period presaged much of what has happened since: the development of the first secure telephone.

Voice scramblers of various sorts had been in existence since the early years of the twentieth century, but none of these systems was in any real sense secure. Scrambling systems were adequate to thwart casual listeners; however, most of them were vulnerable to being unraveled by talented ears—particularly after the development of recording, which permitted the signal to be listened to more than once. The existing scramblers were inadequate to protect high-level traffic, and the US military started development on Sigsaly, the first digital secure telephone.

Sigsaly was in some ways like and in some ways preposterously unlike a contemporary secure phone. Sigsaly was a digital telephone. It used a vocoder developed at Bell Laboratories to convert the speaker's voice into a 2400-bit-per-second stream.[3] The digitized voice was then encrypted with one-time keys stored on phonograph records. The records were shattered after each use to ensure that they could not be reused. Although modern secure phones do not normally use one-time keys, there is nothing bizarre about the practice. Especially remarkable were Sigsaly's size (30 equipment racks, each 7 feet tall) and cost (so high that only two people in the world could afford such installations: Franklin Roosevelt and Winston Churchill—later a few more were installed in high-level military headquarters).

The successes of the American signals intelligence establishment during the war set the stage for further expansion and consolidation.

The Cold War

The late 1940s brought the explosion in technology that has dominated the lives of the Baby Boom generation. Some technologies, most conspicuously atomic energy and radar, were developed as consequences of World War II. Others, including television, had been put on hold by the war and now profited from the overall growth of the electronics industry that the war had produced. In the world of cryptography, it was a great period of conversion from mechanical and electro-mechanical cryptosystems to purely electronic cryptosystems.

Inappropriate as Sigsaly was for any normal military or civilian communication requirement, it showed the possibility of genuinely secure voice communication. However, there were many hurdles to jump. The highest of these was the problem of digitizing the voice. Decades would pass before secure voice equipment at the low data rates of Sigsaly would be widely available. In the meantime, it was necessary to simplify the problem by sending voice at between 6000 and 50,000 bits per second. This created a new cryptographic problem.

Sigsaly had achieved 2400-bit-per-second encryption by using a one-time system to protect its signal. Like all one-time systems, it suffered from inflexibility of key management. This was of little concern when the supply of instruments numbered two, but it would be a limiting factor in wider deployment. What was required was a cryptographic machine capable of operating at voice speeds. Unfortunately, rotor machines, which

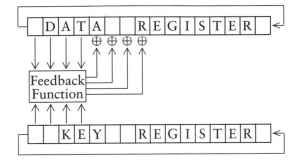

Figure 3.2
Nonlinear shift register.

operated at 10 characters per second, could not come close to keeping up with Sigsaly's 2400 bits (240 characters) per second.[4] The result was the development of shift-register cryptography, which accounts for most of the encrypted bits transmitted today.

During and before World War II, intelligence in general and communications intelligence in particular was attached to military commands. After the war, there was a move to improve the quality of US intelligence by centralizing authority and coordinating the activities of various groups. The year 1947 saw passage of the Central Intelligence Act, which created the Central Intelligence Agency and the position of Director of Central Intelligence (who, at least in principle, coordinated the efforts of the whole US intelligence community). The centralization of signals intelligence took longer and was far less public, but in 1952, after some 5 years of study and reorganization, President Harry Truman signed a secret presidential order creating the National Security Agency. From its creation the new organization sought to capture control of all cryptographic and cryptanalytic work within the United States. Overall, this effort was remarkably successful.

In 1949, Claude Shannon, a professor of electrical engineering at MIT, published a paper entitled "The Communication Theory of Secrecy Systems" in the *Bell System Technical Journal*. In 1967, David Kahn, a journalist who had been interested in cryptography since childhood, published a massive history of cryptography called *The Codebreakers*. In between, no overtly cryptographic work of any significance saw public print in the United States,[5] though various papers whose cryptographic significance is not immediately apparent did.[6]

NSA successfully suppressed cryptographic work in other parts of the US government. Probably as a result of this, application of cryptography to the nuclear command and control system did not get underway until directly ordered by President John Kennedy in the early 1960s.[7]

Horst Feistel's Group at the AFCRC

There was, however, one very significant failure of NSA's territorial ambitions.

Cryptography can play a number of roles that are not explicitly parts of communication. One of these is distinguishing between friends and enemies. During World War II, US warplanes began to carry *Identification Friend or Foe* (IFF) devices to reduce the chance that they would be shot down by "friendly fire." Early IFF devices were analog and were not very secure against replication by opponents seeking to impersonate US forces. IFF without cryptography eventually evolved into a system called the Mark X, which is now an essential component of air traffic control and is used by civilian aircraft, and many military ones, all over the world.

In the early 1950s, the US Air Force recognized the need to improve its IFF equipment and turned to the established cryptographic authorities for help. The Armed Forces Security Agency, NSA's immediate predecessor, had little interest in the problem, and so the Air Force set out on its own. A prototype cryptographic IFF device was built at the Air Force Cambridge Research Center (AFCRC) using the recently developed transistor. Central among this project's objectives was to show that the equipment could be made small enough to fit in a fighter's nose. At this point, the project crossed the path of a man whose role in modern cryptography has been widely underappreciated, despite the fact that he is recognized as the "father of the Data Encryption Standard."

In 1934, the 20-year-old Horst Feistel moved to the United States from Germany. Seven years later, the Japanese attacked Pearl Harbor, the US declared war on Japan, and Germany, in an act of bravado, declared war on the US. Feistel, on the verge of becoming a US citizen, was put under "house arrest"—he could move freely around Boston, where he lived, but was required to report his movements if he went to visit his mother in New York. On January 31, 1944, the restraints were suddenly lifted and Feistel became a US citizen. The following day he was given a security clearance and began a job at the AFCRC.[8]

Feistel, who says that cryptography had interested him since he was in his teens, recalls that when he mentioned his interests shortly after arriving at his new job, he was told that it was not the time for a German to be talking about cryptography. His career in cryptography had to wait until after the hot war against Germany was over and the Cold War against the Soviet Union was underway.

Several years later, Feistel, who had by now built a research group in cryptography at the AFCRC, discovered that the Air Force's cryptographic IFF system was on its way to being put into service without what he considered adequate evaluation. He put his team of young mathematicians, supported by a number of academic consultants, to work analyzing the new system. The group alternately found weaknesses in the original design and discovered how to fix them. Over a period of several years it made a major contribution to modern cryptography, developing the first practical block ciphers.[9]

Although Feistel's group at the AFCRC was in steady communication with NSA and seems thereby to have exerted a profound influence on cryptographic design in that organization, NSA appears eventually to have succeeded in shutting down the Air Force work. In the late 1950s, the group dissolved. Horst Feistel moved, first to MIT's Lincoln Laboratory and then to its spinoff, the Mitre Corporation. In the mid 1960s, Feistel, who devoted himself to one problem throughout his career, attempted to set up a new cryptographic development group at Mitre. He was forced to abandon the project as a result of what was perceived at Mitre as NSA pressure.[10] At this point, Feistel found a new champion, IBM, and moved to the Watson Laboratory in Yorktown Heights, New York.

The 1960s

In 1965, Representative Jack Brooks of Texas authored a thoroughly non-controversial law providing for the purchase and leasing of computer equipment (Brooks Act 1965, 89-306(f)). The law decreed that the Secretary of Commerce was authorized to "make appropriate recommendations to the President relating to the establishment of uniform Federal automatic data processing standards" (ibid., 89-306(f)). Thus the responsibility for setting civilian computer standards fell to the National Bureau of Standards (NBS), which already determined everything from standard US time (based on an atomic clock in Boulder, Colorado) to methods for testing the flammability of fabrics. The new series were called Federal Information Processing Standards; their purpose is to provide incentives for industry to manufacture what the US government needs. Although FIPS are only binding on government organizations, the enormous purchasing power the government brings to bear causes most FIPS to become de facto commercial standards.[11] This assignment of responsibility to NBS was later to prove difficult for the Department of Defense, but there were no objections when the Brooks Act was passed.

In 1967, a non-technical phenomenon had a profound effect on the course of cryptographic technology. Despite NSA's efforts, David Kahn's book *The Codebreakers* was published. Kahn was neither a mathematician nor an engineer, but a historian, and his book focuses far more on the military and diplomatic impact of codebreaking than on its technique. Nonetheless, Kahn explained everything he knew about cryptography and cryptanalysis. All of a sudden, there was a book in print that explained rotor machines and at least mentioned that they had been succeeded by purely electronic devices. Furthermore, what explanation of the technology the book did contain was wrapped in an extensive explanation of why the subject was important. The result was a wave of interest in cryptography that rippled through the engineering world.

IBM was also interested in cryptography in the late 1960s. The company had undertaken to provide automatic teller machines for Lloyds Bank in London. Lloyds, which planned to have several hundred of the machines scattered around London, had a scenario for disaster. It imagined a gang with a confederate inside the post office, which, in Britain, ran the telephone system. The members of the gang would drive around London visiting the teller machines at the start of a bank holiday weekend, when the machines had just been stocked with thousands of pounds of cash apiece. The post office confederate would call each machine in turn and tell it to hand out all its money. At the end of the evening the gang would be several million pounds richer. Nobody else would be likely to notice anything but the fact that some of the machines had run out of money till Tuesday morning, when the bank holiday weekend ended. By then the money would no doubt be sitting comfortably in a bank in Zurich, a city whose bankers had not taken the previous day off.

IBM hired Horst Feistel, who continued with his life's work. Colleagues recall that Feistel had been talking back in the 1950s about much larger systems than could be built at the time. Now, a quarter century later, electronics had caught up with Feistel's imagination. The work resulted in a number of cryptographic systems. One was used in the IBM2984 banking system and was commonly called Lucifer.[12]

The 1970s

The US Data Encryption Standard

In the early 1970s, some people at the National Security Agency and some at the National Bureau of Standards[13] recognized that the Privacy Act of 1974 and other federal laws, together with the increasing use of computers and digital communications by the federal government, would require that approved cryptography be available to government users other than NSA's traditional national-security clients.

NSA was reluctant to provide equipment of its own design to a wider range of users. Its reasoning is easy to imagine. Cryptographic secrecy was part of NSA's security philosophy. The designs of NSA cryptographic systems were uniformly classified SECRET NOFORN,[14] and the equipment in which they were implemented was classified CONFIDENTIAL. The tamper-resistance technology being used today to put SECRET algorithms in UNCLASSIFIED chips did not yet exist. Any equipment that was to be put in the hands of uncleared users would have to embody an unclassified cryptographic algorithm. NSA was afraid, however, that, by making public an algorithm it had designed itself, it would reveal information about its design philosophy and potentially compromise other equipment. A system that was made public would be widely available to attack. If it was broken, and if it resembled other NSA cryptosystems, the attack might work against those as well.

Following the FIPS process, NBS published a solicitation for a standard cryptographic system in the *Federal Register* (USDoC 1973). The process by which the Data Encryption Standard was selected has never been adequately explained, and the identity of the other contestants, if there were any, has not become public. On the face of it, IBM submitted the winning algorithm and NSA, acting as a consultant to the NBS, approved it. In fact it is generally agreed that NSA had a substantial hand in determining the algorithm's final form (Bayh 1978).

Public-Key Cryptography

During the academic year 1974–1975, Whitfield Diffie and Martin Hellman, working at Stanford University, and Ralph Merkle, at the University of California at Berkeley, discovered a new concept in cryptography that was to have profound implications for policy as well as for technology.[15]

The idea was that the capacity to encrypt messages and the capacity to decrypt messages were not, as had always been taken for granted, inseparable. It was possible for one person to have a key that would encrypt messages in such a way that decrypting them required a different key, held by a different person. Diffie, Hellman, and Merkle called the new concept *public-key cryptography*.

Instead of having one key that could both encrypt and decrypt messages, like a conventional cryptosystem, a public-key system had two keys. What was essential was that with one key, called reasonably enough the *public key*, it was very hard to figure out the corresponding *private key* that could be used to decrypt a message encrypted with the public key. Exactly how hard? Just as hard as breaking a cryptogram produced by encrypting something with the public key.

Public-key cryptography has two implications: if I know your public key and want to send you a secret message, I can encrypt it with your public key, and, since only you know your private key, only you will be able to read the message. I don't need to share a secret with you in order to send you a secret message.

Not only did public-key cryptography make a major contribution to key management, it provided a *digital signature* that could be used on electronic documents. Suppose I receive a message from you encrypted with your private key. If I decrypt the message and find that it makes sense, I have evidence that, unless someone else knows your private key, the message must have come from you.

Although the research that led to this new field was carried on in academia, it grew out of two very practical considerations. Diffie had been trying for years to figure out how to build a system that would secure every telephone in the country. Getting away from the traditional need to distribute secret keys from a central facility brought this vision a step closer to reality. By what seems to be coincidence, Diffie had also been thinking about signatures for years. Around the time of Diffie's arrival at Stanford in late 1969, his new boss, John McCarthy, had given a paper on what we would now call Internet commerce, and Diffie had begun wondering what would play the role of the signed contract in the new environment.

In 1976, Diffie and Hellman published a paper entitled "New Directions in Cryptography," which contained a partial solution to the problem of creating a public-key cryptosystem. This system, commonly known as *Diffie-Hellman key exchange*,[16] provides an additional feature with policy implications that took years to be recognized: the creation of truly ephemeral keys.

In early 1977, Diffie and Hellman's paper was read by a team of three young faculty members at MIT, Ron Rivest, Adi Shamir, and Len Adleman. With a far better grasp of number theory than the three West Coast researchers, Rivest, Shamir, and Adleman quickly found a solution to the problem Diffie and Hellman had posed. They named the resulting system after their initials: RSA (Rivest et al. 1978).

The Meyer Affair

Public-key cryptography created quite a splash. In August of 1977 it was described in Martin Gardner's column in *Scientific American*, and requests for the MIT technical report came from around the world. Rivest was scheduled to present the work at an Institute of Electrical and Electronics Engineers (IEEE) meeting in Ithaca, New York, in October. The

IEEE received a letter from one "J. A. Meyer" warning that, because foreign nationals would be present, publication of the result was a potential violation of the International Traffic in Arms Regulations.

This was the first that Rivest and his colleagues had heard of ITAR. Aside from an address in Bethesda, Maryland, J. A. Meyer was unidentified. Nonetheless, the MIT scientists took the warning seriously and halted distribution of their paper. Then an enterprising *Science* journalist, Deborah Shapley, discovered that Meyer worked at the National Security Agency. NSA denied any connection with Meyer's letter, and the MIT scientists decided to resume distribution of the paper (Shapley and Kolata 1977). Rivest spoke on the results at the Ithaca conference, and for the moment the furor subsided.

But 1978 brought new problems. Two inventors filed for patents and found themselves subjected to secrecy orders under a little-known provision of US patent law that permits the federal government to order an inventor to keep secret not only the substance of the invention but the fact that the order has been issued. The first of these inventors was Carl Nicolai, a garage-shop entrepreneur in Seattle who had developed a telephone scrambler based on spread-spectrum technology. The second was George Davida, a professor of computer science at the University of Wisconsin. Davida's invention, more technical than Nicolai's, was a way of combining linear and nonlinear finite automata in a cryptosystem. Most patent secrecy orders are placed on inventions growing out of government-sponsored secret work. Almost all of the remainder are directed at patents filed by large corporations,[17] which have little motivation to argue with the government's decision.[18]

Fights about secrecy orders are rare. But, despite the aspect of the law that makes the order itself secret, both Nicolai and Davida chose to fight. Ultimately, both orders were overturned—Nicolai's ostensibly on the grounds that a mistake had been made and the invention should not have been declared secret in the first place; Davida's on the pretext that, since it had earlier appeared as a Computer Science Department technical report, it could not effectively be kept secret.

Research Funding and Publication Rights

In the late 1970s action developed on another front: funding. Frederick Weingarten was a program officer at the National Science Foundation (NSF). One day in 1977 "two very grim men" walked into his office and informed him that he was "probably" breaking the law by funding cryptography research through the NSF (Weingarten 1997; Burnham 1980, pp. 139–140). He was not, but a new battle ensued.

Len Adleman submitted a research proposal to the NSF, whereupon the MIT scientist found himself in the midst of an inter-agency conflict regarding funding. Because Adleman had proposed research in cryptography, the NSF had sent the application to NSA for review. Now NSA wanted to support Adleman's work. Unwilling to accept funding from NSA for fear that the agency's requirement of prior review could lead to classification of his work, Adleman was caught in a bind: since he had an alternative source of support, the NSF—whose purpose is to support "non-mission-oriented" research—now refused to support him.

Adleman's concerns tied in with another issue that had disquieted the research community: in 1979, the director of NSA, Admiral Bobby Inman, warned that open publication of cryptography research was harmful to national security. Inman threatened that, unless academia and industry could come to a satisfactory compromise with his agency, NSA would seek laws limiting publication of cryptographic research.

Adleman's problem was resolved when it was decided that both NSF and NSA would fund cryptography research. While NSF would make applicants aware of the alternate source of grants, it would not require them to accept NSA support.

Inman's concerns led to the creation of an American Council on Education study panel consisting of mathematicians and computer scientists from industry and academia and two lawyers, including the representative from NSA. The panel recommended a two-year experiment in which NSA would conduct pre-publication reviews of all research in cryptography (ACE 1981). Submissions would be voluntary, reviews prompt. The academic community feared that this process would have a chilling effect on the emerging field, but the experiment proved successful. Concerns eased when relatively few authors were asked to modify their publications. There have been NSA requests that an author not publish, and the agency has made suggestions for "minor" changes in some papers (Landau 1988, p. 11), but the research community reports that such requests have been modest in number. In an ironic twist, there was even an incident in which NSA apparently aided in the publication of cryptography research that the Army had tried to silence.[19]

The 1980s

The conflicts of the 1970s appeared to have abated. Behind the scenes, however, NSA's efforts to limit civilian research in cryptography continued. The result was a protracted delay in any widespread application of cryptography to civilian communications.

For example, in 1982 the NBS issued a *Federal Register* solicitation for algorithms for a public-key cryptography standard (USDoC 1982). RSA Data Security (the corporation formed by Rivest, Shamir, and Adleman) was interested in having the RSA algorithm become a federal standard, but the NSA was not. At the intelligence agency's request, NBS's plan to develop a federal standard for public-key cryptography was shelved (USGAO 1993b, p. 20).

Commercial Comsec Endorsement Program

In the mid 1980s, NSA changed its approach to broadening cryptographic coverage of American communications. Even though the initial promises that DES would be exportable had been broken, NSA was distressed by the algorithm's widespread availability[20] and was looking for a way to put the lid back on the box. Aided by the development of tamper-resistant coatings for chips (Raber and Riley 1989), NSA embarked on a program to supply equipment whose functioning was secret to a much wider user base.

Not only did NSA intend its new Commercial Comsec Endorsement Program (CCEP) to secure a much wider range of American communications, including industrial communications, NSA intended the program to do it with industry money.[21] As announced, the program was open to companies that had SECRET facility clearances and were willing to contribute expertise and funding to the development of secure versions of their products.[22]

Type I (High Grade)	Type II (Medium Grade)	
Winster	Edgeshot	Voice ($\leq$ 100 KB)
Tepache	Bulletproof	Data ($\leq$ 10 MB)
Forsee	Brushstroke	High speed ($\approx$ 100 MB)

Table 3.1
CCEP cryptomodules.

The most significant feature of the new program, however, was that it would provide a new category of equipment certified only for the protection of "unclassified sensitive information" but available without the tedious administrative controls that applied to equipment for protection of classified information. The traditional equipment was termed Type I, the new equipment Type II. Thus NSA was sponsoring the production of equipment directly competitive with DES.

The new undertaking was essentially a marketing effort, and in this situation NSA acted the same way commercial organizations often do: it began to undercut its previous "product line." The agency announced that it would not support the recertification of DES at its next five year review, due to take place in 1988, and told the banking community so in a letter to the chairman of X9E9, the security standards committee.

It didn't wash. The bankers and their DES suppliers, few of whom were members of CCEP, were furious at the attempt to scuttle efforts on which they had been pressed to spend money only a few years earlier. Banking, furthermore, was international and had successfully negotiated special export arrangements in acknowledgement of this fact; secret American technology would not satisfy its worldwide need for security. In the end, NBS showed some backbone; in 1988 it recertified DES over NSA's objections. NSA had second thoughts about the wide availability of Type II equipment and, citing the Computer Security Act of 1987, imposed restrictions on its availability arguably as onerous as those for Type I equipment.[23] As a result, Type II products never approached market success, and few were ever manufactured.

The STU-III

Although technically not a part of CCEP,[24] the third-generation secure telephone unit, STU-III, shared a lot of its technical and administrative approach. The project began in 1983 and, like CCEP, incorporated Type I and Type II devices. The first instruments were delivered in late 1987.[25] Unlike CCEP, they have been a dramatic success, with over 300,000 installed by the mid 1990s.[26]

With CCEP and STU-III, NSA began using public-key cryptography. The exact method, called Firefly, was kept secret[27] but appears to employ the same exponentiation operations used by commercial gear (AWST 1986).

NSDD-145

In September 1984, President Ronald Reagan issued a National Security Decision Directive (NSDD-145) establishing a federal policy of safeguarding "sensitive, but unclassified" information in communications and computer systems—a directive with NSA's fingerprints all over it (Brooks 1992). In 1985 the president's Assistant for National Security Affairs, John Poindexter, sent out a directive implementing NSDD-145 by putting a Defense Department team in charge of safeguarding all federal executive-branch departments and agencies *and their contractors.*[28]

The Poindexter directive, as it came to be known, attracted a lot of attention. Federal executive-branch contractors included a fair number of civilian companies, many of which had little or nothing to do with secret work. Mead Data Central (a supplier of databases, including the Lexis and Nexis systems, which provide law cases and news and magazine stories respectively) was one of the companies affected. Jack Simpson, the president of Mead, told Congress: "We have had a number of visits to inquire about our system, how it works, who uses it, whether we would be amenable to controls or monitors, and whether the Soviets used it. On April 23, 1986, AFMAG [Air Force Management Advisory Group], five people came; September 29, US Government Intelligence Committee, CIA, NSA represented; October 7, FBI; October 21, FBI. Cordial visits, but asking the same questions." (Simpson 1987, p. 328) Cordial though these visits may have been, their effect was chilling. The National

Technical Information Service (NTIS), a database of unclassified federal scientific and technical material, had been part of Mead's information systems. After the visits from representatives of federal agencies, Mead got rid of NTIS. "This may have removed Mead Data Central from concern under NSDD-145," Simpson told Congress. "I guess I wonder about other information providers of NTIS." (ibid.) He got it right.

In 1986, Assistant Secretary of Defense Donald Latham said: "I am very concerned about what people are doing, and not just the Soviets. If that means putting a monitor on NEXIS-type systems, then I am for it." (Schrage 1986) The FBI visited various university libraries, attempting to discover what scientific information foreign students were accessing. Here the government agents ran into an unexpected obstruction: the librarians insisted on subpoenas before they would release information.

A committee of the House of Representatives examined NSDD-145. Legislators saw an inappropriate incursion of presidential authority into national policy, and a turf battle developed. "[T]he basement of the White House and the backrooms of the Pentagon are not places in which national policy should be developed," Representative Jack Brooks declaimed (Brooks 1987, p. 2).

NSA backpedaled. "NSDD-145 in no way sets NSA as a computer czar," Lieutenant General William Odom, NSA's director, told the representatives. "[O]ur role with the private sector is one of *encouraging, advising and assisting* them with regard to their security needs. We view our role, then, as one that is clearly *advisory* in nature. . . ." (Odom 1987, p. 281)

Many in industry and academia beheld NSDD-145 in a different light. "[Latham] is talking about monitoring private computer systems, private information sources, and unclassified data, and we find that incredible," Jack Simpson said before the House Committee (Simpson 1987, p. 328). Cheryl Helsing, chair of the Data Security Committee of the American Bankers Association and a vice president of BankAmerica, told the committee: "NSA's new . . . algorithms . . . absolutely cannot be used by the banking industry. Those conditions might well be appropriate for national defense related security, but are clearly inappropriate for use in our industry." (Helsing 1987, p. 113) Indeed, NSA's encryption algorithms threatened years of development work by the banking industry. Eventually NSA decided to accept the use of old DES-based technology[29] in the financial industry, but in the interim "sixteen months ... elapsed while we worked to educate the NSA about our business," Helsing told Congress (ibid., p. 114).

Shortly after the congressional hearings on NSDD-145 began, the Poindexter directive was withdrawn.[30] "The policy was a good idea, in response to a real security threat," explained a senior Defense Department official. "The problem was that no one thought through all the implications." (Sanger 1987)

The Computer Security Act

The experience with NSDD-145 and NSA's behind-the-scenes actions convinced some US representatives that legislation was needed to reestablish which agency was in charge of assessing the security of civilian computer systems. NSA tried hard to convince the representatives that it was the right agency for the job. "[W]e are beginning to see civil agencies study and understand the usefulness of mechanisms resulting from [NSA's] earlier work," NSA Director Odom testified.

NSA could lead the way; it had "talent": "The [NSA] National Computer Security Center has a staff of more than 300 people," Odom told Congress (Odom 1987, pp. 294–295). He reminded the legislators that NSA already had responsibility for providing security for defense computer systems: "My concern with [the Computer Security Act] in its current form, then, it would create a wasteful, redundant bureaucracy that would busy itself with finding solutions to problems in computer security for the civil and private sector, while another government entity would be busy seeking the same solutions for the defense sector" (ibid., p. 296).

Congress did not buy the NSA director's arguments. The National Bureau of Standards (soon to be renamed the National Institute of Standards and Technology—we will refer to the agency as NIST from here on) was put in charge of developing computer security standards for the civilian sector. The representatives observed that developing civilian standards was a very different game from developing military ones, and that NIST had 22 years' experience with it whereas NSA had none.[31]

The report by the House Government Operations Committee described the concerns about giving such a charge to the intelligence agency: "NSA has made numerous efforts to either stop [work in cryptography] or to make sure it has control over the work by funding it, pre-publication reviews or other methods." (USHR 100-153 *Computer Security Act*, p. 21)[32]

The House committee was explicit that NIST was to be in charge, although NIST was to consult with NSA in the development of computer security standards, including those for cryptography: "By putting NSA in charge of developing technical security guidelines (software, hardware, communications) . . . [NIST], in effect, would on the surface be given the responsibility for the computer standards program with little to say about most of the program—the technical guidelines developed by NSA. This would jeopardize the entire Federal standards program." (USHR 100-153 *Computer Security Act*, p. 26)

The Computer Security Act (Public Law 100-235) was written to ensure that NIST would have responsibility for developing standards for the protection of "sensitive, but unclassified, information." All that remained was to fund the NIST program.

The NSA's Response

The NSA felt it had been had. A TOP SECRET NSA memo described what had occurred as follows:

- In 1982 NSA engineered a National Security Decision Directive, NSDD-145, through the Reagan Administration that gave responsibility for the security of all US information systems to the Director of NSA, eliminating NBS from this.

- This also stated that we would assist the private sector. This was viewed as Big Brother stepping in and generated an adverse reaction.

- Representative Jack Brooks, chairman of the House Government Operations Committee, personally set out to pass a law to reassert NBS's responsibility for Federal unclassified systems and to assist the private sector.

- By the time we fully recognized the implications of Brooks' bill, he had it orchestrated for a unanimous consent voice vote passage.

Clinton Brooks
Special Assistant to the
Director of the NSA
April 28, 1992

Congress legislates, but agencies implement; the ball game wasn't over. Under the Computer Security Act, NIST had been given additional responsibilities, but now it needed funds to go with the new responsibilities. NSA, the largest employer of mathematicians in the United States, had a vast operation working on issues of computer security and cryptography that dwarfed NIST's efforts. In 1987 NSA's *unclassified* computer security program had 300 employees and a budget of $40 million (USHH 102 *Threat of Economic Espionage*, p. 176); NIST's 1987 computer security operation had 16 employees and a budget of $1.1 million (USC-OTA 1994, p. 164). The Congressional Budget Office estimated that implementation of the Computer Security Act would cost NIST $4 million to $5 million dollars annually (USHR 100-153 *Computer Security Act*, p. 43).[33] It was time for Congress to appropriate the funds, but circumstances conspired to make that difficult.

During the Reagan-Bush years, the White House favored funding the Defense Department over funding civilian agencies, and NIST, part of the federal regulatory apparatus, was very much out of favor with a number of Republicans, some of whom were even in favor of eliminating the Department of Commerce. The Gramm-Rudman-Hollings Act[34] severely constrained discretionary funding. Yet by 1990 NIST's operation had a staff of 33 and a budget of $1.9 million.[35]

After the passage of the Computer Security Act, NSA began negotiating with NIST over their respective responsibilities in the development of cryptography. NSA went directly to Raymond Kammer, the acting director of NIST, to discuss drafting a Memorandum of Understanding (MOU) delineating the two agencies' responsibilities under the Computer Security Act. Kammer, the son of two NSA employees, was deeply concerned about protecting national-security and law-enforcement interests in cryptography. His instincts were to defer to the intelligence agency on control over civilian cryptography standards.[36]

The debate surrounding the Computer Security Act, as well as the act itself, had made it clear that NIST was in charge of developing civilian computer security standards. The MOU between NIST and NSA mandated that NIST would "request the NSA's assistance on all matters related to cryptographic algorithms and cryptographic techniques" (US-DoCDoD 1989, p. 2). A Technical Working Group (TWG), consisting of three members each from NIST and NSA, would review and analyze issues of mutual interest, including the security of technical systems, *prior to* public disclosure (USDoCDoD 1989, p. 3). The opportunity to vet proposed standards before any public airing put NSA in a controlling position in the development of civilian computer standards.

In making civilian computer security standards part of NIST's bailiwick, the Computer Security Act had placed decisions regarding the approval of these standards in the hands of the Secretary of Commerce. The MOU changed this so that, although appeals could still be made to the Secretary of Commerce, members of the Defense Department were free to appeal proposed NIST standards to the Secretary of Defense before any public airing. Appeals of TWG disagreements would go to the Secretary of Commerce *or* the Secretary of Defense, and from there to the president and the National Security Council (the same group that had promulgated NSDD-145).

The Government Accounting Office was appalled. Milton Socolar, a special assistant to the Comptroller General, told Congress: "At issue is the degree to which responsibilities vested in NIST under the [Computer Security] act are being subverted by the role assigned to NSA under the memorandum." (Socolar 1989, p. 36) Congress's research arm, the Office of Technology Assessment, described the MOU as "ced[ing] to NSA much more authority than the act itself had granted or envisioned, particularly through the joint NIST/NSA Technical Working Group" (USC-OTA 1987, p. 164). NIST Acting Director Raymond Kammer, who had signed the document, disagreed: "As I've heard people interpreting the . . . memorandum of understanding, it occurs to me that many individuals doing the interpreting are perhaps starting from a perspective I don't share, namely that NSA has had some trouble accepting the act. My experience in the months that I have been negotiating and working with the current management of NSA is that they fully understand the act, and that their understanding and my understanding are very consistent. I have no reservations about their willingness to implement the act as written." (Kammer 1989)

Representatives Jack Brooks and Daniel Glickman viewed the Digital Signature Standard as a test of who was running the show on civilian computer standards;[37] they were proved right.

The Digital Signature Standard

Many breaches of confidentiality are difficult to detect, and even when a breach is clear it is often not at all clear where the problem lies. In light of this, any attempt by a litigant to claim that the insecurity lies with bad cryptography and thus force the protection techniques into the open is likely to fail. On the other hand, digital signatures—as a consequence of their function of resolving disputes between users of electronic networks —are certain to give rise to litigation over the adequacy of the signature methods. This is all the more true because digital signatures are a novel idea in commerce, whereas the notion of protecting messages by encryption is well established even though the details of particular methods may be unfamiliar. It is therefore necessary to provide a public digital signature mechanism that is open to examination by cryptographic researchers and expert witnesses in order to foster the public confidence necessary to achieve acceptance of the new technology.

In the spring of 1989, representatives from NIST and NSA began meeting to develop a set of public-key-based standards. First on the agenda was a digital signature standard to be included in the FIPS series. NIST proposed the Rivest-Shamir-Adleman algorithm (USGAO 1993b, p. 20). During its very public twelve-year lifetime, no cryptanalytic attacks had succeeded in breaking that algorithm. It was an accepted industry standard for digital signatures, and several standards organizations had formally adopted it. But in Technical Working Group meetings the NSA representatives rejected RSA. Presumably they objected to its flexibility, which allowed it to be used for purposes of confidentiality as well as authenticity. Raymond Kammer concurred in the decision, arguing that the infrastructure needed for public key management made RSA an unwieldy digital signature standard. Kammer ignored the fact that any digital signature standard adopted for widespread use would entail a key management facility.

Through 1989 and 1990 the Technical Working Group of NIST and NSA representatives met once a month, and each month no progress was made. "We went to a lot of meetings with our NSA counterparts, and we were allowed to write a lot of memos, but we on the technical side of NIST felt we were being slowrolled on the Digital Signature Standard," recalled Lynn McNulty, NIST's Associate Director of Computer Security, "In retrospect it is clear that the real game plan that NSA had drawn up was the Capstone Chip and Fortezza card—with key escrow all locked up in silicon."[38]

A year after the meetings on digital signatures began, NSA presented its proposal: an algorithm it had developed itself. NSA's justification for its algorithm was a "document, classified TOP SECRET CODEWORD, [that] was a position paper which discussed reasons for the selection of the algorithms identified in the first document" (USDoC 1990b). According to NSA, the proposed algorithm was based on unpatented work done by Taher ElGamal when he was a graduate student at Stanford University under Martin Hellman (ElGamal 1985). Outside the agency even that was questioned. The algorithm had been developed by David Kravitz at NSA (Kravitz 1993), and the technique bore a strong resemblance to one developed by the German mathematician Claus Schnorr, who had patented his algorithm in the United States and in various European countries.[39] Concerned about potential patent conflicts, NIST officials went to negotiate with Schnorr over selling his rights, but the government did not want to pay his asking price (reputed to have been $2 million).

NIST proposed Kravitz's algorithm as the Digital Signature Standard (USDoC 1991b). The computer industry objected to this because the algorithm was not interoperable with digital signatures already in use. The proposed standard had a 512-bit key size, but Bell Labs scientists had already shown that the Kravitz algorithm was not particularly secure with a 512-bit key (LaMacchia and Odlyzko 1991; Beth et al. 1992). Furthermore, it was significantly slower than the RSA algorithm in signature verification, taking roughly 10 times as long on comparable processors.[40]

In abandoning the RSA algorithm in favor of the proposed NSA algorithm, NIST had traveled a considerable distance from the Computer Security Act.

Critics saw the dark hand of NSA behind NIST's bumbles. When questioned by a congressional committee, NIST director John Lyons denied such pressure. "What's your response to charges that NSA, either directly or through the National Security Council, continues to control NIST's computer security program?" Representative Jack Brooks asked Lyons. "My response is that it's not true," said Lyons. "We're running our program. We consult with them, according to the 1987 legislation, but they know and we know that we make these decisions." (Lyons 1992, p. 176) The record, released after Freedom of Information Act litigation, told a different story. A January 1990 memo from the NIST members of the Technical Working Group said: "It's increasingly evident that it is difficult, if not impossible, to reconcile the requirements of NSA, NIST and the general public using the approach [of a Technical Working Group]." (USDoC 1990a) Completely contrary to Congress's wishes, NSA was making the decisions on civilian cryptography.[41]

In its report on the Computer Security Act, the House Government Operations Committee said "NSA is the wrong agency to be put in charge of this important program." (USHR 100-153 *Computer Security Act*, p. 19) Congress concurred and passed the measure. It looked as if the intelligence agency had made an end run around Congress. Under the Computer Security Act, NIST was supposed to develop cryptography standards for the public sector, but the combination of the MOU and NSA's clout prevented such an outcome. Lynn McNulty later commented: "We bent a hell of a lot backwards to meet national security and law enforcement requirements, but we didn't do much to meet user requirements." Various government observers, including the Office of Technology Assessment and the General Accounting Office, concluded that the MOU had put NSA in the driver's seat—not at all the intent of the Computer Security Act.[42]

The proposal for a Digital Signature Standard was put forth in 1991. Public objections resulted in modifications, including a flexible key size (key sizes from 512 to 1024 bits are permitted, in jumps of 64 bits). On May 19, 1994, over strong protests from industry and from academia, the government adopted DSS as Federal Information Processing Standard 186, announcing that the "Department of Commerce is not aware of patents that would be infringed by this standard" (USDoC 1994c).[43]

Ceding Even More Control

While Congress waited to see how NIST would handle implementing a digital signature standard, a transfer of power was occurring behind the scenes. In the drafting of the MOU, NSA had recommended that the FBI be part of the Technical Working Group; NIST staffers had objected, and this clause was dropped (USDoC 1989). But Kammer, the acting director of NIST, was concerned that his agency was not properly equipped to develop civilian cryptography, the job Congress had handed to it. He and Clinton Brooks, advisor to NSA's director, shared their concern with the FBI.

Their initial reception was cool. "The first couple of times [we went there] they said, 'Why are you bothering us?'" recalled Kammer. "They kept giving inappropriate responses; the FBI didn't understand the issue. Cryptography is a somewhat peripheral issue to the FBI." Brooks and Kammer presented the dangers of encrypted telecommunications, but it took the FBI some time to understand. "Ray and I kept encountering lots of blank stares," Brooks said later. "What we were encountering was a lack of appreciation that digital communications was here. Wiretapping was just doing clips, or going to the phone office. But the phone companies had all gone digital. The next step [in understanding] was that encryption was going to exist on the digital lines."[44] To the FBI the cryptographic issues seemed futuristic.

There was a clash of understandings. "A successful FBI agent," Kammer explained, "kicks in the door, arrests the guy, and goes on to the next case." The issues NIST and NSA were raising were more subtle. "A successful NSA man ... well give him a hard problem and the first thing he'll do is sit down and think—sometimes for a very long time."[45] NSA had been thinking about strong cryptography for a long time, but the FBI did not have any experts remotely close to the area. The closest the Bureau had were agents working on defeating electronic locks and alarms.

After a number of visits to the FBI over several months, Brooks and

Kammer encountered James Kallstrom, Chief of the Special Operations Branch of the New York Field Office. "It was obvious," Kallstrom recalls, "that encryption had been around a long time. What was new here was it had never been an issue before for the general public. Old encryption didn't work; it was too bulky, you sounded like Donald Duck. But in the late eighties we could see that it wouldn't be very long before cheap encryption would be around that would put us out of business."[46]

Kallstrom's tenure in New York undoubtedly shaped his viewpoint. Historically, New York State has relied heavily on electronic surveillance.[47] For example, over a third of the 1994 Title III electronic surveillances occurred in New York State.[48] California, whose big cities suffer similar problems of drugs and crime, had one-eighth as many.[49]

Kallstrom could not imagine law enforcement without wiretapping and did not want wiretapping to disappear from law enforcement's arsenal. He went to work: "From the standpoint of this becoming an issue in the government, from the standpoint of law enforcement, we were the user, the customer. An Interagency group was formed; the squeaky wheel was us. We went to both [NIST and NSA]. We have a long-standing relationship with NSA; we have a responsibility for counter-terrorism and intelligence." NSA was immediately part of an interagency group focusing on problems of domestic use of strong cryptography. NIST joined shortly afterwards. "It wasn't a function of official policy. We have always recognized NSA as a premier agency [in intelligence]. NIST was also at the table."[50]

By 1991 the FBI had formulated a policy that included shoring up its ability to perform electronic surveillance, particularly wiretaps, and preventing the establishment of unbreakable cryptography in the public sector. Efforts in support of this policy included the Digital Telephony Proposal and the concept of key escrow, which were introduced to the public in 1992 and 1993 respectively.

In negotiating the MOU, NSA had sought to include the FBI as a full-fledged member of the Technical Working Group (which would have meant that two-thirds of the participants came from either law enforcement or national security). After that effort was rebuffed by NIST scientists, Kammer and Brooks brought the FBI in by different means. The FBI's involvement in encryption issues buttressed NSA's position.

With the end of the Cold War, law-enforcement issues were significantly closer to the public's heart than national-security concerns. By replacing national-security concerns over cryptography with law-enforcement concerns, the FBI succeeded in returning much of the control of civilian cryptography to NSA.

"The whole Digital Telephony [effort] came out of [our meetings]," Clinton Brooks said some time later.[51]

4
National Security

In discussions of cryptographic policy, "national security" is usually shorthand for *communications intelligence*—spying on foreign communications. It is taken for granted that the United States depends on breaking foreign codes for much of its intelligence and that any decline in the success of this activity will make the country less secure. Intelligence, however, is only one of cryptography's roles in national security.

The Concept of National Security

The notion of national security is a relative newcomer to American political iconography. Although the term dates to the early post-World War II era, it does not appear in *Webster's Third International Dictionary*, which was published in 1961 and which sought to capture an up-to-date picture of American English.

The essence of national security is, of course, the protection of the country against attack by foreign military forces. The term is broader than this, but not so broad as to encompass all of the national interest. Its focus is protection of the country, and in particular its government, against threats that are characteristically but not invariably foreign.

National security includes the following:

- Maintenance of military forces adequate to deter attacks on the United States, repel invaders, control domestic unrest, and undertake other military actions that may be in the national interest.

- Provision of intelligence on the capabilities and intentions of all powers, both friendly and hostile, sufficient to inform foreign policy and military action. Such powers are understood to be primarily, but not entirely, national states. They may, in addition, include organizations representing landless peoples, revolutionary movements, terrorist groups, organized crime, trans-national political movements, and multi-national corporations.

- Denying to foreign powers intelligence about the United States that would interfere with American diplomatic, military, or trade objectives.

- Enforcement of certain laws, in particular those governing espionage, terrorism, the integrity of the national-security community itself, and the movements of people and material across borders.

- Maintenance of an industrial base, a resource base, and an infrastructure adequate to support essential government activities, including military forces, intelligence, and relevant aspects of law enforcement.

The set of issues that define the national security is naturally neither free from debate nor immune to change. In the late 1960s and the 1970s, the idea that drug trafficking should be seen as a threat to the national security and approached with military resources and tactics gained substantial ground.[1] Since the end of the Cold War, a quite different constituency has argued for the inclusion of broader economic issues, such as education and competitiveness in the world marketplace.[2]

From the viewpoint of communications security—and its all-important component, cryptography[3]—the relevance of the second and third points —intelligence and security against foreign intelligence—is most apparent. We will examine these first and in more detail, but issues of infrastructure, law enforcement, and offensive capability will also be considered.

The Spectrum of Intelligence

When we speak of intelligence, we will usually mean national intelligence —information obtained by national governmental organizations. The intelligence activities of governments, however, have much in common with those of other organizations. Scholars, reporters, political parties, businesses, criminals, and police all practice intelligence in one form or another. The intelligence-gathering activities of nations are generally more ambitious and include things not accessible to organizations without state power (for example, launching spy satellites), but the similarities outweigh the differences.

The most familiar form of intelligence—so familiar that it is usually not recognized as intelligence—is *open-source intelligence*: information obtained from sources that are not attempting to conceal it. Open-source intelligence is almost the only form of intelligence practiced by scholars, reporters, and business people, but it also plays a major role in national intelligence. In the national case, typical open sources are newspapers, radio broadcasts,[4] foreign government publications, propaganda, maps, and phone books. In industrial intelligence, advertisements and product literature are major sources.

Older open sources have now been joined by the Internet and the World Wide Web. Browsing the Web is practicing open-source intelligence. Google and more specialized search engines give their users access to information on an unprecedented scale.

Operations intelligence is information obtained by observing and recording a target's visible actions and inferring actions that are not visible. Although it is hardly limited to military affairs, a typical example of operations intelligence in a military context was widely touted during the 1991 Gulf War: a pizza parlor near the Pentagon told newsmen that it could always tell when something was about to happen because large numbers of people stayed late at the Pentagon and ordered pizza in the middle of the night.[5]

What most people think of when they think of "spying" is called *human intelligence* (HUMINT), which runs the gamut from interviewing travelers[6] to infiltrating illegal agents and sometimes extends to breaking and entering. In the most basic form of human intelligence, intelligence officers from one country, traveling under diplomatic or journalistic or commercial cover, recruit an agent who has access to secret information. The agent then passes information to the foreign handlers, usually either for ideological reasons or in exchange for money.

In the twentieth century, open-source intelligence, operations intelligence, and human intelligence were joined by a host of new methods having only the barest antecedents.[7] Indeed, David Kahn, cryptology's foremost historian, argues that modern intelligence was created by signals intelligence (Kahn 2006). Generals were unwilling to commit their resources and risk their troops on the words of spies. Only when radio interception gave them access to their opponents communication did they have intelligence they were prepared to believe.

The growth of technology intensive intelligence originated in the use of new technologies to gather intelligence about societies that exercise tight control over the information they release to the outside world and over the movement of people across and within their borders. Their success has pushed back the frontiers of national sovereignty; by limiting the degree to which nations can keep their military preparations secret from each other, it has also become a fundamental stabilizing influence on international relations. For the United States, surprised once by the Japanese at Pearl Harbor and again by al-Qaeda on 9/11, intelligence has become a national obsession.

The techniques of intelligence gathering have also been guided by adaptation to political reality. Throughout the twentieth century, improvements in communication and increases in interdependence have produced the "shrinking of the world" that has changed so much of modern life. This has increased peer pressure among nations, giving rise to the World Court, the United Nations, and other international institutions. In this environment, nations have become more concerned than ever with appearance.

Spying exists, and has perhaps always existed, in a sort of limbo. "Everyone" knows that "everyone" does it, yet it remains frowned upon, hidden, and, under the laws of the nation being spied on, illegal. Most if not all nations use their embassies and consular facilities forintelligence gathering. Some of the activities are aboveboard. Ambassadors, trade representatives, and military attachés all report on both their meetings with representatives of the host country and their observations of life, politics, industry, and military activity. Others are not. Embassy personnel often recruit spies from among the local population or undertake more technical forms of information gathering from the legally protected premises of the embassy (Frost 1994).

Often a host country is aware of the clandestine intelligence activities of foreign diplomatic and consular personnel but finds itself unable to interfere for fear of retaliation against its own diplomats. When an espionage case becomes public, the host country usually feels obliged to put on a show of public indignation, and a scandal ensues. Such was the case when Soviet Colonel Oleg Penkofsky was caught spying for the West in the 1960s, and more recently when CIA officer Aldrich Ames was caught spying for Russia. The embarrassment is most acute when the countries involved are supposed to be friends, as happened in the case of Jonathan Pollard, an American naval intelligence officer found to be spying for Israel. A desire to avoid embarrassments of this sort is one motivation for the development of a variety of new forms of intelligence that do not intrude on the territory of the target country.[8]

Another outgrowth of the "shrinking" is the relationship between the tactical and the strategic. The time-honored practice of climbing a hill to get a look at an opposing army—which, in the past, was of little use except during battle—has evolved into a new field of strategic reconnaissance.

No aspect of modern intelligence is more impressive or more important than *photographic intelligence* (PHOTINT): information from photographs at frequencies both in and out of the human visual range. Although photography dates from the nineteenth century, it did not become a distinctive tool of intelligence until aircraft and later spacecraft gave cameras secure platforms from which to operate—platforms that could observe an opponent's territory from a safe distance.

Today, the most important intelligence photographs are those taken from orbiting satellites. Paradoxically, despite the fact that photo-reconnaissance aircraft fly much closer to their targets than satellites (10–20 miles as opposed to several hundred), the larger cameras carried by satellites produce far more detailed pictures. Images of a Russian shipyard taken by an American KH-11 spy satellite appear to be from a distance of 500 feet rather than the actual 500 miles (Burrows 1987, pp. 166n–166o; Richelson 1990, p. 186).

Photographic intelligence provides high-resolution images of the Earth's surface but is impeded by clouds, sandstorms, and vegetation. The passive form is therefore complemented by the use of radar-imaging satellites, such as the American Lacrosse, which produce lower-resolution images but are unaffected by night and fog and can penetrate trees and even buildings. Orbital lasers open yet other possibilities (AWST 1997b).

Besides cameras and radar, modern intelligence employs a broad range of sensors for *measurement and signatures intelligence* (MASINT), which seeks to characterize objects or events by their observable characteristics and to detect or analyze them by combining information from various sensors. In the late 1940s, the United States began collecting atmospheric samples and testing them for radioactive isotopes in an attempt to discover nuclear tests. It was this technique that made the US aware of the Soviet Union's successful test of a nuclear weapon before it was announced. At about the same time an Air Force activity named *Project Mogul* sought to listen for the sounds of nuclear explosions propagating along the boundaries between layers of the atmosphere.[9]

For decades, the *Sound Surveillance Underwater System* (SOSUS) has tracked the movements of submarines and other ships by means of arrays of microphones lying on the ocean floor. In the 1960s, a family of satellites called Vela-Hotel were put in orbit to watch the earth for nuclear explosions. These satellites exemplify the *signatures* aspect of intelligence, distinguishing nuclear events from other phenomena such as lightning flashes or meteor impacts by characteristics more subtle than the brightness of the flash.[10] More recent satellites called simply Defense Support Program satellites also watch for the infrared signatures that characterize the exhaust plumes of rising ballistic missiles. Satellites were only one part of the wider Vela program for detecting nuclear explosions. Another important element was seismographic. An array of seismometers called *NORway Seismic ARray* (NORSAR) was placed at a location geologically coupled to the area in which the Soviets conducted their nuclear tests. Seismic measurements served to verify compliance with a treaty limiting the yields of underground nuclear explosions.

Measurement and signatures intelligence can be viewed as a refined form of operations intelligence. It seeks out one or more subtle but unavoidable consequences of an event and infers the occurrence and character of that event from the observed phenomena. Its efficacy depends not only on the sensors but on the computing required to draw useful inferences from the data they produce.

Another aspect of modern intelligence that leans heavily on inference may be called *technical intelligence*. As the term suggests, this is the study of an opponent's technology, but the emphasis in this case is on inferences drawn by simulating or duplicating technologies whose existence has been inferred from observations or from information provided by human sources. The British Office of Scientific intelligence made extensive use of such methods during World War II to improve its understanding of developing German weaponry. Accounts of its work convey a novel perspective in which the reports of human agents were essentially regarded as rumors to be confirmed or refuted by technical means (Jones 1978; Johnson 1978).

The various means of gathering intelligence are far from independent. This is true both in the sense that the boundaries are not sharp (it is sometimes difficult to pigeonhole something as photographic intelligence rather than imaging intelligence) and in the sense that frequently information obtained by one technique may be useful or even indispensable in acquiring information by another technique or in interpreting the information acquired by another technique.[11]

Signals Intelligence and Communications Intelligence

We have surveyed a variety of forms of intelligence in an attempt to convey the breadth of modern intelligence work. No one intelligence method exists in a vacuum, and the intelligence analyst draws on information from a wide variety of sources. It is within this context that we now turn to the form of intelligence with which we are most concerned.

Communications intelligence (COMINT) is the practice of extracting information from an opponent's communications. Although, as we shall see, communications intelligence is quite broad, it is embedded within a yet broader category. *Signals Intelligence* (SIGINT) is the information obtained by analyzing signals emitted by a target. When signals intelligence is distinguished from communications intelligence, the broader category includes such electromagnetic phenomena as radar signals, which are not intended to convey information but rather to locate physical objects and measure their movements. The study of radar signals is at the heart of electronic warfare and is prerequisite to all efforts either to jam radars or to evade them by stealth.[12] The areas of signals intelligence other than communications intelligence are collectively called *electronic intelligence* (ELINT) and include *radar intelligence* (RADINT), *telemetry intelligence* (TELINT), and *emissions intelligence* (EMINT).

Although normally categorized as electronic intelligence, some aspects of emissions intelligence can be better regarded as communications intelligence. These include processing ciphertext to extract plaintext signals accidentally encoded "piggyback" and listening to the sounds of electromechanical cryptoequipment as an aid to cryptanalysis (Martin 1980, pp. 74–75; Wright 1987, p. 84; Agee 1975, pp. 474–476). Another aid to communications intelligence is *emitter identification*, the technique of distinguishing individual radio transmitters by minor variations in behavior too small to be eliminated by ordinary quality-control techniques.

One of the most disquieting techniques of emissions intelligence is Rafter, a technique for monitoring the behavior of radio receivers.[13] It is not surprising that it should be possible to exploit the signal of a radio transmitter. That a receiver should reveal the frequency it is listening to is both surprising and frightening. It can be used, for example, to determine who is listening to banned foreign radio broadcasts.

Despite the elaborate taxonomy, the distinctions are not always clear. Telemetry intelligence, for example, is the study of communications between moving platforms (usually aircraft, rockets, or satellites) and their controlling stations. Test firings of ballistic missiles are monitored via radio transmissions from the missile being tested, and interception of these signals permits an opponent to learn almost as much from the test as do the people conducting it.[14] Similarly, communication satellites, spy satellites, and others are controlled from the ground, and interception of the control channel can reveal information about a satellite's attitude, fuel supply, and activities.[15] It is clear that these examples of telemetry intelligence, though commonly classified as signals intelligence, are as much examples of communications intelligence.

In short, although the term SIGINT is sometimes used to distinguish interception of non-communications signals from communications signals, it is also used to encompass both activities. Communications intelligence so dominates signals intelligence that the term "SIGINT" is often used when the narrower term "COMINT" would do.

With the possible exception of human intelligence, communications intelligence exhibits unparalleled breadth and flexibility. Observation of the gross characteristics (often merely the occurrence) of messages can be used to monitor military, diplomatic, commercial, or criminal activity or to detect relationships between persons, organizations, or events that have, to public appearances, no connection.[16] On the other hand, the analysis of carefully selected messages can sometimes reveal the intentions of military or political leaders even more accurately than information obtained by recruiting members of their staffs.[17]

Because it relies primarily on radio reception, with only occasional recourse to transmission or physical taps, communications intelligence rarely results in diplomatic incidents; indeed, rarely is the target aware of being monitored.[18]

Taxonomy of COMINT

Cryptography is often considered, particularly by those primarily concerned with security, to be the only serious barrier to communications intelligence. Histories of the field have generally fostered this impression by painting a picture of war between codemakers and codebreakers. In practice, spying on communications is a multi-stage activity in which each stage plays an essential role. It is entirely possible that the cryptanalysis of a message, once the message has been identified and captured, may be less difficult than acquiring and filtering the traffic to locate it. On balance, the greatest problem in communications intelligence—as in most efforts to learn things—is sorting out the information you are after from the information you are not.

The *sine qua non* of communications intelligence is acquisition of signals. Without communications in the form of radio waves, electrical currents in wires, written materials, or copied disks and tapes, there can be no work for cryptographic or intelligence analysts. The interception of communications presents both a strategic and a tactical aspect.

Strategically, it is crucial to learn as much as one can about an opponent's communications infrastructure. The first step is to come up with the most precise possible description of the target—what the military call the *order of battle*. If the target is a country, it may have millions of residents who in turn make millions of phone calls every day. Most of these calls are not of interest; the people who make them do not work for the government or in critical industries and say little of intelligence value. Describing the target is one of the many areas where *collateral intelligence*—information from sources other than covert interception of communications—plays a vital role. Most of the information about a country and its government can be learned from open sources, such as phone books, newspapers, histories, and government manuals. Some, however, will come from covert sources such as spies, and some will come from communications intelligence itself.

Once the targets have been precisely identified, it is necessary to discover how they communicate with one another. Are their communications carried by high-frequency (HF) radio, by satellite, or by microwave? How accessible the communications are and how they can be acquired is a function of the means chosen. High-frequency radio and satellite transmissions are the most accessible. At the time of World War II, most radio communication and thus most of what was intercepted was HF. Such signals bounce back and forth between the ionosphere and the ground and can travel thousands of miles. This property makes intercontinental radio communication possible; at the same time, it makes it essentially impossible to keep HF signals out of the hands of opponents. Today a large fraction of radio communication is carried by satellite. Satellite downlinks typically have "footprints" thousands of miles across that spread over more than one country.[19] Terrestrial microwave communications are significantly harder to intercept. They travel between towers a few miles or tens of miles apart. Intercept facilities on the ground must generally be located within a few tens of miles of the microwave path and often require facilities in the target country.[20] Terrestrial microwave is nonetheless vulnerable to interception by an astounding, if expensive technique: satellites on the microwave path beyond the receiving antenna. The satellite is in synchronous to orbit remain in the same relative position to its microwave targets. It is placed not over the target country but about a quarter of the way around the Earth from the target (Campbell 1999).

As with the organizational structure, a target's communication practices can often be derived from open sources. Since national and international organizations cooperate in allocating the radio spectrum, it is easier to identify the frequencies used for military, police, or air traffic

Figure 4.1
NSA installations around the world. (Illustration by Roland Silver.)

control communications by consulting regulations and standards than by direct spectrum monitoring.

The output of the strategic or *targeting* phase of communications intelligence is a map of the opponent's communications, which will guide the selection of locations, frequencies, and times of day at which monitoring is conducted. Interception can also be conducted from many sorts of platforms: ground stations, aircraft, ships, embassies, covert locations, and orbiting satellites.

The United States has several major intercept facilities within its borders and a host of others abroad (figure 4.1). Despite attempts to keep these locations secret, many, including Menwith Hill in Britain, Alice Springs in Australia, Alert in Canada, Osburg in Germany, Misawa in Japan, and Shemaya in the Aleutian Islands have been in the news at one time or another (Bamford 1982; Shane and Bowman 1995).

The Soviet Union made extensive use of small ships as collection platforms. Usually operating under very thin cover as fishing trawlers, these boats carried large antennas and were thought to be making their biggest catch in the electromagnetic spectrum. The United States has been less successful with this approach. In the 1960s it commissioned two ships described as research vessels, the *Liberty* and the *Pueblo*, for intercept duty. The *Liberty* was attacked by the Israelis, for no publicly apparent reason, while supposedly intercepting Arab communications in the Eastern Mediterranean during the Six Day War of 1967.[21] A year later, the *Pueblo* was captured by the North Koreans. It turned out to have been carrying many TOP SECRET documents for which it had no apparent need, and most of these fell to its captors. As quietly as it had begun, the United States ceased using small ships as collection platforms.

Airborne collection, by comparison, has been an important component of US COMINT for decades. Boeing 707s, under the military designation RC-135, are equipped with antennas and signal-processing equipment. These aircraft can loiter off foreign coasts for hours at a time. Flying at altitudes of 30,000 feet or higher, they can pick up radio transmissions from well inland.

The use of embassies to do intercept work exemplifies the twilight-zone character of intelligence. Despite widespread "knowledge" that many embassies are engaging in intelligence collection, such activity is a breach of diplomatic etiquette that could result in diplomats' being asked to leave the host country if discovered. All the equipment used must therefore be smuggled in or constructed on the spot and must be made from components small enough to fit inconspicuously in the "diplomatic bag" —a troublesome limitation on the sizes of antennas. Politics and public relations aside, if an embassy is not suspected of interception, it is likely to be more successful. Mike Frost, a Canadian intelligence officer who spent most of his career intercepting host-country communications from Canadian embassies, reported that the Chinese put up a building to block radio reception at the US embassy in Beijing but failed to protect themselves against the Canadian embassy because they did not realize that it too was engaged in interception (Frost 1994).

Interception can also be conducted from covert locations that do not enjoy the legal protection of diplomatic immunity. Britain operated a covert direction-finding facility in neutral Norway during World War I (Wright 1987, p. 9). In the early 1950s, the CIA established a group known as "Staff D" to carry out interception from covert locations.

One of the most ambitious undertakings in communications intelligence has been the development of intercept satellites, which did not arrive on the scene till roughly a decade after their camera-carrying cousins. Low-altitude satellites are not well suited to intercept work. They are relatively close to the transmitter, which is good, but they are moving quickly relative to the Earth, which is not. No sooner have they acquired a signal than they move on and lose it again, because the source has passed below the horizon. The comparison with communications satellites is interesting. The mainstay of satellite-mediated communications has been satellites in synchronous orbits, 22,500 miles up. Only recently have communications satellites been placed in low orbits. Tens of satellites are required so that as soon as one moves out of range of a transmitter on the ground, another comes close enough to take over. Systems of this kind have the advantage that the satellites and the transmitters are cooperating. A system in which the satellites were attempting continuous coverage of uncooperative targets would be far more complex.

Because they are in very high orbits, intercept satellites must carry antennas tens or hundreds of feet across. It is difficult to make an antenna of this size light enough to be lifted into synchronous orbit. In addition, the antenna must be launched in a folded configuration, which adds complexity and detracts from reliability. In sum, communications intercept satellites are more complex and expensive than other types.

Because of its huge size and the low population density of much of its territory, the Soviet Union made more extensive use of radio communications than the United States or Western Europe. Most of the territory of the Soviet Union was far north and not conveniently served by synchronous satellites, so the Soviets developed a family of communication satellites, called Molniya, that move in polar orbits. A *Molniya orbit* passes over the Northern Hemisphere at very high altitude and thus moves quite slowly during this part of its journey. Its perigee, in contrast, is low over the Southern Hemisphere, and that part of the trip goes very quickly. The result is that most of the time the satellite "hangs" above the Northern Hemisphere, where it can be used for high-latitude communications. In order to spy on these communications, the US built satellites, called Jumpseat, that move in Molniya orbits. These satellites are in a position to listen to both radio transmissions from the ground and those from Molniya satellites.

Communications intelligence depends for its success on tactical as well as strategic elements. When an intercept station has been put in the right location, operates at the right time of the day, points its antenna in the right direction, and tunes its radio to the right frequencies, it is rewarded with a flood of traffic too large to record, let alone analyze. The process of examining intercepted traffic to determine what is to be retained and what is not may be as "simple" as detecting which channels within a trunk are active or as complex as recognizing the topic of a conversation. Typical selection processes include active channel detection, called and calling number identification, speaker identification, keyword spotting (in either text or voice), fax recognition, and semantic information processing.

The difficulty of locating and isolating just the right messages is an intrinsic consequence of the volume of traffic in modern communications. Communications intercept equipment must decide in a fraction of a second whether to record a message it has detected or to permit the message to escape. Often it must make the decision to record communications of which it has only one part.[22] If, for example, the two directions of a telephone call are carried on separate facilities, an individual intercept point may have access to only one side of the conversation. Although the entire call may in fact be recorded, so that both sides of the conversation will ultimately be available to an analyst, it will be recorded by two devices acting independently. Should either fail to detect that the call is of interest, and therefore fail to record it, the utility of the other component will be vastly reduced.[23] The problem of identifying traffic of interest among all possible traffic is the problem of *search*.

Communications are organized at many levels. The entities communicating have addresses—in radio these are called *call signs* (commonly known in the case of commercial stations as *call letters*); in the case of telephones they are telephone numbers; in the case of computer networks, they are IP addresses, email addresses, URLs, etc. Messages follow *routes*, which in turn are made up of *links* or *hops* on *trunks*. Within an individual trunk, messages are *multiplexed* into channels, which make up the trunk much as lanes make up a road.[24]

At the lowest level, intercept equipment sits and looks through the space in which messages might be found. At each frequency, or time slot, or code pattern, it listens to see if there is any traffic at all. It may well be the case that most of the channels in a trunk are inactive most of the time.

When intercept equipment detects an active channel, it must decide whether to record what it finds there. This depends on *diagnosis*: characterization of the form and the significance of the signal that has been found. If the channel is a telephone channel, for example, the likely possibilities are voice, fax, and data. The intercept device must try to decide what it is hearing and may then discriminate more carefully depending on the category. The first step will usually be to listen for dial pulses or touch tones and attempt to determine what number is calling and what number is being called. If the call is voice, the device may attempt to determine what language is in use, or even listen for keywords. If the call is fax, it may try to determine whether the transmission is text or pictures. If the call carries data, it will attempt to determine what type of modem is in use and what codes (ASCII, Baudot, EBCDIC) or data formats are present. When text is detected, the equipment may go further and apply semantic processing to determine the subject of the message in much the same way that a search engine tries to locate a topic of interest on the World Wide Web.

One strategy followed by many pieces of intercept equipment should be a caution to anyone using cryptography: if an intercepted message is found to be encrypted, it is automatically recorded. This is possible because at present only a small fraction of the world's communications are encrypted. The first lesson to be drawn from this is that if you encrypt something you had better do it well; otherwise you will only succeed in drawing attention to yourself. The second is that as the use of cryptography increases, the privacy of everyone's traffic benefits.

Once traffic has been diagnosed as interesting, it will be recorded. This is not as simple as it sounds. Typically a signal can be recorded in several different formats, depending on how well it has been understood. It is always possible to make a recording of the waveform being received, but this may turn out to be much bulkier than the message it encodes. For example, recording a modem signal carrying 2400 bits per second of information (about 240 characters a second), without demodulating it, uses up the 48-kilobyte-per-second capacity of a digital audio tape. A direct recording of the signal is thus 20 times the size of the message it contains.

Neither diagnosis, nor recording, nor any form of analysis that may be done on an intercepted signal can be separated from *signal processing*

—study of the signal by mathematical and computational means. Digital signal processing (one of the fastest-growing areas in computing) is revolutionizing communications. The availability of $100 modems is a consequence of the availability of signal-processing chips costing a few dollars apiece.

Demodulating modem signals (which accounts for most of the signal processing in data interception) is far harder for an intercept device than for the modems used by the sender and the receiver. Present-day modems go through a period of training at the beginning of a call during which they study the communications path and "discuss" how best to make use of it. Even if the intercept device is listening to this "conversation," it cannot transmit without revealing its presence, and thus it cannot engage in the negotiations. The signal quality available to the intercept device is therefore rarely as good as that available to the communicating modems.

Only after traffic has been located, demodulated, and recorded do we finally get to the most famous process in communications intelligence, the process of breaking codes: *cryptanalysis*. This book is not the place for a technical discussion of cryptanalysis; such discussions now abound in both the technical and the historical literature of cryptography.[25] It is, however, the place for a discussion of the process of cryptanalysis.

Most of the public literature, both technical and historical, is devoted to *research cryptanalysis*, the process of breaking codes for the first time. This is naturally an indispensable component of any production cryptanalytic organization, but does not account for most of its budget or most of its personnel.[26] The object of "codebreaking" is the development of *methods* that can be applied to intercepted traffic to produce plaintext. In modern cryptanalysis, this is often done entirely by computers, without human intervention.[27]

The process of converting ciphertext to plaintext is called *exploitation*. It follows a process of *diagnosis* closely related to the more general diagnosis of traffic discussed above.

The heart of a communications intelligence organization, however, is not cryptanalysis but *traffic analysis*—study of the overall characteristics (length, timing, addressing, frequencies, modulation, etc.) of communications.[28] Traffic analysis by itself provides a broad picture of the activities of communicating organizations (Wright 1987). One of NSA's most

noteworthy lapses was its failure to detect India's preparations for its nuclear tests in 1998—a failure to detect increased traffic around Pokharan,
southwest of New Delhi, where the tests were conducted.

Moreover, it is essential to assessing the signaling plan, the traffic patterns, and the relationships among communicating entities. Elaborate
databases of observed traffic (Hersh 1986, pp. 258–259) underlie all
COMINT activities.

A last operational point that bedevils communications intelligence is
retention—the preservation of intercepted signals for short or long periods of time until they can be processed, cryptanalyzed, interpreted,
or used. As we have noted, storing a signal that the holder is unable
to restore to its original form typically takes far more memory than
storing an understandable signal. This is justified because, enciphered
messages can be of value even if they are first read only months or years
after they were originally sent. During World War II, Allied cryptanalysts
were sometimes weeks or even months behind on some classes of traffic
(Welchman 1982). Some signals intercepted during the Cuban missile
crisis of 1962 were not read until 2 years later (Hersh 1987). In what is
probably the granddaddy of ciphertext longevity, Soviet messages sent in
the 1940s were still being studied in the 1970s (Wright 1987). Managing
the storage of intercepted material is thus a major problem in all signals
intelligence activities.

After all of the technical processes characteristic of communications
intelligence, the *product* enters into the part of the process common
to information from all intelligence sources: interpretation, evaluation,
dissemination. One process looms larger over COMINT than over perhaps
any other intelligence material: *sanitization*—removal from the intelligence product of information that would reveal its source. Sanitization
to greater or lesser degrees produces intelligence of varying levels of classification.[29]

Secrecy in Communications Intelligence

It is impossible to exaggerate the importance of security to every phase
of communications intelligence. In other areas of military activity, secrecy plays an important role but is rarely indispensable to success. A
superior army often vanquishes its adversary despite lacking the element

of surprise. Even in the area of nuclear weapons (where it abounds), secrecy serves primarily to prevent proliferation. If all of America's nuclear secrets were to be published tomorrow, nuclear weapons would remain just as destructive as they are today and almost as effective as weapons of war. In contrast, communications intelligence would be rendered significantly less effective by disclosure of its techniques and capabilities. Even a credible warning to an opponent that its communications are being intercepted and exploited can result in the opponent's taking action to restore the security of its communications and can destroy the results of many years of intelligence work.[30]

Once traffic has been identified and recorded, shipping it home for further analysis presents security problems of its own. If the intelligence is needed promptly, telecommunication channels must be used. The traffic is, of course, encrypted to conceal from the opponent the details of what is being recorded, if not the fact of interception itself. The circumstances, however, give the opponent a measure of control over what is transmitted on the channel and may provide the opportunity for a chosen-plaintext attack (see chapter 2) on the cryptography. Moreover, unless careful measures are taken to counter traffic flow analysis, correlation is likely to reveal much detail of the interceptors' activities to the opponent.[31]

Current Status of the COMINT Product

Communications intelligence is enjoying a golden age.[32] The steady migration of communications from older, less accessible media—both physical and electronic—has been the dominant factor. The loss of information resulting from improvements in security has been consistently outweighed by the increased volume and quality of information available. As a result, COMINT has been improving for more than 50 years and has become a growth industry.

Even 50 years of success has not made the supporters of COMINT confident that the success will continue, however. From the beginnings of the multinational arms buildup that followed World War II, there have been repeated warnings that improvements in cryptography would bring about the demise of communications intelligence. After the emergence of a public cryptographic technology in the late 1970s, these warnings became especially shrill and were joined by self-confident predictions from the academic and commercial cryptographers that they could produce unbreakable systems and that this would put NSA and its cousins out of business.[33]

The independent cryptographers may well have been correct in their technical bravado but entirely wrong in their view of its consequences. Equating unbreakable cryptography with the security of communications is like equating cryptanalysis with signals intelligence.

It is often said that the intelligence agencies of the major powers can no longer break each other's high-grade systems and must subsist on reading the traffic of Third World countries (Simmons 1986). Although the intelligence community itself has done all it can to foster this view, the steady expansion of COMINT facilities[34] suggests it is too modest.

The status of cryptanalysis in the contemporary world is hard to determine, owing to pervasive secrecy. Oddly enough, although the "Russian Project" is the most secret of NSA's secrets, the fortunes of an activity this important are hard to conceal. What evidence there is makes it plausible that high-grade Russian traffic continued to be read at least until the early 1980s and may still be accessible today.

In its early years, the Soviet Union, like most of the world at that time, relied on code books to secure its military and diplomatic communications. This practice appears to have come to an end in 1927 with MI5's raid on the London offices of the All-Russian Cooperative Society and with the prime minister's admission in Parliament that Britain had been reading Soviet messages for years. It is presumably at that point that the Russians began the extensive reliance on one-time systems that was long characteristic of their operations. In the 1930s and the 1940s their use in Soviet diplomatic communications seems to have strained the facilities for key production to the breaking point, and they began to reuse keying material. Despite the subtle worldwide pattern of the reuse, it resulted in some of their most sensitive messages' being read (Wright 1987). Discovery of this fact after World War II must have led to a broad program to improve the security of Soviet communications.[35]

The Russians were undoubtedly aware of rotor machines and other mechanical cipher equipment as early as the 1920s, but they seem not to have made much use of this awareness before the end of World War II. With their capture of the eastern part of Germany and the acquisition of many of the papers of Pers Z (probably the best of the German crypt-analytic organizations; see Kahn 1967), the awareness must have been enhanced, and perhaps their interest was piqued.

In the late 1940s a cryptographic laboratory was established at Mar-fino, in the suburbs of Moscow. The focus of its efforts was secure tele-phones, of which it produced several, some analog and some digital.[36]

If developments in the Soviet Union followed a course similar to those in the West, rotor machines could comfortably operate at teletype speeds of 50–110 bits per second (bps), but could not keep up with the 2400 bps and higher needed for digitized voice. This led to the development of purely electronic shift-register systems, although rotor machines re-mained in use for text traffic for many years thereafter.

In the late 1950s, according to Peter Wright (1987, p. 148), NSA and its British counterpart, the Government Communications Headquar-ters (GCHQ), jointly mounted an attack on a Russian machine they called "Albatross." Development cycles in cryptography are long, and at that date this was probably a rotor machine. Wright makes no con-crete statement about the success or failure of the project, but the self-congratulatory tone in which he describes pushing the endeavor suggests success.[37]

Traffic encrypted by Soviet cipher machines was also read by the Amer-icans during the 1960s. The messages, encrypted in a Soviet cryptosys-tem which NSA code-named *Silver*, played a prominent role in a 25th-anniversary post mortem of the Cuban missile crisis, held at Harvard University, at which it was revealed that for several hours the Cubans had taken control of a Soviet military base and of some of the nuclear missiles. NSA was not able to read the traffic at the time it was sent; it only became aware of this critical new dimension of the crisis when the messages were first read in 1964.

In their analysis of a number of spy cases from the 1970s, Corson et al. (1989, pp. 94–95) refer to an "NSA intercept from the Soviet Embassy in Washington in April 1977." They go on to say: "The cable was sent by Ambassador Anatoly Dobrynin to the Foreign Ministry in Moscow. It referred to advice Henry Kissinger had given Dobrynin on how to deal with the new Carter administration in the ongoing SALT II negotiation." (ibid., 1989, pp. 94–95) It strains credulity to suppose that such a telegram would have been sent in clear. If the telegram was intercepted by NSA, it must have been cryptanalyzed. The process by which the authenticity of the cable was established lends further weight to this view. The CIA officers involved are quoted as saying that "the only way to confirm the authenticity of the cable was to go out to NSA, pull the transcripts of other cables sent from the Soviet Embassy, and compare the style, content, and timing." As a result, "the experts at NSA concluded that the cable was real and not a Soviet disinformation effort" (ibid., pp. 97–98). This information is all the more persuasive because the authors mention "intercepts" without appearing to have given any thought to cryptography. Their concern is entirely with the content of the cable and its implications about the propriety or impropriety of Kissinger's relationship with Dobrynin.

Evidence of still more recent US success in reading high-level Soviet traffic arises in connection with the September 1983 destruction of Korean Airlines 007. Seymour Hersh's book on the subject describes the interception of a call from Khabarovsk to Moscow placed via the Soviet Raduga satellite and intercepted by the US Jumpseat satellite—which had been placed in a similar orbit for just that purpose. Hersh (1986, p. 232) quotes an unnamed NSA official as saying that "the cipher signal snapped on and some long-precedence message was sent." He remarks that the "NSA officials would not say anything further about the message." Others were more forthcoming, including a senior US intelligence officer who "vividly recalled his reaction well after the shootdown of Flight 007, upon being shown a copy of the deputy commander's intercepted and decoded message to Sakhalin."

The most recent evidence of the continuing success of cryptanalysis involves Iranian communications. Ahmad Chalabi, a Shiite member of the Iraqi government and a founder of the Iraqi National Congress, was accused of leaking to the Iranian government the fact that the US was able to read diplomatic traffic between Teheran and its embassy in Baghdad. No credible explanation of how Chalabi would have known this with any certainty has been put forward, and it appears more likely that the real leak came from the US government. In an effort to discredit Chalabi, who had fallen out of favor in its eyes, the US released an intercepted diplomatic cable quoting Chalabi's warning—solid evidence that, whether the

Iranians had previously known it or not, the US was reading Iranian traffic (Galbraith 2006, p. 30). [Disclosure: Chalabi and Diffie studied mathematics together at MIT from 1961 to 1965.]

Non-Cryptographic Impediments to Interception

If cryptography has not stopped communications intelligence, other developments must at least have slowed it down. In recent decades, the loss of intelligence resulting from the use of cryptography to protect communications appears to have been eclipsed by losses due to other developments not intended primarily for security. These include optical fiber, high-speed modems, and dynamically routed communications.

Between World War II and the appearance of optical fiber, the major developments in transmission technology had the effect of rendering communications more vulnerable to interception. Microwave relays were more accessible than the copper wires they replaced, and satellite channels were more accessible still. Optical fiber, on the other hand, is directly competitive with these radio technologies in cost and bandwidth, and immeasurably more secure. Although undetectable taps on unprotected fiber circuits are possible, they always require physical contact, which is often infeasible. Owing to its economic advantages, optical fiber has been used to reduce the vulnerability of US communications and those of other nations around the world.[38]

A more interesting signal-acquisition problem has arisen out of improvements in modem technology. For decades, Telex and similar low-speed data-communication facilities were the backbone of both commercial and government communications in most of the world. Data rates increased gradually from 50 bits per second to 75 to 110 to 150, and finally to 300. Around 1980, the speeds of inexpensive modems jumped to 1200 bps. Today, they are 28,800 bps.[39] Since the older modems acted essentially independently, each using a phase-locked loop to interpret a set of data pulses in relation to a predictable timing pulse, an intercept modem had no difficulty in doing exactly the same thing.

The new modems not only indulge in initial training to optimize their use of particular communication circuits; they also employ auto-cancellation: both modems transmit simultaneously on the same set of frequencies, and each subtracts its own transmission from the signal it is receiving.[40] Even at 2400 bps, this presents serious difficulties for a passive intercept device attempting to separate the two halves of the signal. At 4800 bps, 9600 bps, and higher, the problem becomes progressively more difficult. Furthermore, it appears to be, in a sense, intrinsic. If the intruding modem can separate and interpret the two data streams, it is receiving information twice as fast as the "legitimate" modems. This suggests that a modem using the same techniques as the intercept device could operate twice as fast. In many cases, the development of the technology of communications and that of communications intelligence proceed independently or even synergistically. In the case of modems, improving technology works directly, if unintentionally, against interception.

The increasing difficulty of acquiring modem signals goes hand in hand with another trend in modern communications: better modems have led to an explosion in the use of dialed-up point-to-point connections to replace leased lines. Private networks often use the same circuits month after month or even year after year. Once such a network has been mapped and access points located, the same intercept facilities can be employed for long periods of time. Furthermore, the ownership of such a net typically determines much about the traffic it carries, which drastically reduces the need for further filtering. In contrast, dialed-up point-to-point connections must be identified within the larger traffic volume of a common carrier's network. This is complicated by *dynamic routing*.[41] Even after it has been determined that a high fraction of the traffic between two particular telephone numbers is worth targeting, it may be difficult to acquire this traffic because different circuits are established on different calls.

The impact of dynamic routing has in some measure been mitigated by commercial developments. The Internet today is intended less to survive a nuclear attack than to serve the needs of millions of customers moving trillions of bits. For the most part, its facilities are owned by a small number of large communications carriers who handle packets by a strategy known as *hot-potato routing*: when you get a packet you try to hand it to the network that owns the destination as quickly as possible. In practice this means that communication between parties on two different networks will be carried on one of two channels, depending on which direction it is going. An intercept facility placed in an appropriate position will have access to a large body of material and will not have to contend with packets following a wide variety of paths.

A related development in switching systems, common-channel signaling (the practice of sending signaling information out of band in a separate digital signaling channel), can be both a blessing and a curse to the interceptor. It is a blessing in that it gathers together in one place the calling number, the called number, and the way the call is to be handled and routed. It is a curse because the common channel can be routed more securely—through copper or fiber, or on an encrypted channel.[42] If this is done, the call itself carries no identifying information and becomes difficult for an opponent to locate. This characteristic makes it possible to upgrade an existing wire-line communication system to a radio-based system of a much higher capacity with little loss of security. All signaling is routed through the pre-existing (and more secure) wires to minimize the vulnerability of the radio circuits.

The Impact of Encryption on Communications Intelligence

Although the spread of encryption technology is not at present the most serious cause of lost communications intelligence, its potential impact on intelligence activities should not be underestimated. Many of today's secure telephones require the users to secure the call as a distinct action from making the call. The process takes 10–20 seconds—long enough to be a deterrent to doing it at all. The digitized voice is of lower quality and may exaggerate other unpleasant phenomena, such as line noise. The callers are likely to say at least a few words to each other before initiating security. If the message is short enough and seems innocuous, they may not bother with security at all. All this leaves room for various sorts of information leakage.

The Integrated Services Digital Network (ISDN)—a set of telephone standards for direct digital telephone service—potentially alleviates the problems of POTS-oriented secure phones.[43] The time required to initiate a secure call drops to under a second and encryption has no effect on voice quality since the signal is digitized in any case. Should ISDN fulfill its promise to permit digital end-to-end negotiation before the called phone rings, the need to initiate security explicitly will be eliminated and the result will be a form of secure caller ID.

The future of voice telephony is Internet telephony, Voice over IP, which lends itself even more readily to full automation than ISDN. Skype, one popular VoIP system, automatically encrypts all calls between Internet clients.[44]

Extensive use of link encryption can also have devastating effects on intelligence gathering. When link encryption is applied to microwave beams and to satellite channels, it conceals everything passing over them; the intruder sees nothing but a steady flow of random data that does not even reveal whether real communication is taking place. Typically, however, link encryption cannot be applied by the users and must be supplied by the carrier. Link encryption will therefore provide users with protection against some spies but not others. In a world with an ever-growing number of interconnected and competing communications carriers, this opens numerous opportunities to couple communications intelligence with human intelligence and network penetration.

Despite the possibilities, the vast majority of the world's traffic is currently in plaintext.[15] This makes it feasible to sort traffic in real time to determine which messages are of interest and which are not. On circuits where the fraction of ciphertext is not too high, the fact of encryption itself provides a valuable clue to the potential significance of intercepted material.

Combined with the limited use of encryption is the diversity of cryptographic products in use throughout the world. The relatively small fraction of traffic that is encrypted today is encrypted in a wide variety of cryptographic systems. This enables interceptors to recognize traffic by identifying the encryption techniques or equipment used. This *diagnosis* of cryptosystems need not require cryptanalysis or cryptomathematical statistics. Distinct cryptosystems typically employ different data formats that can easily be distinguished, and it is desirable from the COMINT viewpoint to preserve those characteristics of communications that permit the filtering of traffic and the selection of messages.

The rise of international encryption standards, even de facto standards, may make this task immeasurably more difficult. We will have more to say about twenty-first-century cryptography in chapters 10 and 11.

One of the distinguishing characteristics of cryptography is that it is robust. Much cryptographic equipment is located close to its users and is likely to survive any attack that does not destroy the users themselves. Cables and optical fibers, like roads and railways, are vulnerable to attack all along their lengths.

The first British military action of World War I was the cutting of an undersea cable, which forced the Germans to use radio for messages to North America and made their communications vulnerable to interception (Kahn 1967, p. 266). Similar scenarios were played out during the Normandy invasion in World War II[46] and at the start of the first Gulf War.[47]

The impact of encryption (and other technical developments) on the interceptor depends very much on the interceptor's position. If the surveillance is entirely external, pointing even the fanciest antennas at the target, a comprehensive program of radio encryption will defeat it. If the surveillance is internal, built into the communications infrastructure for any of a variety of possible purposes, it will be inside this layer of encryption and little affected by it. This has two important consequences. First, it is very difficult for any individual or group within a society to protect its communications comprehensively. It can make use of end-to-end encryption but this will leave the pattern of communications visible. Any greater degree of protection, such as anonymity services, requires the society's cooperation or at least tolerance. Second, it points up the sensitivity of any monitoring capability built into a communication system. By design, the monitoring system will bypass most of the investment in security against external opponents. It will itself become the target and will be especially vulnerable to insider attacks. Security must therefore be a primary consideration in the construction of any such system.

Information Warfare

The meaning of the term "information warfare" is far from settled and the term is applied to subjects that range from modern but established military practice to complete science fiction. In one of its solider embodiments, information warfare is the management of information in warfare. In World War II, pilots would receive intelligence information in a pre-flight briefing; during the mission they would get no new information except for what they could see with their own eyes and an occasional radio message. Today, however, fighter and bomber pilots are assisted from takeoff to landing by the products of a real-time intelligence machine that integrates information from signals intelligence, satellites, surveillance aircraft, and other combatants. It will tell them whether the targets for which they set out have already been destroyed, whether interceptors have scrambled to meet them, or whether previously concealed anti-aircraft batteries have become active and present a threat. It is one of the major objectives of the modern military to close its information-processing loop, bringing observation, decision, and action closer together. The first Gulf War was both a test bed for and a triumph of this approach, which is now solidly established in American military doctrine (Campen 1993a).

Where information processing is an essential military tool, it will naturally be subject to attack. Radar installations are now vulnerable to missiles that follow a radar beam and destroy its source,[48] and much recent military thinking has gone to improving strategies for attacking communication facilities and surveillance aircraft. The possibilities include frying computers with high-power microwaves and shorting them out with carbon fibers.[49] Some attacks on information resources are meant not to destroy them but merely to render them temporarily ineffective. This aspect of information warfare is an outgrowth of the established field of electronic warfare, in which radio and radar are pitted against jamming,[50] decoys, and chaff.

Discussions of this sort are real enough and genuinely high-tech, but in a sense unimaginative. The heart of information warfare today is the notion of attacking the enemy with information alone. This idea is not entirely new. In classical warfare it is called propaganda and disinformation. A less classical antecedent is the practice of communications deception: making use of the opponents' own signals intelligence activities to fool them.[51]

The present-day concept is rooted in the essential role of information not just in battle but in all aspects of society. An opponent who is critically dependent on information will be catastrophically vulnerable to corruption of that information. The notion has been enveloped in an apocalyptic aura by the development of *viruses* and *worms*[52]—malignant forms of software that reproduce within an opponent's computers and eventually cause them to malfunction. Computer viruses originated as a malicious prank and are now a widespread hazard of the computer world.[53] The military vision is that by the application of millions of dollars and hundreds of people far more subtle forms of viruses, suitable as weapons in military conflicts, can be developed.[54]

The impact of such invaders has already been quite noticeable. One incident brought down a telephone "loop carrier" switching system, disabled the tower at the Worcester Airport and shutting down the airport for six hours (Festa 1998). An attack on a sewage treatment plant in Maroochy Shire, Australia resulted in the release of thousands of gallons of untreated sewage (Shea 2003). A safety monitoring system at the Davis-Besse nuclear power plant was disabled by the Slammer worm. Fortunately, the plant was off at the time and there was no immediate hazard. The worm had bypassed the plant's firewall by entering through a machine on the unsecured network of a contractor (Poulsen 2003).

Computer Intelligence

One aspect of information warfare that is unquestionably real, though how much of it is occurring is hard to assess, is the practice of obtaining information by active intrusion into a target's computers or networks. We shall call this field *computer intelligence*.

Both the strengths and the weaknesses of communications intelligence derive from the fact that it is passive. On one hand, its passive character means that communication spies are rarely caught. On the other, its passivity deprives it of the chance to go after particular pieces of information and restricts it to listening to what opponents decide to transmit. This raises the cost of interception by obliging the interceptors to winnow through vast quantities of traffic in order to find what they want to know. A passive eavesdropper must wait for some legitimate user to access the information and then record the result; an active one can go to a database and extract a particular piece of information.

Intrusions into American computers by a group in Germany with ties to the KGB are described in a 1989 book by Clifford Stoll. An operation in Tripoli by the Israeli Mossad provides an interesting example of the intersection between human intelligence and the low-tech end of network intelligence. Using a phone line that actually originated in Israel but appeared to originate in France,[55] and masquerading as French shipping insurers, the Mossad recruited the harbormaster in Tripoli and "ran" him for more than 2 years (Ostrovsky 1990, chapter 16). With the worldwide linking of computers through the Internet, new techniques for extracting information by active penetration are at the frontier of intelligence research (Schweizer 1993, pp. 158–163) and are being developed all over the world.

At a meeting on information warfare at Stanford University, members of the President's Commission on Information Warfare and Critical Infrastructure Protection acknowledged that there has not yet been an example of information warfare in its pure form. No nation has attacked another nation's computers using information. Nor is it believed that a politically motivated attack on computers using information alone has been made by terrorists or other non-national groups. Nonetheless, information warfare is very real, and very alive as a subject of military speculation, planning, and development. Not a month passes without a conference, meeting, or war game devoted to the subject.

In the late 1990s, these issues appeared to be largely theoretical. They are no longer. It is clear that the Chinese government has "invested significantly in cyberwarfare training and technology" (Kaplan 2005, p. 54). Japan has already suffered a number of attacks originating in China and South Korea (Faiola 2005). Japan is not alone. The US has also been targeted.

"China has downloaded 10 to 20 terabytes of data from the NIPR-Net [The Department of Defense's Non-Classified IP Router Network]," Major General William Lord, director of information, services and integration in the Air Force Office of Warfighting Integration and Chief Information Officer, reported in 2006.[56] We have evidence of clear and highly targeted attacks. For example, the following set of attacks sought military computers that had specific known vulnerabilities:

- "At 10:23 P.M. PST, [attackers] found vulnerabilities in computers at the US Army Information Systems Engineering Command at Fort Huachuca, Arizona.

- At 1:19 A.M. PST, they found the same hole in computers at the military's Defense Information Systems Agency in Arlington, Virginia.

- At 3:25 A.M. they hit the Naval Ocean Systems Center, a defense department installation in San Diego, California.

- At 4:46 A.M. PST, they struck the United States Army Space and Strategic Defense installation in Huntsville, Alabama." (Thornburgh 2006a)

Of course, we do not know for sure that these files were stolen by the Chinese military. But what we do know would surely indicate that. We know the files were "zipped" and immediately transmitted to computers in South Korea, Hong Kong, or Taiwan, and then to the People's Republic of China. Attacks were fast: in and out of the targeted computers in 10–30 minutes. And, most telling to government investigators, "these guys never hit a wrong key" (Thornburgh 2006b).

We also know some things that have been taken: specifications for the aviation mission-planning system for Army helicopters from the Army Aviation and Missile Command and Falconview 3.2, the flight-planning software used by the Army and Air Force (Espiner 2005).

On balance, the Department of Defense takes the security of its networks—even its unclassified networks—more seriously than do most corporations. If penetration on this scale could happen to a military network, it seems prudent to assume that it is also happening to civilian networks.

The relevance of information warfare to cryptographic policy is twofold and straightforward. The major worry of most pundits is that critical elements of national infrastructures such as transportation systems and power grids are being connected to control systems that communicate via the Internet. Much of the plausibility of this concern lies in the lack of authentication in current computer networks. Viruses might get in because new versions of programs are loaded over the Internet, and there is no easy way of telling a genuine program from an alternate one prepared by intruders. Furthermore, the information that opponents would need to mount an attack is available as a result of the general lack of security in communications. Widespread deployment of cryptography in the "command and control" of the civilian infrastructure would solve both problems.

The Relationship of Security and Intelligence

In loose correspondence with the various categories of intelligence are security measures intended to counter them and limit their effectiveness. Thus, for example, human intelligence can be countered by limiting information access to vetted personnel, photographic intelligence can be countered by camouflage, and open-source intelligence can be countered by restricting public access to information or mixing false information with genuine.[57] Cryptography is the centerpiece of communication security, the countermeasure to communications intelligence.

As with the various aspects of intelligence, security measures are far from independent. For example, good personnel security is essential to communication security, and communication security can in turn make a major improvement in operations security.

The Security of Communications in the United States

No nation in the world is more dependent on electronic communications than the United States. As a result, no nation is more vulnerable to subversion of its commerce, its money supply, and its civic functions by electronic intruders. Attempts to address this vulnerability take two forms:

- Protecting American communications by government action in the same way that the country as a whole is protected by defense and law-enforcement agencies.

- Leaving the protection of most communications to the private sector and encouraging such protection by such measures as standards, incentives, and regulation. This parallels the way in which physical security is provided by locks and alarm systems in civil society, often in consideration of reduced insurance premiums.

In practice, any comprehensive solution must have elements of both.

In the 1970s, the US government made its first attempts to secure broad segments of American communication rather than narrow classes of military, diplomatic, and intelligence traffic. Some communications (in Washington, New York, San Francisco, and other areas that harbored Soviet diplomatic or consular facilities) were routed through underground cables rather than over microwave relays, analog and digital encryption devices were developed for the protection of telephone trunk lines,[58] the security of common-channel interoffice signaling was improved,[59] and telephone satellite channels were encrypted.

With the demise of the Soviet Union, whose hostility to the United States was supported by a massive intercept capability, and with the migration of more and more of our critical infrastructure to Internet- and Web-based mechanisms, the focus of our concerns has shifted from passive intercept to active attack. In the process, national security and commercial security have become intertwined.

For most purposes, the Internet is the most effective and economical communications medium in the world, and businesses have been quick to improve their functioning and lower their expenses by taking advantage of it. The flexibility and worldwide ubiquity of the Internet have also made it an ideal culture medium for a variety of activities that threaten the security of both critical and commercial infrastructure.

The threats can be loosely categorized into a half a dozen forms:

- Break-ins to websites—Most businesses have customer-facing websites that advertise their wares, allow communication with their employees, and perform other functions. Competitors, detractors, and customers may all find ways of interacting with the website that are not what the provider had hoped for. Extraction of more information than the provider intended to provide frequently goes unnoticed, but defacements or perversions of function can cause the website provider embarrassment and financial loss.

- Viruses and worms are an automated form of computer penetration that can be spread by almost any form of computer communication and have shown tremendous destructive potential.

- Denials of service—When opponents cannot break into a website, they may still be able to mount an attack that prevents it from functioning correctly. Such attacks quickly developed the sophisticated technique of capturing less-well-protected computers and turning them into *zombies*. The zombies are woven together into a *botnet* and used to attack particular targets, a technique called *distributed denial of service*.

- *Spam*—Email on the Internet is billed not by the message or by the bit but by the month. There is no disincentive to sending lots of mail. This is analogous to—but even more extreme than—the artificially low bulk mailing rates that support the clutter in our physical mailboxes. Unwanted mail that varies from uninvited to repulsively unwelcome also serves to support other forms of computer malfeasance. Spam can be used to spread viruses and worms, gather information about active email accounts, and commit fraud.

- *Phishing*—Spam uses a number of mechanisms to trick the recipient into providing information that can be used for *identity theft* or other nefarious purposes.

- *Spearing*—Targeted attacks encouraging a small, carefully selected group to install a patch in their security systems. The patch is in fact a vulnerability.

The threats, coupled with the staggering commercial importance of the Internet, have created a new security industry with revenues in the tens of billions of dollars a year. The new industry is more complex than its "purely national security" predecessor. The only defenses against Soviet eavesdropping were proactive. If we failed to prevent them from getting useful information, there was rarely anything we could do after the fact. Commercial security employs a combination of preemptive measures—firewalls, intrusion detection systems, encryption—with forensic and investigative techniques that deter opponents who are more subject to legal retribution than was the Soviet Union.

Our efforts to date, however, fall far short of providing the degree of protection desired in a communications infrastructure that has become indispensable to American prosperity and security.

Federal Policies and Programs

The challenge, from the national-security viewpoint, is to achieve a two-fold objective:

- Improve the security of communications and computing within the United States and for US government and commercial activities abroad.

- At the same time, attempt to minimize the impact both on US intelligence activities and on domestic security that could result from having the country's own technology used against it.

This objective is difficult, if not impossible, to achieve by a reactive strategy of permitting events to unfold as they will and responding to them piecemeal. Threats to American intelligence capacity, both domestic and foreign, can be anticipated, and policies can be developed to nullify them. Only a misplaced sense of fair play would demand that threats to American well-being should be allowed to develop freely when the means to control them are at hand.

Export Controls

Most of the federal activities discussed so far do not affect the public directly. For one thing they are secret. When foreign policy is successful, people who give the subject some thought may attribute a share of the success to intelligence. When the United States is surprised by something —like the taking of the hostages in Iran, the Iraqi invasion of Kuwait, or the attacks of 9/11—poor intelligence is likely to be blamed. Intelligence has, however, no visible day-to-day impact on the lives of most Americans.

There are, however, federal activities in support of intelligence that affect many people—usually, as those people see it, adversely. These are the export-control laws. Although the US Constitution prohibits export tariffs, it does not prohibit an outright ban on exporting particular things to particular countries.

All exports from the United States are regulated under one of two laws: the Arms Export Control Act (22 U.S.C. 2571–2794) and the Export Administration Act (50 U.S.C. App. 2401–2420). The Arms Export Control

Act takes precedence over the Export Administration Act and confers on the Department of State the authority to regulate the export of anything it deems to be a weapon of war (or, as the export laws term it, a *munition*). Items ruled to be munitions require individually approved export licenses designating the customer, the application, and often conditions for the handling or redeployment of the item.

Things that are not munitions but that may have military applications are called *dual-use items*. If the Department of State decides that something is a dual-use item, it transfers jurisdiction over its export to the Department of Commerce, which administers the Export Administration Act. Under the Export Administration Act, exporters can receive licenses to export to broad classes of customers in broad regions of the world. In the area of cryptography, for example, equipment using the Data Encryption Standard to authenticate bank-to-bank wire transactions was allowed to be exported to banks in most countries in the world even when export of comparable equipment for other applications was not. Under the Export Administration Act, furthermore, the Department of Commerce is obliged to take into account the foreign availability of equivalent products[60] in deciding whether to grant or deny an export permit—that is to say that it can block exports only where there is evidence that such action is actually likely to prevent a foreign customer from acquiring a product with equivalent capabilities. Under the Arms Export Control Act, no such test of foreign availability is required. All cryptographic devices that do not fall into certain narrow categories are regulated as munitions and require individually approved licenses.

Many of the actions of the export-control authorities seemed ludicrous and inspired widespread resentment. In 1994, Philip Karn, a security engineer at the cellular telephone maker Qualcomm, applied for a license to export a copy of Bruce Schneier's popular book *Applied Cryptography*. The license was granted, and the accompanying letter stated that the Department of State did not have authority over published material—a view commendably in accord with the First Amendment. Karn then applied for an export permit for a small part of Schneier's book—an appendix containing source code for cryptographic algorithms—transcribed onto a floppy disk, rather than on paper. That application was denied. This case, which is working its way through the federal courts, has made the export-control regime an object of ridicule, but the cryptographic export policies of the United States may appear less foolish and irrational when examined in light of communications intelligence practices.

One natural objective of cryptographic export control is to limit foreign availability of cryptographic systems of *strategic capability*—those capable of resisting concerted cryptanalysis by US intelligence agencies. Were this the only objective, export control in the cryptographic area would be much like export control in other areas—items that have only military uses or have been explicitly adapted to military applications would be treated as munitions, others would not.[61]

Probably the most important objective of the export-control regime in the area of cryptography is to slow the widespread deployment of cryptographic systems of sufficient strength to present a serious barrier to traffic selection. Rather than limiting the export of cryptosystems whose traffic would take weeks, months, or years to break, the objective is to prevent the export of cryptosystems that cannot be broken in real time by intercept equipment in the field. This is a far lower bar, and it precludes the export of any system that could reasonably be said to provide acceptable security for most commercial applications.[62]

It also appears to have been an objective of export control—and, if so, one that had remarkable success—to prevent widespread adoption of standard cryptographic systems. The development of standards would be expected to have two effects from an intelligence viewpoint. It would expand the use of cryptography, thereby complicating both traffic selection and exploitation. It could also result in a uniform appearance of broad categories of messages, making the problem of selection harder still.

More recently, US policy has shifted from suppressing to promoting standard cryptosystems. This change will be explored in subsequent chapters.

A final objective of export control goes virtually unnoticed. It is to maintain an ongoing assessment of the quality, availability, and functioning of commercially supplied cryptographic equipment. Would-be exporters are required to disclose the details of their products to the government on a routine basis. Even if they are not obliged to modify their products in order to get export approval, this guarantees that NSA will have the details of each product's functioning on file. The process of acquiring information on how cryptographic products work is thereby separated from any actual occasion on which their traffic is being intercepted, thus contributing to security. From this point of view, a product exported under an export-control permit is entirely different from and far preferable to one exported without any permit or any reporting requirement.[63]

By limiting the strength of exportable cryptosystems to well below what the users felt they needed, export control created a direct conflict between the needs of the government and the needs of commercial and private cryptographic users. It is an oft-expressed opinion that commercial communications do not require the same level of protection as military communications. This is probably more a reflection of the fact that the military are aware of who their opponents are and of the level of effort that these opponents put into attacking them than a reflection of the value of the communications. The communications of commercial organizations are often worth hundreds of millions of dollars,[64] and many industrial secrets, along with much personal and personnel information, have long lifetimes. Air traffic control, power grid regulation, and control of communication networks are essential to the working of society; their disruption would expose participating corporations to immense liabilities and might cost lives as well as dollars.

Cryptographic keys are often held to be the most sensitive of all secrets, because anyone who has access to the keys can gain access to all other secrets (Clark 1986, p. 11-1313). In a "flipside" to this vision, controlling the export of cryptography was seen as essential to controlling the export of information in general. With the increasing importance of intellectual property to modern commerce, it was thought that if smugglers had access to encrypted communications, the export of any form of information would become impossible to regulate and the United States would lose all control of its "electronic borders." Cipherpunk talk of crypto-anarchy did little to allay the government's fears.

Fortunately, the attitudes toward cryptography that characterized the Cold War and its immediate aftermath have begun to change. As we will examine in the latter chapters, export controls have been relaxed, and high-grade cryptography has been adopted in national standards. These developments hold promise of the more harmonious relationship between national security and commercial security that the modern world requires.

5

Law Enforcement

The Function of Law Enforcement:
Solution versus Prevention

The purpose of law enforcement is to prevent, interdict, and investigate crimes and to prosecute criminals. There is a certain logic to the order in which these objectives are presented; it goes from the most anticipatory to the most reactive. The closest relationship is between investigation and prosecution, which are by and large the most visible and best known of law enforcement's functions. We will examine these first, then turn to prevention, of which a major component, deterrence, is closely related to the success of investigation and prosecution. Finally we will examine interdiction, which is the area that most often brings law enforcement into conflict with civil liberties.

If asked to name a crime, most people would pick a typical crime: murder or robbery or rape or fraud. Everybody agrees that these are crimes (although there is often disagreement about whether a particular event is an instance of the crime), and they are crimes in which the victims are identifiable. Except (of course) in the case of murder, police generally begin their investigation by questioning the victim to get a description of what has happened and to get "leads."

The investigation of crimes has not changed since the rise of the police in the mid nineteenth century, but the way in which they go about it has changed a great deal.

In the nineteenth century, police work was largely a matter of interpersonal relations. The policeman walking a beat depended as much on rapport with the people of the neighborhood as on a gun or a nightstick. In the absence of a radio or even a call box, he did not have the option —the first resort of police today—of communicating with the station before taking action of any kind. If he encountered trouble, his choices were to deal with it himself or to turn and go for help. The actions of the police in investigating crimes reflected the same skills and resources they used on patrol. They relied less on forensic evidence, record keeping, and communications and more on talking to people and knowing the community. Police action after a crime would rely heavily on an expertise in community functioning acquired continually as the police developed informants or kept watch on markets known to include stolen goods among their wares.

Since then, the character of policing has changed dramatically, and today the police depend as heavily on technology as on their skill in dealing with people. Nowhere is this more apparent than in detective work. The invasion of technology began around 1900 with the development of fingerprinting[1] and forensics. Forensics, as popularized in the Sherlock Holmes stories, has become a mainstay of detective work, and today crime laboratories maintain vast files of common items (such as paper, automobile paint, and duct tape) that enable them to identify and track these items when they are found at crime scenes (Fisher 1995, pp. 159–193).

When police investigate the scene of a crime today, their first action is to seal off the scene and take numerous samples—fingerprints, cartridge cases, clothing fibers, traces of DNA, and so on. In conjunction with the lab analysis of these samples, they will make extensive use of their records on people (suspects or witnesses) and on cars, guns, jewelry, and other objects that may have been involved.

The act of investigating a crime consists most conspicuously of attempting to discover who committed the crime, but it may also involve determining the nature and extent of the crime and whether there was a crime committed at all. If this process is successful, the next step is prosecution. Once a suspect has been identified, the police must supply the state's attorney with sufficient evidence to prove the suspect's guilt. Sometimes this process is barely distinguishable from investigation. The police may, for example, observe a suspect's movements until they have a persuasive case that he was in a position to commit the crime and was the only person in such a position. In other cases, particularly in crimes such as burglary or fraud that are likely to be repeated, the process of developing evidence takes on a life of its own and looks quite different. The police may watch someone, whom they suspect of committing a crime but against whom they lack sufficient proof, in hopes that the suspect will grant a repeat performance. They may go so far as to create attractive conditions for the commission of the crime and lie in wait for the suspect to take the bait.

The prevention of criminal activity relies on a combination of security and deterrence. Security measures, ranging from locks and fences to surveillance cameras and patrols, make it more difficult to get away with crimes. Deterrence is in part a result of security and in part a result of the investigation and prosecution of crimes. It persuades a would-be criminal that even success in committing a crime will not mean he has gotten away with it. Security and deterrence are not always easy to distinguish. Police walking beats or patrolling in cars convey the impression that a crime might be observed and stopped outright; they also serve to remind the citizens that the police are present and are likely to find and prosecute the perpetrators of crimes.

The important factor unifying security and deterrence is that neither is aimed at particular individuals. Both are intended to prevent the commission of crimes—to communicate to everyone that criminal efforts are unlikely to succeed and likely to be discovered and punished.

Interdiction of crimes has a different flavor. It is aimed not at everyone, but at particular people who have been detected in the process of planning or preparing to commit criminal acts. Unlike deterrence, it involves concentration of police attention and possibly police action on people who have yet to commit any crime. Any police program of interdiction must, therefore, involve watching people who, at least under US law, are entitled to a presumption of innocence.

A Brief History of the Police

The police as we know them today—an armed force maintained by the state to perform the functions described in the previous section and paid a salary rather than a share of fines[2]—are a rather recent phenomenon. Police appeared in France as one of the products of the revolution, but police in the United States stem more from the British tradition.

In 1829, Home Secretary Robert Peele established the Metropolitan Police in London. Initially this force was limited to uniformed patrolling, but in 1842 it was expanded to include the Criminal Investigation Division, responsible for detective work. London's salaried force became the organizational model for police throughout Britain and subsequently the United States. Police forces were formed in all major US cities during the latter half of the nineteenth century.

At the time of their founding, the fundamental mandate of the police was a combination of deterrence (through their visible presence on patrol) and investigation. Under earlier British law, there was less provision than there is today for discovering the perpetrators of a crime when their identities were not obvious; it was the responsibility of the injured parties to investigate and "solve" the crime. Over time, however, investigating and solving crimes has evolved into one of the most important functions of the police.

At the time of their founding, the police were viewed not as an outgrowth of the state's ability to make law but as a manifestation of its ability to use violence. As time went by, they gradually became essentially the only body entitled to employ violence in anything other than self-defense and narrowly construed instances of defense of property. They also came to be seen as the lower rungs of a ladder whose upper steps are the prosecutors and the courts.

In the original British conception, police powers of arrest were limited, and police were supposed to have little or no influence in whether cases were brought to trial. Over time, however, the influence of the police over such decisions has increased dramatically,[3] as has their influence over the making of laws.[4]

The Use of Wiretaps in Law Enforcement

Both before and after the development of telecommunication, a central element of police work has been the acquisition of information about criminals' plans and conversations without their knowledge or cooperation. There are two fundamental ways in which this can take place:

- through conversations in which a criminal is talking to someone who, usually unbeknownst to the criminal, is an undercover police officer or is providing information to the police

- through conversations between criminals being overheard by the police or by their agents.

The development of telecommunication has had an impact on both of these scenarios, but it has affected the latter far more than the former.

Our first scenario occurs widely in traditional police work—for example, when a victim of extortion or blackmail relates the threats he or she has received to the police, or when the police employ a stool pigeon. Telecommunication has left many of these practices unaffected while opening new opportunities for others. Demands by kidnappers and blackmailers are often delivered by phone and may readily be taped at the victim's end for use as evidence or for other investigative purposes. Such taping, legally and practically distinct from a wiretap, is called a *consensual overhear*.

Our second scenario is the home territory of electronic surveillance.

The crimes we have called typical are investigated largely at the interface between the criminal and non-criminal worlds. But neither the typical crimes nor the broader class of acts generally accepted as crimes and having identifiable victims exhaust the criminal repertoire. Many activities that are criminal under law do not have readily identifiable victims. Not coincidentally, the criminal status of such activities—which include such classic activities of organized crime as prostitution, gambling, and drugs—is often controversial.

Under these circumstances, the police cannot proceed from questioning an aggrieved victim to investigating and prosecuting an offender. They must instead attempt to infiltrate the criminal activity—to intrude on the interactions of a group of people who are, by and large, either satisfied purveyors or satisfied customers in the illegal trade.

For many crimes, including conspiracies to fix prices, bribe government officials, or commit terrorist acts, such infiltration is a difficult procedure.

The participants are wary of newcomers, and normal routes of investigation are sealed; it may even be hard to determine what crimes are occurring. It is in such environments that electronic surveillance comes into its own, providing information about the crime being committed and insight into the structure of the criminal organization.

Rarely is it possible to develop or plant an informer high in a syndicate. Information from surveillance enables law enforcement to develop a coherent view of a conspiracy. In the Illwind investigation of corruption in the procurement of weapons for the US military, FBI agents were afraid to run credit checks on suspects for fear of alerting them. Instead they used wiretaps and electronic bugs (which were even more productive) to assemble a picture of the conspiracy. Hundreds of agents were involved in the surveillance (Pasztor 1995, pp. 190–191). Wiretaps can also provide mundane details—the suspect's daily schedule (Weiner et al. 1995, p. 223), or the personal relationship between the suspect and his co-conspirators—that can vastly simplify an investigation.

Though equivalent to electronic bugs from a legal viewpoint, wiretaps are generally easier and less dangerous to install. They usually do not provide the same quality of information, however. Criminals are typically considerably less forthcoming over wiretapped telephones than in bugged rooms.[5] Microphone rather than wiretap evidence was responsible for convicting John Gotti, head of the Gambino crime syndicate; information obtained by means of a bug in a Gotti hideaway convinced underboss Sammy Gravano to testify against Gotti.[6] Electronic bugs were also responsible for the conviction of John Stanfa, a Philadelphia crime boss (Anastasia 1994; Hinds 1994).

Even when wiretaps do not directly provide evidence, they can be useful. Wiretaps informed US agents about meetings between the spy Aldrich Ames and his Soviet handlers (Weiner et al. 1995, p. 246).[7] Wiretaps also apprised agents of the fact that Ames's wife had aided him in his illegal activities. Investigators used this knowledge to extract a more detailed confession from Ames in exchange for a reduced sentence for his wife (ibid., pp. 261–262 and 288).

In general, wiretaps appear to be of greater value in gathering intelligence than in developing evidence. They can expose the connections within an organization, and they may reveal events before they occur. Not to be undervalued is the fact that knowledge of the possibility of wiretapping renders the telephone far less useful to criminals.

Wiretaps and Their Relatives

How Wiretaps Work

Words spoken into a telephone begin their journey at the microphone of the handset, which may be connected to the phone by wire or (in the case of a cordless phone) by radio. From the phone, the signal is passed down a line cord to a wall socket. If the phone is in an office building, its next stop is typically a phone closet where the wires from many phones on the same floor come together. If the phone is in a private home, the signal generally goes directly to a junction box on an outside wall and then by wire to a pole; it is likely to run on poles for only a short distance before it joins wires from other houses in a junction box (similar to the office phone closet) and is routed underground. At this point, the signal takes a fairly long hop and makes its way to the local telephone exchange, where it is received on a frame.

By comparison with the previous elements, the frame plays a more profound role in the call's progress. Up to that point, phone lines were arranged in a basically geographical way, lines coming from nearby houses being routed close together. The frame rearranges the lines to put them in numerical order by phone number. From the frame the lines go into a telephone switch, the formerly mechanical and now electrical equipment that routes calls from one telephone to another. From this point on, no path is permanently bound to any particular phone.

From the viewpoint of security, a phone call is vulnerable to wiretapping at every point along its path. A wiretap may be placed in or close to the phone.[8] It may be placed in a phone closet or in a junction box. It may be placed on the frame or inside the telephone switch. From the viewpoint of an intruder, however, things look entirely different and each of the possible opportunities for wiretapping has serious disadvantages.[9] If he tries to put the tap in the phone, he may get caught in the act of planting the device.[10]

Tapping intermediate junction boxes, whether in phone closets, on poles, or elsewhere, presents both the danger that the installation will be observed and the danger that the tap will later be found by maintenance personnel. Precisely because a junction box serves a number of different clients, it is subject to unpredictable visits by installers and repairmen connecting, disconnecting, or troubleshooting the phones of subscribers other than the target. Any of these maintenance personnel may mistake the tap for a wiring error and disconnect it or take some action that will reveal the tap to its target.

One traditional point for attaching wiretaps was the frame, where an incoming line can readily be connected to an outgoing one in such a way as to make the call accessible at an additional external location. Occasionally this is done for quality-control purposes in the normal course of telephone operations. Like a junction box, however, a frame is tended by numerous people from whom the tappers are trying to keep the tap concealed.

Until the advent of digital telephone switches, the phone, the junction boxes, and the frame were essentially the only points at which a tap targeting an individual line could be placed. It was within the ability of intelligence agencies to target multiplexed flows of traffic further removed from the targets, but only because they had both the budget and the legal mandate to listen to whatever they could find that was of value. Police wiretapping, by comparison, was targeted at individuals and had to be conducted on a limited budget with limited cooperation from the telephone companies.

Digital telephone switches, such as AT&T's ESS series and Northern Telecom's DMS-100, introduce a new (and, from the police viewpoint, far superior) way of wiretapping. The new technique is to make use of the switch's ability to create conference calls. (A tapped call is, after all, simply a conference call with a silent and invisible third party.) Conference calling, however, was not designed with the wiretapper in mind. The technology was intended primarily for creating conference calls in which all connected phones were active and cooperating participants. The new switches had secondary applications to debugging and quality-control monitoring that were closer to the tapper's desires, but a switch's capacity to provide these services was typically limited to a few lines at a time. Although this was generally sufficient for law enforcement use in the past, law-enforcement agencies have recently begun pushing for vastly expanded wiretapping facilities.

Pen Register and Trap and Trace

As we noted in the previous chapter, the backbone of a communications intelligence organization is not its ability to read individual messages, valuable as that is, but its ability to keep track of a broad spectrum of communications through traffic analysis. Although this is less true of law-enforcement intercepts, a sort of traffic analysis, generally targeted at individual lines, plays an important role here.

A log of the numbers of all phones called by one particular phone is called a *pen register*.[11] In the United States the ability to record called numbers has been an essential component of billing for a long time and thus has been built into telephone equipment for a long time.[12] Call logging can also be carried out by equipment on a subscriber's premises (Jupp 1989).

The inverse activity—taking note of the numbers of all phones that call a particular phone—is called *trap and trace*. Unlike logging the numbers of outgoing calls, this was very difficult before the changes made during the past two decades in telephone signaling.[13] In the era of electro-mechanical telephone switching, call tracing took many minutes, during which the caller had to be kept on the line.[14] With digital switching, however, tracing information is almost always available until at least the last switch through which the call passes, and it is often available to the receiving telephone in the form of Caller ID.[15]

As in intelligence work, analysis of the patterns of telephone calls can reveal the structures of organizations and the movements of people. Where billing records preserve such information, it can be employed after the fact, even though the need to do so was not anticipated.[16]

Electronic Surveillance in Context

To be understood correctly, electronic surveillance must be seen in the context of the use of technology by police.

Police have often been at the forefront of the use of new technologies.

Scotland Yard was connected to the district police stations of London by telegraph in 1849, only 5 years after the telegraph was invented. In 1878 the Chicago police introduced telephone boxes along policemen's routes. "Wirephoto," credited with the capture of a criminal as early as 1908, became widespread after World War I.

Some elements of today's police technology are entirely new; others have clearly recognizable roots but have changed so much that they are only barely recognizable in relation to their ancestral forms. The most basic of these are the technologies that shape the operations of a police organization: record keeping and what we will call by the military term "command and control."

Police records serve both a strategic and a tactical function. Strategically, they allow police to decide how to deploy their forces. Tactically, they provide information on persons, property, and events connected with the investigation of individual crimes. Police records were once local to cities or districts, and sharing of information between police was a slow, uncertain process. Today, however, a computerized National Crime Information Center provides information to federal, state, and local police forces throughout the United States. This source of information is augmented by police access to commercial credit databases, national telephone directories, and other online information services.

The utility of records is closely connected to another mainstay of police technology: communications. Today most on-duty officers, whether patrolling in cars or walking beats, are in immediate radio contact with their stations. Police cars in some cities carry data terminals that allow direct access to printed records; in other places, officers on patrol must make voice contact with a dispatcher who has access to databases. Police forces today not only have access to nationwide (and often worldwide) records, but much of that access is directly available to officers in the field.

Radio communication also gives the dispatcher immediate access to officers on patrol. To utilize this effectively, the dispatcher must have information about where the officers are and what they are doing. At regular intervals, officers report their locations and activities, and the information they provide is entered on maps and status boards. In the most advanced setups, tracking systems such as Teletrack automatically report the location of each patrol car and display it on the dispatcher's map. Coupled with databases of locations developed to support the 911 emergency system,[17] this permits the dispatcher to recognize where a telephone call is coming from and identify the nearest available emergency personnel. Within a patrol car, it may provide automatic directions to the place the car has been told to go.

The same technologies that permit police forces to track their officers permit them to track and watch individuals and goods with unprecedented ease. Many truck, bus, and automobile fleets use tracking systems to monitor the movements of their vehicles. This allows vehicles to be tracked if they are stolen and makes it difficult for the drivers to use them for purposes other than those the employer intended. Surveillance is also facilitated by computerized road tolls, optical character recognition of license plates, and specialized networks for tracking "bumper-bugged" vehicles (Burnham 1996, p. 138).

Police track more than vehicles. It is a rare person in the modern world who can avoid being listed in numerous databases. From a police viewpoint, the process of identification is a process of matching a person with society's records about that person.[18] Until recently, fingerprints were of more use in confirming identity than establishing it. Even long after the availability of computers, the search of a large fingerprint file was a slow process requiring expert human labor.

Even though fingerprint records have been computerized, fingerprints are not an ideal way of tracking the movements of people, and they are rapidly being supplanted or augmented by other technologies.

Video cameras are now ubiquitous. It is popularly believed that such cameras run "tape loops" containing, for example, the last half-hour's view of people entering and leaving a bank. In fact, videotape is cheap, and many cameras record images for much longer periods. After the bombing of the Murrah federal building, the FBI collected videotapes from all over Oklahoma City, synchronized them using the shaking that resulted from the blast, and watched the movements of people before and after the explosion.[19] One developing technology allows automatic identification of people from their videotaped images (Busey 1994). Another new form of technology that allows "surreptitious fingerprinting" is infrared imaging of the veins in the face. Like fingerprints, these are unique to individuals. Unlike fingerprints, these veins can be detected by hidden infrared cameras installed in airports and other public places.

Today almost every American adult carries a driver's license[20] and other forms of identification that are hard—and illegal—to counterfeit. The near ubiquity of identification cards has given them wide social acceptability. Many communities require hotel registrants to show identification, presumably as an anti-prostitution measure. The 1995 decision by the Federal Aviation Authority that air travelers could be required to show government-issued identification when checking in at airports and more drastic post-9/11 security measures have made anonymity impossible for law-abiding travelers. The Supreme Court ruling that the police cannot require a person to show identification (*Kolender, Chief of Police of San Diego, et al. v. Lawson* 461 US 352 (1983)) has brought no noticeable change in police practice.

Possibly the most important way of tracking individuals is through credit cards. Since credit cards are the easiest way of making most purchases and are essentially required of persons renting cars or checking into hotels, the databases used for billing and credit verification contain good pictures of most people's movements.

Material objects too are tracked; they are also examined. Fear of terrorism initiated the development of a broad range of devices intended to search for guns and bombs. The capabilities of the most recent equipment go far beyond that, however. Some baggage x-ray machines can be programmed to look for sharp things, for guns, for drugs, for precious metals, or for fruits and vegetables. Similar devices can look at an entire truck and detect drugs right through its aluminum skin. There are detectors based on magnetometry, on vapor analysis, on neutron activation analysis, and on nuclear magnetic resonance. Detection of mass concentrations from their gravitational effects is in an experimental stage.

Evaluating the numerous technologies that have appeared over the past 100 years for their law-enforcement potential versus their criminal potential produces a preponderance of cases that favor law enforcement. On the criminal side, chemistry has given us a variety of new drugs that have been condemned by society but enjoy good sales anyway. Improvements in manufacturing and competition in the international arms market have given us "Saturday night specials." Photography has contributed to blackmail. Computers and communications have brought us a new kind of crime—theft of computer and communication services—even while contributing to the operations of both criminals and police. It is hard to see much that microscopy, x-rays, database technology, microbiology, infrared imaging, MRI, or numerous other technologies have contributed to criminal enterprises; they have, however, given the police a host of techniques for tracking, identifying, and monitoring both people and physical objects. On balance, the impact of technology is so weighted on the side of law enforcement as to make it remarkable that crime has survived at all.

Blurring the National Security/Law Enforcement Distinction

In general, "national security" refers to the government's operations outside the borders of the United States and "law enforcement" to its domestic operations. However, at times—most dramatically in the 1960s and the early 1970s—the distinction has been blurred.

In the mid 1970s, the Church Committee, a year-long Senate investigation of US intelligence operations, found evidence of "domestic intelligence activities [that] threaten to undermine our democratic society and fundamentally alter its nature" (USS 94d, p. 1). The Church Committee recommended that the CIA, NSA, the Defense Intelligence Agency, and the armed services be precluded, with narrow and specific exceptions, from conducting intelligence activities within the United States, and that their activities abroad be controlled so as to minimize impact on the rights of Americans (ibid., p. 297). Out of the Church Committee report grew a sharp delineation between laws governing (domestic) law-enforcement investigations and those governing (foreign) national-security ones. National-security investigators were allowed to operate with considerably more latitude outside the borders of the United States than within. This sharp delineation was in line with the 1878 Posse Comitatus Act, post-Civil War legislation that prohibited Army involvement in domestic arrests or searches and seizures.[21]

Within a decade of the Church Committee's recommendation, the line began to blur. In 1981 President Ronald Reagan declared that international drug trafficking posed a threat to national security (McGee 1996c), and the Military Cooperation with Civilian Law Enforcement Agencies Act sharply increased the Army's role in anti-drug efforts. The military's responsibilities in anti-drug efforts grew. The 1989 Defense Authorization Act put the Department of Defense in charge of applying US command, control, communications, and intelligence assets to monitor illegal drugs.

The National Guard does not face Posse Comitatus restrictions unless it is on federal duty, and it has been given a greater role in drug interdiction. In addition, Army Special Forces and the Marines patrol the Southwest and California for drug smugglers. Military intelligence officers also watch for gang and criminal activity in a number of US cities (McGee 1996b,c). Law enforcement and the military have become closely linked in their anti-drug activities. The scale of military participation is evident in the fact that since 1989 the military has spent over $7 billion on anti-drug operations (McGee 1996b).

Aside from drug trafficking, various international events have been cited as evidence of a need for closer coordination between national security and domestic law enforcement. One such event was the 1991 collapse of the Bank of Credit and Commerce International. According to a 1992 report by Senators John Kerry and Hank Brown, the CIA had discovered the essentials of the bank's criminal activities by 1985 but had never properly conveyed these facts to law-enforcement agencies.[22] Proponents of closer cooperation between intelligence and law-enforcement agencies maintain that globalization and the end of the Cold War make separation between national security and domestic law unrealistic.

A series of laws passed during the 1980s made various acts occurring outside the borders of the United States criminal acts prosecutable within the United States if they involved American citizens. These laws included the Omnibus Diplomatic Security and Antiterrorism Act of 1986 (Public Law 99-399), which established jurisdiction over violent terrorist acts against Americans overseas; the Act for the Prevention and Punishment of Hostage-Taking (18 USC §1203 (1988)); the Aircraft Sabotage Act (Public Law 98-473, 18 USC 63.2, 40 USC App. 1301, 1471 1972, (Supp. v. 1987)); and Public Law 101-519, which established US district court jurisdiction over suits to recover damages in international terrorism cases. The passage of these laws left unclear how they were to be implemented, and, in particular, what role national security should play in international law enforcement.

The Clinton administration initiated internal discussions on these issues. Then the 1997 Intelligence Authorization Act (§814) stated the following:

> ... elements of the Intelligence Community may, upon the request of a United States law enforcement agency collect information outside the United States about individuals who are not United States persons. Such elements may collect such information notwithstanding that the law enforcement agency intends to use the information collected for purposes of a law enforcement investigation or counterintelligence investigation.

This wording carefully steers clear of permitting the intelligence community to spy on Americans directly, but opens the way for unprecedented collaboration between the intelligence and law-enforcement communities. Since 9/11, however, the wall between intelligence and law enforcement has been eroded and cooperation between intelligence and law enforcement has been expanded on many fronts. This is discussed further in chapter 11.

It has long been recognized that agents of foreign powers—we used to talk about spies, now we talk about terrorists—may commit crimes in the US but do not behave like ordinary criminals and are not readily controlled by ordinary law-enforcement activities. To start with, they often have the support of foreign intelligence or covert operations services that provide them with equipment, training, or information not available to common criminals. Not only may they be provided with money, expertly forged credentials, and weapons smuggled in in the diplomatic bag, they may be provided with intelligence about the actions of the law enforcement agencies that are pursuing them. More fundamentally, foreign agents are not part of American society and do not care about its censure. They may accept going to prison as a risk of the work or even look forward to prison as a badge of honor. In prison, they may act on training about how to behave as a prisoner, how to escape, how to continue to work for the home country, etc. Their spirits may be buoyed by the knowledge that they may be traded for American agents captured by their own side.

These considerations are considered adequate justification for special laws dealing with agents of foreign powers and special agencies for carrying out those laws. In the United States the Foreign Intelligence Surveillance Act, the Classified Information Procedures Act, and more recently the USA PATRIOT Act have been passed with this in mind. Such laws diminish the presumption-of-innocence-based protection that the Constitution guarantees to ordinary citizens suspected of crimes and allow the government to act "more expediently," most particularly, in secret.

Unfortunately, there is an opposite side of this coin that generally faces down and goes unobserved. If the Constitution's guarantees of due process, presumption of innocence, and the right to be confronted by one's accusers are to be upheld, laws aimed at agents of foreign powers must be circumscribed so that they cannot be broadly applied. Unfortunately, this is not done as often as it should be and prosecutors dealing with ordinary criminal activities are quick to make use of the new "tools" to improve their conviction rates.

6

Privacy: Protections and Threats

Protecting the national security and enforcing the laws are basic societal values. Often they stand in competition with another basic societal value: privacy. The competition is hardly an equal contest. National security and law enforcement not only have political constituencies, they are represented by major societal organizations. Privacy has no such muscle behind it. As a result, although an attachment to privacy endures and at times grows, privacy is often violated.

The Dimensions of Privacy

Our focus throughout this book is on privacy in communications. However, it is valuable to draw back and view privacy in a broader context.

Two hundred years ago, if you chose to speak to a colleague about private matters, you had to do it in person. Others might have seen the two of you walk off together, but to overhear your conversation an eavesdropper would have had to follow closely and would likely have been observed. Today, the very communication links that have made it possible to converse at a distance have the potential to destroy the privacy such conversations previously enjoyed.

From video cameras that record our entries into shops and buildings to supermarket checkout tapes that list every container of milk and package of cigarettes we buy, privacy is elusive in modern society. There are records of what we do, with whom we associate, where we go. Insurance companies know who our spouses are, how many children we have, how often we have our teeth cleaned. The increasing amount of transactional information—the electronic record of when you left the parking lot, the supermarket's record of your purchase—leaves a very large public footprint, and presents a far more detailed portrait of the individual than those recorded at any time in the past. Furthermore, information about individuals is no longer under the control of the person to whom the information pertains; such loss of control is loss of privacy.

Privacy as a Fundamental Human Right

Privacy is at the very soul of being human. Legal rights to privacy appeared 2000 years ago in Jewish laws such as this: "[If one man builds a wall opposite his fellow's] windows, whether it is higher or lower than them . . . it may not be within four cubits [If higher, it must be four cubits higher, for privacy's sake]." (Danby 1933, p. 367) The Talmud explains that a person's neighbor "should not peer and look into his house."

Privacy is the right to autonomy, and it includes the right to be let alone. Privacy encompasses the right to control information about ourselves, including the right to limit access to that information. The right to privacy embraces the right to keep confidences confidential and to share them in private conversation. Most important, the right to privacy means the right to enjoy solitude, intimacy, and anonymity (Flaherty 1989, p. 8).

Not all these rights can be attained in modern society. Some losses occur out of choice. (In the United States, for example, candidates for office make public much personal information, such as tax and medical records, that private citizens are allowed to keep private.) Some losses are matters of convenience. (Almost no one pays bills in cash anymore.) But the maintenance of some seclusion is fundamental to the human soul. Accordingly, privacy is recognized by the international community as a basic human right. Article 12 of the 1948 Universal Declaration of Human Rights states:

> No one shall be subjected to arbitrary interference with his privacy, family, home or correspondence, nor to attacks upon his honour and reputation. Everyone has the right to the protection of the law against such interference or attacks. (Academy on Human Rights 1993, p. 3)

The 1967 International Covenant on Human Rights makes the same point.[1]

The Soviet Union, East Germany, and other totalitarian states rarely respected the rights of individuals, and this included the right to privacy. Those societies were permeated by informants,[2] telephones were assumed to be tapped and hotel rooms to be bugged: life was defined by police surveillance. Democratic societies are supposed to function differently.

Privacy in American Society

Privacy is essential to political discourse. The fact is not immediately obvious, because the most familiar political discourse is public. History records political speeches, broadsides, pamphlets, and manifestos, not the quiet conversations among those who wrote them. Without the opportunity to discuss politics in private, however, the finished positions that appear in public might never be formulated.

Democracy requires a free press, confidential lawyer-client relations, and the right to a fair trial. The foundations of democracy rest upon privacy, but in various democratic societies the protection of privacy is interpreted in varying ways. Britain, for example, has much looser laws regarding wiretaps than the United States. A number of European nations extend more protection to individuals' data records than the US does.

Privacy is culture dependent. Citizens of crowded countries such as India and the Netherlands hold very different views of what constitutes privacy than citizens of the United States. The American concept developed in a land with a bountiful amount of physical space and in a culture woven from many disparate nationalities.

In the 1970s, as rapid computerization brought fear of a surveillance society, some nations sought to protect individuals from the misuse of personal data. Sweden, Germany, Canada, and France established data-protection boards to protect the privacy and the integrity of records on individual citizens. When the US Congress passed a Privacy Act with a similar goal,[3] President Gerald Ford objected to the creation of another federal bureaucracy, and no US data-protection commission was ever established (Flaherty 1989, p. 305). A number of states have data-protection and security-breach notification laws and California includes a right to privacy in its state constitution. It would, however, be a mistake to view the lack of a major regulatory apparatus in the United States as a complete lack of legal protection of privacy.

Privacy Protection in the United States

Most Americans believe that privacy is a basic right guaranteed by the Constitution. The belief has some truth to it, but not nearly as much as some believe. Nowhere is the word privacy mentioned in the Constitution, nor is a right to privacy explicit in any amendment. Privacy is nonetheless implicit to the Constitution.

The First Amendment protects the individual's freedoms of expression, religion, and association. The Third Amendment protects the private citizen against the state's harboring an army in his home, the Fourth against unreasonable search or seizure. The Fifth Amendment ensures that an individual cannot be compelled to provide testimony against himself. The Ninth Amendment reserves to "the people" those rights that are not enumerated in the Constitution. And "the Fourteenth Amendment's guarantee that no person can be deprived of life, liberty or property without due process of law, provides an additional bulwark against governmental interference with individual privacy" (USSR 93 *Federal Data Banks*, p. ix).

The Early Years

Three hundred years ago, the colonists' portion of the North American continent consisted of farms, small towns, and a few small cities. Eavesdroppers were easily avoided by walking to a place where one could not be overheard. Although the colonists' application of English common law provided for the punishment of eavesdroppers (Friedman 1973, p. 254), more typically such crimes were handled by their discoverers (Flaherty 1989, p. 89).

The problem of the mail being less than private was less easily disposed of, for mail delivery entailed dependence on others. In colonial America mail delivery was haphazard; letters from Europe would be left by a ship's captain at a local tavern, awaiting pickup by their intended recipients but meanwhile open to inspection by all passersby. Within the colonies mail was delivered by travelers and merchants, or by special messengers, and privacy was likewise not assumed. In 1710, in creating a postal delivery system in the colonies, the British government established privacy protection similar to what existed in England; at least in law, the opening of letters and other forms of tampering was forbidden without authorization by the secretary of state (Seipp 1977, p. 11).[4]

Despite the law regarding privacy of the mails, the British government maintained a "secrets office ... whose staff routinely opened correspondence that the government considered potentially subversive" (John 1995, p. 43). Letters from London and the countryside were opened in the Secretary's private office, while Irish correspondence were opened by a clerk in Dublin Castle (Ellis 1958, p. 64). Foreign mail was opened in a special office whose names over the years were the "Private Foreign Office," "Secret Department," and "Secret Office" (ibid., p. 65). The size of this office varied; by 1810, there were ten on staff (ibid., p. 69). The French maintained a *cabinet noir* for a similar purpose (John 1995, p. 43).

Before the establishment of the United States, Postmaster-General Benjamin Franklin believed that his own mail was being opened and read (Franklin 1907, pp. 461–462). Later, Thomas Jefferson held a similar concern.[5]

The political revolutionaries who established the United States thus had a visceral understanding of the importance of the postal privacy. In the Postal Act of 1792, Congress did three important things: postal officials were prohibited from inspecting the contents of mail (unless the mail was undeliverable); the mailing of newspapers was at a very low cost, thus encouraging political discourse; and Congress was given the right to determine postal routes (John 1995, p. 31). The latter two had surprising consequences. Because the legislature had—and used—the ability to expand postal routes, including into locales in which the postal routes were not self-supporting, the US postal service expanded rapidly. By 1828, the United States had twice as many post offices as Britain and five times as many as France (ibid., p. 5).[6] Because of the democratizing effects of the Postal Act, fears about a strong federal postal system did not emerge. Rather, the postal service was one of the few strong federal institutions early in the nation's history (Starr 2003, p. 3).

In 1825 Congress addressed the problem of spying in the mails with the Postal Act (4 Stat. 102, 109), which prohibited prying into another person's mail. Practice did not necessarily follow the law. During the 1850s there were complaints about the "greedy fingers" through which the mail passed,[7] and during the Civil War there were government attempts to open private civilian mail.[8] In 1878 the Supreme Court ruled that the government could not open first-class mail without a search warrant (*Ex Parte Jackson*, 96 US 727, p. 733).

The invention of the telegraph led to a new way of communicating and two new ways of eavesdropping: one could tap the wire[9] or one could read the messages later from copies kept by the telegraph companies. The latter was the search method preferred by the government.

At the beginning of the Civil War, the government took control of the telegraph wires and seized copies of all telegrams sent within the previous 12 months (Plum 1882, p. 69). For the duration of the war, the federal government censored all dispatches emanating from Washington (Randall 1951, pp. 482–483). But the War Department did not have full success in controlling the medium. In 1864, attempting to track down the source of a false newspaper story that the president planned to call up an additional 400,000 men, the government sought copies of all telegrams sent out from Washington, but company operators refused to cooperate. They were arrested and held for several days until Army investigators uncovered the perpetrator—a stock manipulator (Bates 1907, pp. 228–243).

Some of the complexity of the fight over the privacy protection afforded to telegraphs was due to the fact that, unlike the mails, this new communication medium was controlled by private enterprise. The government sought two seemingly contradictory goals: to protect privacy of communications from the prying eyes of operators (Seipp 1977, pp. 83–95) and to establish broad search privileges for itself.

Telegraph companies sought to assure the public that communications would be private (ibid., pp. 89–90). The government's determination to obtain copies of telegrams pressed against this, and the conflict came to a head in 1876 with the contested presidential election between Hayes and Tilden. During a fight over electoral votes from Louisiana and Oregon, a House committee questioned the manager of Western Union's New Orleans office about telegrams. Under orders from Western Union's president, the manager refused to reveal the contents of the disputed dispatches. The House held the manager in contempt. Another Western Union manager in Oregon faced a similar situation with the Senate. The company responded that it would henceforth destroy copies of telegrams as quickly as account keeping would allow (*New York Tribune* 1876). However, the policy was never carried out, and on January 20, 1877, with the manager of the New Orleans office in custody at the Capitol and the company's president ill, Western Union gave in to congressional pressure, and responded to the subpoena (Seipp 1977, p. 54). Western Union prepared a bill for Congress in which telegrams were provided the same legal protections as US mail (*New York Tribune* 1880). Although the bill was reported favorably out of committee, it was not heard of again.

Several court decisions put a partial closure on the issue. One of these involved a St. Louis grand jury investigation of a gambling ring that allegedly involved Missouri's governor and the police commissioner of St. Louis. When the manager of the Western Union office refused to hand over copies of telegrams (*New York Times* 1879), the company appealed and won a partial victory; the Missouri Supreme Court ruled that, although telegraph messages were not accorded the privacy protection the company sought (the same as for US mail), any request for copies of telegrams had to include both the date and the subject of the message (*Ex Parte Brown*, 72 Mo. 83 95 (1880), in Seipp 1977, p. 58). A number of other decisions around that time reached the same conclusions (ibid., p. 59).[10]

After the Civil War the United States experienced rapid industrialization and urbanization. Small towns have never been much known as respecters of privacy, but in such locales the ubiquity of gossip was mitigated by the fact that news did not travel far. As the invention of rotary presses, linotypes, and automatic folders led to a sharp increase in the number of newspapers and to the advent of "yellow journalism," that ceased to be true. Against such a backdrop Samuel Warren, a socially prominent paper manufacturer, and his former law partner Louis Brandeis, later to become a Supreme Court Justice, wrote an article on privacy for the 1890 *Harvard Law Review*. Warren and Brandeis argued that the changes caused by technology called for a response in law:

> That the individual shall have full protection in person and in property is a principle as old as the common law; but it has been found necessary from time to time to define anew the exact nature and extent of such protection.... [I]n very early times, the law gave only for physical interference

with life and property ... Later, there came a recognition of man's spiritual nature, of his feelings and his intellect. Gradually the scope of these legal rights broadened; and now the right to life has come to mean the right to enjoy life,—the right to be let alone.... (Warren and Brandeis 1890, p. 193)

Instantaneous photographs and newspaper enterprise have invaded the sacred precincts of private and domestic life; and numerous mechanical devices threaten to make good the prediction that "what is whispered in the closet shall be proclaimed from the housetops." (ibid., p. 195)

The success of the US Postal Service in contributing to the development of the country in the nineteenth century rested on two characteristics not shared by other postal systems of the time. On one hand, it promoted rural growth by subsidizing rural mail; the cost of sending a letter within the US was the same regardless of where it went. A letter that had to be carried miles to a postbox on a country road cost the same as one that was delivered to an address near the central post office of a big city. The other was one not shared by European mail systems: the US mail was not the agent of government spying. Postal patrons could send letters secure in the knowledge that their mail was private.[11]

By comparison, the French mail was very much a tool of surveillance.

Privacy as a Principle of Law

Law review articles may shape legal philosophy, but it is statutes and court decisions that determine the law. In the first round between telephone technology and privacy, privacy lost.

In 1928 Prohibition was still in force and illegal liquor distribution was big business. Among the "businessmen" were Roy Olmstead and his partners, who had a large liquor importing and distribution operation in Seattle. The enterprise employed 50 people, including salesmen, dispatchers, deliverymen, bookkeepers, and even an attorney. Outside the city they had a ranch with an underground hideaway for storage, and throughout the city they had smaller caches. Their headquarters, in a large downtown office building, had three phone lines in the main office. In addition, Olmstead and several of his associates had office lines at home.

Federal agents installed wiretaps in the basement of the headquarters building and on lines in streets near the conspirators' homes. For months four prohibition officers eavesdropped on the conspirators as they ordered cases of liquor, arranged bribes for the Seattle police, and reported news of seized alcohol. The wiretap evidence played a crucial role in Olmstead's conviction, but should the evidence have been admitted? No warrants were sought, but neither did the agents enter private homes or offices to place the taps (*Olmstead* 1928). Because the evidence had been obtained from warrantless wiretaps installed by the government, the defendants claimed that the wiretapped evidence had been obtained in violation of the Fourth Amendment. Use of the evidence also violated the Fifth Amendment, the defendants argued, because they had unwillingly become witnesses against themselves.

The US Supreme Court did not buy these arguments, and five of the justices voted to uphold a lower court's conviction. In an irony of history, the most famous opinion to come out of the case is Justice Louis Brandeis's dissent, which includes these passages:

> When the Fourth and Fifth Amendments were adopted, "the form that evil had heretofore taken" had been necessarily simple. Force and violence were then the only means known to man by which a Government could directly impel self-incrimination. It could compel the individual to testify—a compulsion effected, if need be, by torture. It could secure possession of his papers and other articles incident to his private life—a seizure effected, if need be, by breaking and entry. Protection against such invasion of "the sanctities of a man's home and the privacies of life" was provided in the Fourth and Fifth Amendment by specific language.... But "time works changes, brings into existence new conditions and purposes." Subtler and more far-reaching means of invading privacy have become available to the Government. Discovery and invention have made it possible for the Government, by means far more effective than stretching upon the rack, to obtain disclosure in court of what is whispered in the closet. (Brandeis 1928, p. 473)

> Unjustified search and seizure violates the Fourth Amendment.... [I]t follows necessarily that the Amendment is violated by the officer's reading the paper without a physical seizure, without his even touching it; and that use, in any criminal proceeding, of the contents of the paper so examined ... any such use constitutes a violation of the Fifth Amendment.
>
> The protection guaranteed by the Amendments is much broader in scope. The makers of our Constitution undertook to secure conditions favorable to the pursuit of happiness. They recognized the significance of man's spiritual nature, of his feelings and his intellect.... They sought to protect Americans in their beliefs, their thoughts, their emotions and their sensations. They conferred, as against the Government, the right to be let alone—the most comprehensive of rights and the right most valued by civilized man. To protect that right, every unjustifiable intrusion by the Government upon the

privacy of the individual, whatever the means employed, must be deemed a violation of the Fourth Amendment. And the use, of evidence in a criminal proceeding, of facts ascertained by such intrusion must be deemed a violation of the Fifth. (ibid., pp. 477–478)

A decade later, in the Nardone cases (*Nardone v. United States*, 302 US 379 (1937) and 308 US 338 (1939)), the Supreme Court avoided constitutional questions and used the Federal Communications Act as a basis for making warrantless wiretapping illegal. In effect, though not in law, this supported the principles Brandeis had espoused in *Olmstead*. In later rulings the Court outlawed the use of warrantless electronic surveillance of any type. Other cases before the Court further helped to define the parameters of privacy.

A 1958 ruling held that the First Amendment right to free association included the privacy of such association. In an effort to curb the political activities of the National Association for the Advancement of Colored People, Alabama filed suit against that organization, which had helped organize the Montgomery bus boycott and secure the admission of blacks to the segregated state university. The NAACP was ordered to produce a number of records, including its membership list. In the climate of the times, this would have been dangerous to the individuals named, and the NAACP refused to comply. The Court sided with the defendants:

[O]n past occasions revelation of the identity of its rank-and-file members has exposed these members to economic reprisal, loss of employment, threat of physical coercion, and other manifestations of public hostility.... (*NAACP v. Alabama*. 357 US 449, 1958, p. 462)

[I]mmunity from state scrutiny of membership lists that the Association claims on behalf of its members is here so related to the rights of the members to pursue their lawful private interests privately and to associate freely with others in so doing as to come within the protection of the Fourteenth Amendment.[12] (ibid., p. 466)

Seven years later, in *Griswold v. Connecticut*, the Supreme Court greatly expanded privacy protections by including rights that were not specifically enumerated within the Bill of Rights. Connecticut's Birth Control Act outlawed the sale and use of contraceptives, and the state had prosecuted the executive director of Planned Parenthood and a New Haven doctor who had given out contraceptive information. The

Supreme Court ruled that the state had no business legislating on matters having to do with what goes on in the privacy of the bedroom. In so doing, it developed the theory that the protections afforded by the Bill of Rights cast wide shadows:

> This law ... operates directly on an intimate relation of husband and wife and their physician's role in one aspect of that relation.
> The association of people is not mentioned in the Constitution nor in the Bill of Rights. The right to educate a child in a school of the parents' choice —whether public or private or parochial—is also not mentioned. Nor is the right to study any particular subject or any foreign language. Yet the First Amendment has been construed to include certain of those rights. (*Griswold v. Connecticut*, 381 US 479, 1965, p. 482).

> In *NAACP v. Alabama*, 357 US 449, 462, we protected the "freedom to associate and privacy in one's associations."... In other words the First Amendment has a penumbra where privacy is protected from governmental intrusion....
> [S]pecific guarantees in the Bill of Rights have penumbras, formed by emanations from those guarantees that help give them life and substance.... (ibid., p. 483)

Two years later, the Supreme Court ruled that telephone calls, including those made in public places, were private. Charles Katz had placed a $300 bet from a phone booth in Los Angeles. An FBI team investigating interstate gambling had, without benefit of a search warrant, installed an electronic bug in the phone booth and picked up Katz's portion of the conversation. The Court ruled:

> [T]he Fourth Amendment protects people, not places. What a person knowingly exposes to the public, even in his own home or office, is not a subject of Fourth Amendment protection.... But what he seeks to preserve as private, even in an area accessible to the public, may be constitutionally protected. (*Charles Katz v. United States*, 389 US 347, 1967, p. 311)

Katz's conviction was overturned. However, the Supreme Court was not heading down a one-way path to greater privacy protection, as Mitchell Miller, who used his checking account for an illegal moonshine business, discovered.

In December 1972, a deputy sheriff in Houston County, Georgia, stopped a suspicious van driven by Miller's business partners. The van was carrying materials for making a still. Several weeks later, during a fire at a warehouse Miller had rented, firemen and police discovered a distillery and whiskey, on which no taxes had been paid. The Bureau of Alcohol, Tobacco, and Firearms issued a subpoena to Miller's bank for his account records, which the bank gave to federal agents. A grand jury subsequently indicted Miller on various moonshining charges. Arguing that because the seized material constituted self-incriminating material the subpoena was an illegal search and seizure, Miller tried to suppress the account information. The Court did not support him:

> We find there is no intrusion into any area in which respondent had a protected Fourth Amendment interest....
> On their face, the documents subpoenaed here are respondent's "private papers".... (*United States v. Miller*, 425 US 435, 1976, p. 440)

> [I]f we direct our attention to the original checks and deposit slips, rather than to the microfilm copies actually viewed and obtained through the subpoena, we perceive no legitimate "expectation of privacy" in their contents. The checks are not confidential communications, but negotiable instruments to be used in commercial transactions. All of the documents obtained, including financial statements and deposit slips, contain only information voluntarily conveyed to the banks and exposed to their employees in the ordinary course of business. (ibid., p. 442)

Here the Court deemed only that which has an "expectation of privacy" worthy of Fourth Amendment protection. Such a decision is very much a double-edged sword for privacy. On the one hand, it forms the basis for such privacy rulings as the Kyllo decision (*Kyllo v. United States*, 533 US 27, 2001), in which the Supreme Court ruled that the warrantless use of a thermal-imaging device to determine "hot spots" in a private house (and thus the likelihood that marijuana was being grown) was a violation of Fourth Amendment rights.[13] On the other, the lack of privacy in Internet communications leaves that domain highly unprotected under the "expectation of privacy" standard. This is why privacy protection for communication is often established by legislative enactment. Statutes define what the expectation of privacy is for new technological environments.

Privacy Threatened

In the name of efficiency, the US government and many businesses have amassed huge databases containing profiles of their "customers." The gains in efficiency are accompanied by losses of privacy.

Credit cards leave a trail of where their holders travel, where they shop, and what they buy. Studying billing patterns, the European branch of American Express tailors the promotional material it sends its customers to their individual purchasing patterns, and plans are afoot to offer the same "service" in the United States. "'A Northern Italian restaurant opens in midtown Manhattan,' [the CEO of American Express] says, by way of an example. 'We know from the spending of card members which of them have eaten in Northern Italian restaurants in New York, whether they live in New York or Los Angeles. We make the offer of a free or discounted meal available only to those card members.'" (Solomon 1995, p. 38) The long-distance telephone company MCI offers its users discounts on calls to the ten numbers they call most often, and the carrier identifies those numbers for the customers. These companies are not unique in the amount they know about their customers' purchasing patterns, though they are perhaps more public about it.

Although there has been no public protest of the two transactional data collections described above, there are periodic episodes in which the public says to a company "Leave us alone." In the early 1990s the software firm Lotus and the Equifax credit bureau were developing a CD-ROM that would contain the names, estimated incomes, and purchasing habits of 120 million Americans. The companies received 30,000 letters opposing the project, and it was killed (Piller 1993, p. 11). Similarly, for exactly 10 days Lexis-Nexis, a leading information broker, offered an online service that included access to Social Security numbers. An outpouring of objections from customers caused the company to discontinue the service (Flynn 1996).

On occasion the federal government has stepped in to safeguard privacy. In 1970 Congress passed the Fair Credit Reporting Act, which ensures that individuals have rights in connection with credit records maintained by private databases. The publicity surrounding Judge Robert Bork's rental habits[14] led to the Video Privacy Protection Act of 1988, which prohibits release of video rental records. The Federal Privacy Act places limits on the types of information federal agencies may collect on individuals. California, Connecticut, the District of Columbia, Illinois, New Jersey, and Wisconsin all have laws that protect cable TV subscribers' records against release. Almost all states recognize the doctor-patient privilege (*Privacy Journal* 1992). Legislative protections of privacy are sparse but not unknown.

In the United States of the mid 1990s, there were over 500 commercial databases buying and selling information (Piller 1993, p. 10). Despite this proliferation of private purveyors of information, US citizens fear their government more. A Harris survey showed that in response to the question "Which type of invasions of privacy worry you the most in America today—activities of government agencies or businesses?" Fifty-two percent of the respondents answered that government agencies were their greater worry; 40 percent said business (Center for Social and Legal Research, p. 7).

Data collection by the US government dwarfs that by private enterprise. It is a rare American indeed who is not in a government database; most are in many. In the early 1990s, the agencies of the federal government supported 910 *major* databases (USGAO 1990, p. 2) containing personal data: 363 in the Department of Defense, 274 in the Department of Health and Human Services, 201 in the Department of Justice, and 109 in the Department of Agriculture (ibid., p. 41). Fifty-six percent of these systems allowed access by other federal, state, and local agencies; some even allowed access by private organizations (ibid., p. 18). Computer matching promotes efficiency and curtails fraud and duplication.[15] We no longer know how many federal databases there are; in 1996, the reporting requirements of the 1974 Privacy Act[16] were diminished and the data are effectively no longer made public.[17] Other things changed too.

The 1974 Privacy Act requires data minimization (only data "relevant and necessary" to the task should be kept) and puts into place rules about data sharing between agencies (with a clear description of the records to be shared and not without a written agreement describing the purpose and justification). There is, however, no set of requirements on what data the government is allowed to purchase from databrokers; such companies did not exist in 1974, when the Privacy Act was passed. While in the early 1990s data collection by the US government dwarfed that by private enterprise, that is no longer the case. Inexpensive storage[18] and fast matching algorithms have made commercial databrokers a profitable—and almost completely unregulated—business, completely changing the meaning of "know your customer." This revolution has also changed the information available to government.

After the 9/11 attacks, the US government turned to databrokers to expand its information on the US population. One of these is Acxiom, an Arkansas company with "information on almost 200 million people living in 110 million households" (O'Harrow 2005, p. 61). Acxiom's information on American households includes not only "names, birth dates, genders, addresses," but "number of adults [in the household], the presence of children, their genders and ages and school grades," house assessment and market value, the families' occupations and net worth and estimated income (ibid., p. 49). It appears that there is not much that Acxiom doesn't know. Another large databroker is ChoicePoint, a Georgia company marketing to state police, with over seventeen billion online public records—and forty thousand more records added each day (ibid., p. 145). It is estimated that ChoicePoint has records on more than 220 million people (ibid., p. 145).

The lack of regulation delimiting government acquisition of information collected by private databrokers is a serious hole in government privacy law, but one seemingly difficult to change at a time when fear of terrorism drives the political agenda. In the 1950s and the 1960s, when fear of communism was paramount, the government response included investigating and disrupting political activities and attempting to discredit civil-rights leaders, including Martin Luther King Jr. The government's power is immense, and has often been used to invade citizens' privacy. In the 1950s and the 1960s, those invasions were a matter of policy. Now technology has been developed to enable such invasions of a far grander scale. Changing public policy is rarely easy. Changing policies that are buttressed by widespread commercial practices and large investments in equipment is rarely possible.

Privacy Lost

Invasions of privacy have occurred despite legal provisions to the contrary. The debate about cryptography is a debate over the right of the people to protect themselves against government surveillance. But privacy intrusions are difficult to uncover. Many are suspected but few are proven. Those that are discovered are frequently exposed only as a result of years of litigation or through a major investigation like the Church Committee hearings of the 1970s.[19] Thus in order to judge the citizenry's need for protection against government surveillance, we take the long view and we look more closely at the government invasions of personal privacy over the last 50 years.

The 1940s

Article 1 of the US Constitution requires that a census be taken every 10 years to determine proper representation in Congress. In the first 50 years of the republic census questions were limited to numbers and age (Wright and Hunt 1900, pp. 132–133), but by 1840 the census included questions about the employment of family members (ibid., p. 143). Public objection led the Census Bureau to tell the public that the public statistical tables included only aggregate data and to remind the census takers that "all communications made to [you] in the performance of this duty … [are] strictly confidential" (ibid., p. 145). By 1880 the questionnaire had become considerably more detailed, asking about employment status, education level, and place of birth. The 1890 census showed a similar interest in personal detail, and people objected. In New York City alone, 60 people were arrested for refusing to answer census questions (*New York Tribune* 1890).

For the 1900 census, Congress made unauthorized disclosure of information about individuals a misdemeanor; for the 1920 census, a breach of confidentiality was made a felony (Davis 1973, p. 200). The current law, passed in 1929, explicitly states that no one other than "sworn employees of the Census Office"[20] shall be able to examine individual reports.

In 1980 the Census Bureau advertised that even during World War II, when there had been fears of a fifth column among Japanese-Americans living on the West Coast, the Census Bureau never gave out information about individuals (Okamura 1981, p. 112). The Census Bureau lied. A 1942 War Department report described the role of the Census Bureau in the roundup of Japanese-Americans:

> The most important single source of information prior to the evacuation was the 1940 Census of Population. Fortunately, the Bureau of the Census had reproduced a duplicate set of punched cards for all Japanese in the United States shortly after the outbreak of war and had prepared certain tabulations for the use of war agencies.... These special tabulations, when analyzed, became the basis for the general evacuation and relocation plan. (USDoW 1943)

These people lost their incomes, their property, and their rights as citizens. The Census Bureau, whose data about individuals, under law, was not to be released, supplied the information that helped the military to round up the Japanese-Americans.[21]

Other Americans lost privacy rights during the war too. Expressing concern about Communist influence, the FBI wiretapped the Congress of Industrial Organizations' Council and Maritime Committee; the Tobacco, Agricultural, and Allied Workers of America; the International Longshoremen's and Warehousemen's Union; the National Maritime Union; the National Union of Marine Cooks and Stewards; and the United Public Workers of America (Theoharis and Cox 1988, pp. 10 and 438). It bugged the United Automobile Workers and the United Mine Workers.

The privacy of the mails was also invaded. Beginning in 1940 and continuing until 1973, FBI and CIA agents read the private mail of thousands of citizens.[22] The justification was the Cold War, but Senate investigators later called this "essentially domestic intelligence" (USSR 94 *Intelligence Activities: Staff Reports*, p. 561). Without warrants and without congressional or clear presidential authority, intelligence agents opened and perused the mail of private citizens, senators, congressmen, journalists, businessmen, and even a presidential candidate. Domestic peace organizations, such as the American Friends Service Committee, had their mail opened, as did scientific organizations, including the Federation of American Scientists. So did the writers Edward Albee and John Steinbeck. Americans who frequently visited the Soviet Union or corresponded with people there were singled out, as were educational, business, and civil-rights leaders (USSR 93 *Electronic Surveillance for National Security*, pp. 574–575). In one program, more than 200,000 pieces of mail were illegally opened.

Telegrams were even less private. The National Security Agency requested copies of telegrams to and from certain foreign intelligence targets, and Western Union complied with the request. RCA Global and ITT went further, giving NSA access to the "great bulk" of their telegrams. In the early stages of this program, the government received paper tapes, and the volume of material precluded sorting on content. In the early 1960s, RCA and ITT World Communications switched to magnetic tape for storing telegrams, and then the two companies shipped copies of *all* overseas telegrams to NSA. The agency read all telegrams to or from people on the "watch list," which included both foreigners that the government had decided were attempting to exert control on US policy and Americans citizens engaged in peaceful protest against government policy (USSR 94 *Intelligence Activities: Rights of Americans*, p. 108).

The mail and telegram searches were largely untargeted invasions of privacy. There were also highly targeted ones. For example, President Harry Truman did not trust Thomas Corcoran, a Washington lobbyist who had been a close confidant of President Franklin Roosevelt. Truman asked the FBI for a telephone tap on Corcoran's phone. Conversations were recorded in which Corcoran and Supreme Court justices Hugo Black, William O. Douglas, and Stanley Reed discussed the pros and cons of various possible nominees for chief justice (Charns 1992, p. 25). After considering sitting justices William O. Douglas and Robert Jackson, Truman went outside the Court for his nominee, choosing Treasury Secretary Fred Vinson, his friend and poker buddy. However Truman made his choice for chief justice, he is known to have discarded certain nominees after seeing the Corcoran transcripts. We do not know if the wiretapped conversations influenced Truman in his decision, but we do know that the other participants had not intended Truman to be a silent participant.[23]

The 1950s

The opening of mail continued after World War II and after the Korean War. Various programs to photograph mail remained in effect, and photos were indexed, filed, and stored.

Throughout the 1950s the FBI aggressively pursued Communists wherever it thought they might be found. By the mid 1950s even many FBI agents did not view the American Communist Party as a serious threat,[24] but the bureau continued to investigate the party and related groups, conducting searches that were far out of proportion to the perceived threats. For example, the Socialist Workers Party was the subject of 20 years of wiretaps, burglaries, and bugs by the government investigators, even though the FBI knew that the group did not advocate violence (Shackelford 1976). Through warrantless break-ins the bureau photographed such papers as "Correspondence identifying contributors to SWP election campaign fund," "Correspondence re arrangements for [name deleted] to debate at Yale University," and "Letter ... detailing health status of ... Nat'l Chairman."[25] Eventually the Socialist Workers Party sued the US government for violations of constitutional rights, winning more than $250,000 in damages.[26]

The FBI's pursuit of Communists led to investigations of such subversive domestic organizations as parent-teacher associations, civil rights organizations, and various racial and religious groups (USSR 94 *Intelligence Activities: Rights of Americans*, p. 67). The FBI launched probes of the Ku Klux Klan and the John Birch Society. It also investigated the Southern Christian Leadership Conference (SCLC), the Congress of Racial Equality, and the NAACP, despite the fact that these three were all staunch advocates of non-violence. FBI agents overstepped many bounds in these inquiries, including those of personal privacy.

The 1960s

In all the inappropriate investigations, one victim stands out, a man whose privacy was repeatedly and egregiously violated by the FBI: Martin Luther King Jr. For a number of years the FBI had been investigating King's alleged ties to the Communist Party. King's closest connection to the Communist Party was through his friend and advisor Stanley Levison. But Levison had left the Party in 1954. The FBI was well aware of this:

in 1960 it had attempted to recruit Levison in order to have him rejoin the Party as an FBI spy (Friedly and Gallen 1993, p. 24).

In 1963, during debates on a pending civil rights bill, Senator James Eastland of Mississippi, using information given to him by the FBI, charged that King and the SCLC were being advised by Communists. Attorney General Robert Kennedy authorized wiretaps on King (USSR 93 *Electronic Surveillance for National Security*, pp. 111–112). The FBI decided that the bugging of King was also authorized (ibid.). No evidence of Communist influence was discovered, but King remained a subject of wiretaps and bugs. The surveillance had uncovered other information too valuable for FBI Director J. Edgar Hoover, who hated King, to ignore.

Electronic surveillance in hotel rooms King stayed in when traveling picked up the minister telling bawdy stories and raucously partying (Darrow 1981, p. 109). Although this information had nothing to do with Communist activity in the civil rights movement, the FBI used the tapes in an attempt to discredit King. A version of the tapes was sent to his wife, and Hoover played the tapes for President Lyndon Johnson.

In a further attempt to discredit King, William Sullivan, Associate Director of the FBI, wrote a monograph on Communist influence in the civil rights movement that was then made available to the heads of intelligence agencies, the Department of State, the Department of Defense, and the United States Information Agency; it included a salacious section on King's personal life, apparently based on information from the bugs (USSR 94 *Intelligence Activities: Staff Reports*, pp. 173–174). The impetus for this siege was King's having received the Nobel Peace Prize. Enraged by this international honor, Hoover tried to spread the news of King's personal life anywhere honors for the civil rights leader were being contemplated.

There was also political espionage against King. At the request of the White House, the FBI sent 30 special agents to the 1964 Democratic National Convention in Atlantic City. Their job was to assist the Secret Service and to make sure that the convention was not disrupted by civil disturbances. The FBI interpreted the latter charge broadly enough to cover political activities. Johnson was especially concerned about a challenge to the seating of the regular Mississippi delegation by the Mississippi Freedom Democratic Party. From an FBI wiretap of King's hotel room in Atlantic City,[27] Johnson's staff discovered the efforts of King and his associates to unseat the Mississippi delegation. A Senate investigating committee later observed that "an unsigned White House memorandum disclosing Dr. King's strategy in connection with a meeting to be attended by President Johnson suggests there was political use of these FBI reports."[28]

The investigations of King were not based on national-security considerations. In the case of the Democratic National Convention, the purpose of the wiretap was to provide political intelligence for the president. As was well known to the FBI at the time, the other wiretaps and bugs had no legitimate purpose.

Where many saw peaceful dissent in the civil rights movement and in the protests against the Vietnam War, the FBI saw domestic upheaval and began a massive program of surveillance. Riots in Los Angeles and other cities in the summer of 1965 led to heavy FBI surveillance of black neighborhoods. In 1966, FBI field offices were told to begin preparing semimonthly summaries of "existing racial conditions" and of the activities of all civil rights organizations (USSR 94 *Intelligence Activities: Rights of Americans*, p. 71). By 1972 the bureau had over 7400 informants in the ghettos, including, for example, "the proprietor of a candy store or barber shop" (ibid., p. 75). The informants were to attend meetings held by "extremists," to identify them when they came through the ghetto, and to identify persons distributing extremist literature (Moore 1972). Since the FBI's definition of extremists included such advocates of nonviolence as Martin Luther King and Ralph Abernathy,[29] it is not surprising that so many informants (one of every 3300 black Americans) were employed by the FBI during this period.

Fear of potential Communist infiltration led the FBI to investigate the anti-war movement (USSR 94 *Intelligence Activities: Rights of Americans*, p. 49), and agents conducted their probes as if these apprehensions had a factual basis. For example, during the spring of 1968, the FBI sought to wiretap the National Mobilization Office, which was planning to hold demonstrations during the Democratic National Convention in Chicago. In this case, despite several requests by the FBI, Attorney General Ramsey Clark refused to give the bureau permission to wiretap (Clark 1968).

The FBI also investigated the women's liberation movement and various university, church and political groups opposed to the Vietnam War (USSR 94 *Intelligence Activities: Rights of Americans*, p. 167). A woman whom the FBI had hired to report on the group Vietnam Veterans against the War later told a Senate committee: "I was to go to meetings, write up reports ... on what happened, who was there, ... to try to totally identify the background of every person there what their relationships were, who they were living with, who they were sleeping with. ..." (USSH 94 *Intelligence Activities: FBI*, p. 111)

A Senate subcommittee estimated that between 1967 and 1970 the US Army maintained files on at least 100,000 Americans (USSR 92 *Army Surveillance*, p. 96), including Joan Baez, Julian Bond, Rev. William Sloane Coffin Jr., Arlo Guthrie, Jesse Jackson, Martin Luther King Jr., Representative Abner Mikva, Dr. Benjamin Spock, and Senator Adlai Stevenson III.[30]

Much of the information contained in the Army's files was personal. "For example, the profile of a well-known local civil rights leader reported that he had four brothers, five sisters, and a widowed mother. An entertainer was recorded as married and the father of six children. ... Another sketch stated that due to flat feet and torn ligaments, ___ failed to pass Selective Service physical examinations, and his draft board classified him 1-Y." (USSR 93 *Military Surveillance*, p. 56) Though these people had not served in the Army, the Army's files contained information about their financial affairs, sex lives, and psychiatric histories (USSR 92 *Army Surveillance*, p. 96).

The military also kept files on the American Friends Service Committee, Americans for Democratic Action, the Congress of Racial Equality, Clergy and Laymen Concerned about the War, the NAACP, the National Mobilization Committee to End the War in Vietnam, the SCLC, Veterans and Reservists to End the War, and Women's Strike for Peace, among others (USAINTC 1969). Army agents reported on such subversive activities as labor negotiations conducted by the sanitation workers' union in Atlanta and actions by Wisconsin welfare mothers who wanted higher payments (Stein 1971, p. 274). As a Senate subcommittee observed, "considerations of privacy, relevance, and self-restraint were cast to the winds" (USSR 93 *Military Surveillance*, p. 56).

The 1960s were also a time when President John Kennedy used national security as a pretext for FBI wiretaps on opponents of his administration's Sugar Bill (see chapter 6) and on several journalists,[31] and when the FBI kept President Johnson supplied with political intelligence through biweekly summaries of contacts between foreign officials and various senators and congressmen (USSR 94 *Intelligence Activities: Rights of Americans*, pp. 119–120).

Despite his use of the King wiretaps and his reliance on the FBI for political surveillance of congressional opponents to the Vietnam War, President Johnson turned against the use of wiretaps.[32] Johnson called the Title III provisions for wiretapping undesirable even as he signed the Omnibus Crime Control and Safe Streets Act of 1968 (*Congressional Quarterly Weekly* 1968b, p. 1842).[33] Richard Nixon took his oath of office as president 7 months later. In his first full day as president, Nixon told a reporter that his attorney general, John Mitchell, would govern wiretapping "with an iron hand" (White 1975, p. 125). Within 4 months, and without Mitchell's approval, Nixon's administration began wiretapping for political purposes.

The political wiretapping began after the *New York Times* reported that the United States had been bombing Cambodia for some time. Henry Kissinger, Nixon's national security advisor, asked J. Edgar Hoover to have the FBI find out who had leaked this information to the *Times*. Seventeen people, including several journalists, were wiretapped in this investigation.

Two of those tapped, John Sears and James McLane, were domestic-affairs advisors with no access to classified national-security material (USSR 94 *Intelligence Activities: Staff Reports*, p. 337). White House speechwriter William Safire was also wiretapped; he had been overheard on an existing tap promising a reporter background material pertaining to a presidential address on revenue sharing and welfare reform (USSR 94 *Intelligence Activities: Staff Reports*, p. 337). No national-security leaks were ever uncovered by this investigation.

Joseph Kraft, a columnist, was another target. Kraft had once been a favorite of Nixon's, having coined the term "Middle America" to describe Nixon's supporters. But in the spring of 1969 Kraft criticized Nixon's peace efforts, and in June John Ehrlichman, Nixon's counsel, arranged for Kraft's phone to be tapped. A "security consultant" to the Republican National Committee installed the warrantless wiretap (USHH 93 *Impeachment Inquiry*, p. 150). Kraft flew to Paris a week later to cover the Vietnam peace talks, and the tap was removed. But the Assistant Director of the FBI, William Sullivan, followed Kraft to Paris and arranged for further electronic surveillance (Sullivan 1969). The switching system of the hotel made it impossible to install a phone tap, so a microphone was placed in Kraft's room (USSR 94 *Intelligence Activities: Staff Reports*, p. 323). No evidence was ever found to support Ehrlichman's claim (Ehrlichman 1973) that Kraft was tapped for reasons of "national security." Indeed, in 1974, William Ruckelshaus, formerly a Deputy Attorney General and now Acting FBI Director, told Congress: "The justification would have been that [Kraft] was discussing with some —asking questions of some members of the North Vietnamese Government, representatives of the government. My own feeling is that this is just not an adequate national security justification for placing any kind of surveillance on an American citizen or newsman."[34] (Ruckelshaus 1974, pp. 320–321)

These wiretaps were ordered for national-security reasons, but summaries forwarded to the White House included such information as this:

"meat was ordered [by the target's family] from a grocer," that the target's daughter had a toothache, that the target needed grass clippings for a compost heap he was building, that during a conversation between his wife and a friend, the two discussed "milk bills, hair, soap operas, and church" (FBI 1975b)

Although the FBI observed that this was "non-pertinent information" (USSR 94 *Intelligence Activities: Staff Reports*, p. 344), agents transcribed the conversations and faithfully sent the contents on to the White House. At least once the agents exercised some discretion. A wiretap on the home line of Henry Brandon, a *London Sunday Times* correspondent, picked up a conversation between the journalist's wife and her close friend Joan Kennedy, wife of Senator Ted Kennedy, in which Mrs. Kennedy discussed "problems with Teddy." The agent in charge said "I knew what those people would do with this stuff," and he destroyed the transcript (Hersh 1983, p. 324). President Nixon described the information he did receive as "gobs and gobs of stuff: gossip and bull" (USHR 93 *Statement of Information*, p. 1754).

Under Attorney General Mitchell, the FBI was empowered to collect political intelligence on the anti-war movement. On November 15, 1969, some 250,000 people marched on Washington in what the next day's *New York Times* described as "a great and peaceful army of dissent" (Herb 1969). Nine days earlier, Mitchell had approved an FBI request to wiretap the march's organizers.[35]

The excesses of the Nixon era were not limited to wiretaps. Tom Huston, a White House staffer, had been put in charge of developing a report on the connection between domestic unrest and foreign movements and had devised a plan in which the resources of the Central Intelligence Agency, the Defense Intelligence Agency, the Federal Bureau of Investigation, and the National Security Agency were to be pooled to fight domestic unrest. NSA—contrary to law—would intercept communications of US citizens using international facilities. Rules regarding mail interception, electronic surveillance, and surreptitious entry would be relaxed. "Covert [mail] coverage is illegal, and there are serious risks involved," wrote Huston. "However, the advantages to be derived from its use outweigh the risks." (Huston 1970, p. 194) Regarding surreptitious entry, Huston wrote: "Use of this technique is clearly illegal: it amounts to burglary." (ibid., p. 195) Yet President Nixon approved the plan. Five days later—before it took effect—he rescinded his approval. That the Huston plan came within a hairsbreadth of being national policy shocked the nation when it was revealed several years later.

The 1970s and the 1980s

The purpose of the FBI's Library Awareness Program, conducted between 1973 and 1988 at a number of public and university libraries,[36] was to investigate their use by foreigners. The FBI wanted to know about "Soviets [who] have come ... [with] unusual database requests" (Foerstal 1991, p. 56), about "computerized literature searches performed for library patrons" (ibid., p. 69), and about "reading habits of a visiting Russian student ... and anyone else 'similarly suspicious in nature'" (Robins 1988).

The FBI ran into trouble when it attempted to enlist librarians in surveillance. Librarians have a deep and abiding commitment to disseminating information, not protecting it. Librarians also strongly defend the privacy of their patrons. When the FBI sought to discover who was borrowing "unclassified, publicly available information relating more often than not to science and technical matters" (FBI 1987), many librarians would not comply with the informal requests and refused to release information about patrons without subpoenas. The American Library Association publicly opposed the Library Awareness Program. By 1989, most states had adopted laws making borrowers' records confidential (Foerstal 1991, p. 133).

When the Reagan administration came to power, a cornerstone of its foreign policy was support of the Duarte regime in El Salvador. The Committee in Solidarity with the People of El Salvador (CISPES) was a group of Americans who supported the opposition movement. In 1981 the FBI began an investigation of the Dallas chapter of CISPES to determine whether the group was in violation of the Foreign Agents Registration Act.[37] It was not.

In 1983, using information supplied by an informant, the FBI began a second investigation of CISPES, this time focusing on the group's alleged connections with international terrorists. "The FBI undertook both photographic and visual surveillance of rallies, demonstrations, etc., in its investigations of CISPES," the bureau later reported. "This technique involved the taking of photographs during demonstrations, surveillance of rallies on college campuses, and attendance at a mass at a local university. The purpose of taking photographs during demonstrations was for use or future use in identifying CISPES leaders." (FBI 1989, p. 2) The bureau kept files on more than 2300 individuals and 1300 groups (ibid.). One source "provided the FBI a copy of another person's address list by gaining unconsented access to the desk where the address list was located." In another case, "FBI agents posing as potential home buyers toured the home of a subject of the investigation with a real estate agent" (ibid., pp. 5–6). The FBI expanded the investigation using long-distance telephone records of the Dallas chapter; agents added 13 CISPES offices to the list of those being probed, and the investigation grew.

There was no justification for these actions. The investigation was based on the words of an unreliable informant. CISPES was not a terrorist organization, nor was it allied with one. Testifying before Congress in 1988, FBI Director William Sessions said: "The broadening of the investigation in October 1983, in essence, directed all field offices to regard each CISPES chapter, wherever located, as a proper subject of investigation. Based on the documentation available to the FBI by October 1983, there was no reason ... to expand the investigation so widely." (Sessions, p. 122) Once again the privacy of Americans engaged in political protest had been violated.

The Senate investigating committee observed: "The CISPES case was a serious failure in FBI management, resulting in the investigation of domestic political activities that should not have come under governmental scrutiny. It raised issues that go to the heart of this country's commitments to the protection of constitutional rights. Unjustified investigations of political expression and dissent can have a debilitating effect upon our political system. When people see that this can happen, they become wary of associating with groups that disagree with the government and more wary of what they say or write. The impact is to undermine the effectiveness of popular self-government." (USSR 101-46 *FBI and CISPES*, p. 1)

The 1990s

Those words describe what happened to a group of seven Palestinians and one Kenyan living in the Los Angeles area in the 1980s, a group that became known as the L.A. 8. Beginning in 1987, the FBI, working with the Immigration and Naturalization Service, tried to deport the eight. At issue were the L.A. 8's support for the Popular Front for the Liberation of Palestine (PLFP), an organization that is included in the US State Department's list of terrorist organizations (USDoS 2005, p. 183).

It is undisputed that in 1986 the group of seven men and one woman had helped organize a fundraiser for the Palestinian cause; what constitutes the cause was, and remains, the issue. The fundraiser was a very public event—1200 people attended (King 2005a)—and was a festival to celebrate the eighteenth anniversary of the founding of the PLFP. As the *Los Angeles Times* reported many years later, "The preparations seemed fairly unremarkable. Posters were taped to walls. Palestinian magazines, including copies of a PFLP publication, *Al Hadaf*, were arranged on tables. A troupe of amateur dancers practiced a folk dance known as the dabka." (King 2005a) The fundraising, according to the L.A. 8, was for orphanages and hospitals; the FBI held otherwise. But did the government even believe its own case?

At first the L.A. 8 were accused of violating the McCarran-Walter Act, a McCarthy-era law which made the support of groups that advocated the "doctrines of world communism" a deportable offense. Even then the FBI knew the charges were harassment: just months after the arrest of the eight, FBI Director William Webster testified to Congress that "if these individuals had been United States citizens, there would not have been a basis for their arrest" (USHR-101 *Webster Nomination*, p. 95). When in 1989 a federal court declared the McCarran-Walter Act charges unconstitutional in *American-Arab Anti-Discrimination Comm. v. Meese* (714 F. Supp. 1060, C.D. Cal. 1989), the eight were charged with visa violations (Dempsey and Cole 1999, p. 35).[38] Congress then repealed those provisions of the McCarran-Walter Act, and the seven men and one woman were instead charged by the US government with providing material support to a terrorist organization (ibid., pp. 35–36), although *not* to a terrorist activity. This is a distinction with a difference: many organizations on the US terrorist list support social and civil infrastructure such as schools and hospitals separately from their support of military and terrorist activity. Had the FBI believed that the L.A. 8's fundraising was actually for terrorist activities, the L.A. 8 would have been so charged. When asked if the government won the case, whether the L.A. 8 would be deported, a government spokesman answered "Probably not. . . . Obviously, if we get information that suggests one of the others did something. . . ."[39] (King 2005b) The situation became ever more Kafkaesque after the terrorist attacks of 9/11, when first the PATRIOT Act[40] and then the Real ID Act[41] were retrospectively applied to 1986 activities of the L.A. 8 (ibid.).

As of this writing, the L.A. 8 have been in legal limbo for 19 years, creating lives for themselves in the United States yet not daring to leave the country for fear of not being readmitted. Their legal situation has not gone unnoticed by the Arab-American community, a community whose support is critical in the U.S. effort against terrorism. The case against the L.A. 8 was just one of a number of FBI cases investigating First

Amendment activity by Arab and Palestinian groups in the United States. For example, from 1979 to 1989 the FBI investigated the General Union of Palestinian Students, a college organization for social and political activities. While the reasons for beginning the investigation are unclear (Dempsey and Cole 1999, p. 44), there were even fewer reasons for continuing the investigation for 10 years.

Other breaches of privacy have also occurred despite safeguards. A recent government report noted that employees of the Internal Revenue Service's Data Retrieval System browsed the agency's database for information on accounts of "friends, neighbors, relatives and celebrities" (Edwards 1994).[42] Authorized users of the FBI's National Crime Information Center similarly misused that system (USGAO 1993a). These, of course, were only the activities that were discovered.

Why Privacy?

Despite strictures to prevent abuses, the US government has invaded citizens' privacy many times over the last 50 years, in many different political situations, targeting individuals and political groups. Politicians have been wiretapped, and lawyers' confidential conversations with clients have been eavesdropped upon by FBI investigators.[43]

Sometimes invasion of privacy has been government policy; sometimes a breach has occurred because an individual within the government misappropriated collected information. The history of the last five decades shows that attacks on privacy are not an anomaly. When government has the power to invade privacy, abuses occur.

Conflict between protecting the security of the state and the privacy of its individuals is not new, but technology has given the state much more access to private information about individuals than it once had. As Justice Louis Brandeis so presciently observed in his dissenting opinion in *Olmstead,*

> "in the application of a constitution, our contemplation cannot be only of what has been but of what may be." The progress of science in furnishing the government with means of espionage is not likely to stop with wiretapping. Ways may some day be developed by which the Government, without removing papers from secret drawers, can reproduce them in court, and by which it will be enabled to expose to a jury the most intimate occurrences

of the home. Advances in the psychic and related sciences may bring means of exploring unexpressed beliefs, thoughts and emotions. . . . Can it be that the Constitution affords no protection against such invasions of individual security? (Brandeis 1928, p. 474)

Preservation of privacy is critical to a democratic political process. Change often begins most tentatively, and political discussion often starts in private. Journalists need to operate in private when cultivating sources. Attorneys cannot properly defend their clients if their communications are not privileged. As the Church Committee observed:

> Personal privacy is protected because it is essential to liberty and the pursuit of happiness. Our Constitution checks the power of Government for the purpose of protecting the rights of individuals, in order that all our citizens may live in a free and decent society. Unlike totalitarian states, we do not believe that any government has a monopoly on truth.
>   When Government infringes those rights instead of nurturing and protecting them, the injury spreads far beyond the particular citizens targeted to untold numbers of other Americans who may be intimidated. (USSR 94 *Intelligence Activities: Rights of Americans*, p. 290)

> Persons most intimidated may well not be those at the extremes of the political spectrum, but rather those nearer the middle. Yet voices of moderation are vital to balance public debate and avoid polarization of our society. (ibid., p. 291)

What type of society does the United States seek to be? The incarceration of Japanese-Americans during World War II began with an invasion of privacy and ended in the tyrannical disruption of many individual lives. Could the roundup of Japanese-Americans have occurred so easily if the Census Bureau's illegal cooperation had not made the process so efficient? The purpose of the Bill of Rights is to protect the rights of the people against the power of the government. In an era when technology makes the government ever more efficient, protection of these rights become ever more important.

Citizens of the former Eastern Bloc countries attest to the corruption of society that occurs when no thought or utterance is private. No one suggests that people living in the United States face imminent governmental infringements of this type, but in 1972 Congressional staffers wrote that "what separates military intelligence in the United States from its counterparts in totalitarian states, then, is not its capabilities, but its intentions" (USSR 92 *Army Surveillance*, p. 96). Electing officials we believe to be honest, trusting them to appoint officials who will be fair, and insulating the civil service from political abuse, we hope to fill the government with people of integrity. Recent history is replete with examples of abuse of power. Relying solely on intentions is dangerous for any society, and the Founding Fathers were careful to avoid it.

The right to be let alone is not realistic in modern society. But in a world that daily intrudes upon our personal space, privacy and confidentiality in discourse remain important to the human psyche. Thoughts and values still develop in the age-old traditions of talk, reflection, and argument, and trust and privacy are essential. Our conversations may be with people who are at a distance, and electronic media may transmit discussions that once might have occurred over a kitchen table or on a walk to work. But confidentiality—and the perception of confidentiality —are as necessary for the soul of mankind as bread is for the body.

7

Wiretapping

Wiretapping is the traditional term for interception of telephone conversations. This should not be taken too literally. The word is no longer restricted to communications traveling by wire, and contemporary wiretaps are more commonly placed on radio links or inside telephone offices. The meaning has also broadened in that the thing being tapped need no longer be a telephone call in the classic sense; it may be some other form of electronic communication, such as fax or data.

Compared with the more precise but more general phrase "communications interception," the word "wiretapping" has two connotations. Much the stronger of these is that a wiretap is aimed at a particular target, in sharp contrast to the "vacuum cleaner" interception widely practiced by national intelligence agencies. The weaker connotation is that it is being done by the police.

The history of wiretapping in the United States is in fact two histories intertwined. It is a history of wiretapping *per se*—that is, a history of the installation and use of wiretaps by police, intelligence agencies, honest citizens, businesses, and criminals. It is also a history of society's legal response to wiretapping by these various groups.

The origins of wiretapping lie in two quite different practices: eavesdropping and letter opening. "Eavesdropping," although once more restricted in meaning,[1] has come to describe any attempt to overhear conversations without the knowledge of the participants. "Letter opening" takes in all acquisition, opening, reading, and copying of written messages, also without the knowledge of the sending and receiving parties. Telecommunication has unified and systematized these practices.

Before the electronic era, a conversation could only be carried on by people located within earshot of each other, typically a few feet apart. Neither advanced planning nor great effort on the part of the participants was required to ensure a high degree of security. Written communications were more vulnerable, but intercepting one was still a hit-or-miss affair. Messages traveled by a variety of postal services, couriers, travelers, and merchants. Politically sensitive messages, in particular, could not be counted on to go by predictable channels, so special couriers were sometimes employed.

And written messages enjoyed another sort of protection. Regardless of a spy's skill with flaps and seals, there was no guarantee that, if a letter was intercepted, opened, and read, the victim would not notice the intrusion. Since spying typically has to be done covertly in order to succeed, the chance of detection is a substantial deterrent.

Electronic communication has changed all this in three fundamental ways: it has made telecommunication too convenient to avoid; it has, despite appearances, reduced the diversity of channels by which written messages once traveled; and it has made the act of interception invisible to the target.

Conversation by telephone has achieved an almost equal footing with face-to-face conversation. It is impossible today to run a successful business without the telephone, and eccentric even to attempt to do without the telephone in private life. The telephone provides a means of communication so effective and convenient that even people who are aware of the danger of being overheard routinely put aside their caution and use it to convey sensitive information.

As the number of channels of communication has increased (there are now hundreds of communication companies, with myriad fibers, satellites, and microwave links), the diversity of communication paths has diminished. In the days of oxcart and sail, there was no registry of the thousands of people willing to carry a message in return for a tip from the recipient. Today, telecommunications carriers must be registered with national and local regulatory bodies and are well known to trade associations and industry watch groups. Thus, interception has become more systematic. Spies, no longer faced with a patchwork of *ad hoc* couriers, know better where to look for what they seek.

Perhaps more important, interception of telecommunications leaves no telltale "marks on the envelope." It is inherent in telecommunication— and inseparable from its virtues—that the sender and the receiver of a message have no way of telling who else may have recorded a copy.

Any discussion of wiretapping, particularly a legal discussion, is complicated by the fact that electronics has not only made interception of telecommunications possible; it has also made it easier to "bug" face-to-face conversations. Bugging would be nearly irrelevant to the central subject of this book—cryptography and secure telecommunications— were it not for the fact that bugs and wiretaps are inseparably intertwined in law and jurisprudence and named by one collective term: *electronic surveillance*.

Wiretaps and bugs are powerful investigative tools. They allow the eavesdropper to overhear conversations between politicians, criminals, lawyers, or lovers without the targets' knowing that their words are being shared with unwanted listeners. Electronic surveillance is a tool that can detect criminal conspiracies and provide prosecutors with strong evidence—the conspirators' incriminating statements in their own voices —all without danger to law-enforcement officers. On the other hand, the very invisibility on which electronic surveillance depends for its effectiveness makes it evasive of oversight and readily adaptable to malign uses.[2] Electronic surveillance can be and has been used by those in power to undermine the democratic process by spying on their political opponents. In light of this, it is not surprising that Congress and the courts have approached wiretapping and bugging with suspicion.

Today, communication enjoys a measure of protection under US law, and neither government agents nor private citizens are permitted to wiretap at will. This has not always been the case. The current view—that wiretaps are a kind of search—has evolved by fits and starts over a century and a half. The Supreme Court ruled in 1967 that the police may not employ wiretaps without court authorization. Congress has embraced this principle, limiting police use of wiretaps and setting standards for the granting of warrants. The same laws prohibit most wiretapping by private citizens.

The rules against unwarranted wiretapping are not absolute, however. For example, the courts ruled in 1992 (*United States v. David Lee Smith*,

978 F. 2nd 171, US App) that conversations over cordless phones were not protected and that police tapping of cordless phones did not require a search warrant. A 1994 statute (Communications Assistance for Law Enforcement Act of 1994, Public Law 103-414, §202) extended the warrant requirements of the earlier law to cover cordless phones. The law also makes some exceptions for businesses intercepting the communications of their own employees on company property.

Constitutional Protection

Protection for the privacy of communications stems primarily from the Fourth Amendment to the US Constitution. Epitomizing the underlying principle of the Bill of Rights—that individual citizens need protection against the power of the state—the Fourth Amendment asserts "the right of the people to be secure in their persons, houses, papers, and effects, against unreasonable searches and seizures."

The Fourth Amendment was a response to the British writs of assistance, which empowered officers of the Crown to search "wherever they suspected uncustomed goods to be" and to "break open any receptacle or package falling under their suspecting eye" (Lassen 1937, p. 54). In framing the laws of the new nation, the founders sought to avoid creating such unrestricted governmental powers. However, they also recognized that since criminals try to hide the evidence of their crimes, law-enforcement officials must have the power to conduct searches. The Fourth Amendment thus allows searches, but restricts them, specifying that "no Warrants shall issue, but upon probable cause, supported by Oath or affirmation, and particularly describing the place to be searched and the persons or things to be seized."

As written, the Fourth Amendment protects citizens against invasions of property. Although strongly suggested by the phrase "secure in their persons, houses, papers, and effects," the word 'privacy' is never used. The view that the Fourth Amendment protects things less tangible than property has taken more than a century to develop.

Wiretaps of the Nineteenth Century

When the telegraph joined the covered wagon and the stagecoach as a channel of long-distance communication, wiretapping followed quickly. During the Civil War, General Jeb Stuart traveled with his own tapper.[3] In California in 1864, a former stockbroker obtained market information by intercepting telegraph messages; he was prosecuted for violating a surprisingly early California statute against wiretapping (Dash et al. 1959, p. 23).

The convenience of voice communication made it obvious that intercepted telephone calls would be a rich source of information. In 1899 the *San Francisco Call* accused a rival, the *San Francisco Examiner*, of wiretapping conversations between the *Call* and its reporters and stealing the *Call*'s exclusives. In 1905, the California legislature responded by extending an 1862 law prohibiting telegraph wiretapping to telephones (ibid., pp. 25–26).

The first tapping of telephones by police occurred in the early 1890s in New York City. An 1892 New York State law had made telephone tapping a felony, but New York policemen believed that the law did not apply to them and employed wiretaps anyway (ibid., p. 35). In 1916 the mayor of New York was found to have authorized wiretapping of some Catholic priests in connection with a charity-fraud investigation, despite the fact that none of the priests were suspected of participating in the illegal activity (*New York Times* 1916). The state legislature discovered that the police had the ability to tap any line of the New York Telephone Company. Using this power with abandon, the police had listened in on confidential conversations between lawyers and their clients, and between physicians and their patients. The *New York Times* reported that "in some cases the trunk lines of hotels were tapped and the conversations of all hotel guests listened to" (ibid.).

During this period, the federal government played no legislative role— except during World War I, when concern that enemy agents would tap phone lines led to a federal Anti-Wiretap Statute (40 Stat. 1017, 1918). After the war, law-enforcement agencies began to find wiretapping increasingly valuable. Wiretapping is said to have been the most frequently used tool for catching bootleggers during Prohibition (Dash et al. 1959, p. 28).

The *Olmstead* Decision

As was mentioned in chapter 6, Roy Olmstead was caught running a $2 million-a-year bootlegging operation during Prohibition. Convicted partially on the basis of evidence obtained from warrantless wiretaps installed by federal agents, Olmstead appealed, and the case made its way to the US Supreme Court.

Though closely divided, the Court ruled that the evidence obtained by tapping the defendants' phone calls had not involved any trespass into their homes or offices. According to the Court: "There was no searching.... The evidence was secured by the use of ... hearing and that only...." (*Olmstead v. United States*, 277 US 438, 1928, p. 464) Five justices agreed that the Fourth Amendment protected tangibles alone, that conversation was an intangible, and that therefore using the evidence from the wiretaps did not constitute unreasonable search and seizure.

In his dissenting opinion in the *Olmstead* case (which has become one of the most quoted of judicial opinions), Justice Louis Brandeis argued that protections provided by the Bill of Rights should operate in a world of electronic communications. In 1928 the telephone was already necessary for commerce and was rapidly becoming an integral part of daily life. Brandeis described the threat that wiretapping posed to privacy:

> Whenever a telephone line is tapped, the privacy of the persons at both ends of the line is invaded, and all conversations between them upon any subject, and although proper, confidential and privileged, may be overheard. Moreover, the tapping of one man's telephone line involves the tapping of the telephone of every other person whom he may call, or who may call him. As a means of espionage, writs of assistance and general warrants are but puny instruments of tyranny and oppression when compared with wire-tapping. (Brandeis 1928, pp. 475–476)

Despite his eloquence, Brandeis's viewpoint did not carry the Court. There was widespread consternation over the Court's ruling, but although there were several attempts to do so, Congress did not make wiretapping illegal.

The *Nardone* Cases

In 1934, Congress passed the Federal Communications Act (FCA), which placed jurisdiction over radio and wire communications in the hands of the newly created Federal Communications Commission and established a regulatory framework that has dominated American telecommunications ever since. The FCA prohibited the "interception and divulgence" of wire communications. Although its wording was quite similar to that of the Radio Act of 1927 (the law in effect at the time of the *Olmstead* case), the Supreme Court used the FCA to reverse *Olmstead.*

In *Olmstead*, the Court had ruled on constitutional grounds and had found warrantless wiretaps legal. A decade later, it considered the case of Frank Carmine Nardone, another accused bootlegger. This time, the Court ruled on the basis of the new law and held that information from wiretaps placed by federal agents was not admissible as evidence (*Nardone v. United States*, 302 US 379, 1937).

Two years later Nardone was back. Having been convicted in a new trial that used evidence derived from the warrantless wiretaps, Nardone appealed, arguing that the evidence should not be admissible. The Court concurred and held that information even indirectly derived from wiretaps could not be used as evidence (*Nardone v. United States*, 308 US 338, 1939). The same day, in a different case, the Supreme Court ruled that the FCA applied to federal wiretapping of intrastate as well as interstate communications (*Weiss v. United States*, 302 US 321, 1939). In response to these decisions, Attorney General Robert Jackson ordered a halt to FBI wiretapping (Gentry 1991, p. 231).

These decisions appeared to prohibit wiretapping by federal law-enforcement agencies, but the prohibition was overtaken by events and quickly eroded.

Evidence versus Intelligence

With the start of World War II, FBI Director J. Edgar Hoover, citing the danger of spies and other subversives, pressed to have Attorney General Robert Jackson's anti-wiretapping order overturned (Morgenthau, May 21, 1940). In view of the Supreme Court's *Nardone* decisions, this would take some fancy footwork. But, under Jackson, the Department of Justice had interpreted the *Nardone* decisions to mean that it was unlawful to both "intercept" *and* "divulge" communications, and had decided that it was not unlawful to intercept communications as long as the contents were kept within the federal government (USSR 94 *Intelligence Activities:*

Staff Reports, p. 278). Hoover urged President Roosevelt to authorize wiretapping for what we have come to call national-security purposes. The president acceded to Hoover's request, but his order stopped short of giving the FBI blanket approval to wiretap:

> I am convinced that the Supreme Court never intended any dictum in the particular case which it decided to apply to grave matters involving the defense of the nation....
>
> You are therefore authorized and directed in such cases as you may approve, after investigation of the need in each case, to authorize the necessary investigative agencies that they are at liberty to secure information by listening devices ... of persons suspected of subversive activities against the government of the United States, including suspected spies. You are requested furthermore to limit these investigations so conducted to a minimum and to limit them insofar as possible to aliens. (Roosevelt 1940)

Jackson, who wanted no part in wiretapping, made a fateful decision in response to the presidential directive: he instructed Hoover to maintain the records of all wiretaps, listing times, places, and cases (Theoharis and Cox 1988, p. 171). In so doing, Jackson effectively permitted the FBI to use wiretapping, free of Department of Justice oversight.

In 1941, Francis Biddle, the succeeding Attorney General, wrested back control of wiretaps from Hoover and turned down applications he felt were unjustified (Biddle 1941; USSR 94 *Intelligence Activities: Rights of Americans*, p. 37). In 1952, Attorney General J. Howard McGrath, in a letter to Hoover supporting the use of wiretaps "under the present highly restrictive basis," made explicit the requirement that all FBI wiretaps required the attorney general's prior approval (McGrath 1952). However, McGrath did not undo the custom that surveillance orders operated without time limits. Only in 1965, when Attorney General Nicholas Katzenbach recommended that authorizations be limited to 6 months (Hoover 1965), did the FBI change its practice.

In 1940 and 1941, several bills that attempted to establish a legal basis for electronic surveillance were introduced in Congress. One of these was endorsed by Roosevelt and Jackson (USSR 94 *Intelligence Activities: Staff Reports*, p. 280), but Hoover opposed any legislation requiring warrants for wiretapping (Hoover 1941) and no legislation passed.

When the war ended, Hoover sought and received continued wire tapping authority from President Harry Truman. Indeed, the power was broadened. In his reauthorization request to Attorney General Thomas Clark, Hoover omitted the final sentence of Roosevelt's original memo, which had required that electronic surveillance be kept to a minimum and limited "insofar as possible to aliens" (Gentry 1991, p. 324). Clark forwarded the amended memo and urged Truman to approve wiretapping in cases "vitally affecting domestic security, or where human life is in jeopardy," and Truman apparently signed it without being aware of the shift from earlier policies (Clark 1946).

In time Truman's aides discovered the change. A 1950 memo to Truman from George Elsey, assistant counsel to the president, reported that "not only did Clark fail to inform the President that Mr. Roosevelt had directed the FBI to hold its wiretapping to a minimum ... he requested the President to approve very broad language [that was] a very far cry from the 1940 directive" (Elsey 1950). Truman, however, took no action to rescind the expanded authority. The new authorization meant that national security was no longer the sole justification for wiretaps (Clark 1946).

By adding "domestic security" to the list of reasons for which wiretapping could be employed, Clark's memo substantially broadened Roosevelt's directive. Developing intelligence is quite different from collecting evidence. In particular, intelligence investigations have less narrowly defined goals than criminal investigations. Developing intelligence is neither attempting to find evidence of a specific crime nor developing a case against a specific suspect. Rather, it is attempting to discern a pattern of behavior: What is the structure of an organization? What are its goals? What are its methods?

Those who work in intelligence emphasize the degree to which they operate under legal restraints intended to protect the rights of Americans. Yet the very character of intelligence work makes it unlikely that these restraints weigh as heavily on them as on criminal investigators. Intelligence officers don't go into court to face opposing attorneys. Criminal investigations are unsuccessful without convictions. Intelligence investigations—even in counterintelligence, where prosecution is sometimes appropriate—can be deemed successful even if no prosecutions occur. The changes Clark made were thus quite significant. They moved wiretap investigations into a shadowy area where, by and large, they were protected from public scrutiny.

Since the two *Nardone* rulings made both evidence from federal wiretaps and evidence tainted by federal wiretaps inadmissible in court, the Department of Justice believed that if wiretapping had been used in a case, it could not prosecute. This eventually gave rise to an elaborate FBI methodology for concealing evidence of wiretapping.

In 1949, Judith Coplon, a Department of Justice employee, was caught as she was about to hand over 28 confidential FBI documents to Valentin Gubitchev, a Soviet employed by the United Nations. Hoover sought to prevent the documents—which revealed FBI wiretapping of Coplon—from appearing at Coplon's trial, but the judge ruled that the government had to release copies to the court. Fearing public disclosure of the FBI's wiretapping practices, Hoover tried to persuade the attorney general to drop the charges against Coplon, but without success. Hoover responded with new procedures regarding wiretapping. FBI reports of "highly confidential" sources, including wiretaps, would not be included in the main case files; instead they would be kept in especially confidential files (FBI 1949).[4]

A second trial revealed that Coplon herself had been wiretapped, despite denials by an agent who turned out to have read the transcripts. Hoover then went to even greater lengths to hide the paper trail between wiretaps and their logs. Agents working on a case were kept in the dark about any associated wiretaps so that they could not even accidentally reveal wiretap information in court. Hoover accomplished this by disguising information derived from wiretaps when it appeared in the case files. He was largely successful in this strategy, but he hoped for looser rules regarding wiretaps.

In 1954, President Dwight Eisenhower's attorney general, Herbert Brownell, was pressing for legislation to permit warrantless wiretapping as an aid in prosecuting communists. "You can't trust a Communist to tell the truth on the witness stand," Brownell (1954a, p. 202) argued, "and you can't trust the courts not to leak information about wiretap applications." Brownell was frustrated because there was "evidence in the hands of the Department as a result of investigations conducted by the FBI which would prove espionage in certain ... cases." He argued that a change in the wiretap law would enable the Department of Justice to prosecute certain cases that it was currently prevented from pursuing (ibid., p. 203).

The House Judiciary Committee accepted Brownell's argument and recommended passage of legislation permitting warrantless wiretapping, but the full House of Representatives disagreed (USSR 94 *Intelligence Activities: Staff Reports*, p. 284). The legislation was amended to require a warrant, but this version did not receive support from the Department of Justice and died (ibid.). Although Congress periodically considered legislation on wiretapping, there were no new federal laws on the subject until the 1960s.

Meanwhile, as a result of the Federal Communications Act's prohibition against "interception and divulgence" of wired communications, wiretaps were useless as court evidence. As Attorney General put it to Congress in 1965:

> I think perhaps the record ought to show that when you talk national security cases, they are not really cases, because as I have said repeatedly, once you put a wiretap on or an illegal device of any kind, the possibilities of prosecution are gone. It is just like a grant of immunity.... I have dismissed cases or failed to bring cases within that area because some of the information did come from wiretaps. But here we feel that the intelligence and the preventive aspect outweigh the desirability of prosecution in rare and exceptional circumstances. (USSH 89 *Invasions of Privacy*, p. 1163)

In 1950, and again in 1953, presidential directives authorized the FBI to investigate "subversive activity" (USSR 94 *Intelligence Activities: Rights of Americans*, p. 45). Since these directives failed to provide guidelines for such investigations, the FBI took a very broad view of what constituted subversive activity. Surveillance, often including wiretapping and electronic bugging, was not limited to those suspected of crimes or even to detecting suspected criminal activities.

Though many other types of domestic activities were the targets of surveillance, Hoover stressed the danger of Communist subversion, and the FBI collected intelligence about the influence of Communists in a variety of categories, including "political activities, Negro question, youth matters, women's matters, farmers' matters, veterans' matters" (FBI 1960a).

Believing that the growing civil rights movement was Communist inspired and would lead to violence, Hoover kept it under careful observation. In 1956 he briefed a cabinet meeting about the Communists' "efforts" and "plans" to influence the movement. This briefing demonstrates how far the FBI overreached its mandate in internal security investigations. Not limiting his discussion to the possibility of violence, Hoover went on to present to the cabinet the legislative strategy of the NAACP and "the activities of Southern Governors and Congressmen on behalf of groups opposing integration peacefully" (FBI 1956).

Permission to conduct investigations of domestic "subversive activity" without restraint gave the FBI free rein in collecting wide-ranging domestic intelligence. The FBI's choice of electronic surveillance as the means for doing this allowed it to keep its activities hidden and uncontrolled for a long time. An official of the Nation of Islam was wiretapped for 8 years without any efforts to prosecute him for illegal activities (Hoover 1956), and the Socialist Workers Party was the target of FBI wiretaps and bugs for 20 years.

The extent of the FBI's wiretapping under J. Edgar Hoover has never been clear.[5] Although Congress received annual testimony from Hoover, it was unable to discover how much wiretapping was actually occurring, since the figures Hoover gave did not include the taps installed by field agents on their own or the taps installed by local police at the FBI's request.

Hoover kept the transcripts of wiretaps—many of them hidden in obscure files—even when they revealed no evidence of criminal activity. Since wiretaps on suspected spies and organized crime figures sometimes picked up conversations with politicians or other influential people, Hoover developed a mass of material with great political value. This is how he ended up with transcripts of intimate conversations between John Kennedy and Inga Arvad, a Danish reporter and former Miss Europe who had visited Germany for social functions with high Third Reich officials, including Hitler. The FBI was investigating allegations that Arvad was a German spy (Anderson 1996a, pp. 48–52). The investigation itself may have been legitimate. The recordings were made during the World War II, while Kennedy was a Naval officer. Nonetheless, although the investigation produced no evidence of espionage, recordings of his pillow talk with Arvad were still in the FBI files when Kennedy became president nearly 20 years later.

It is believed that, without any pretense of investigating criminal activities, Hoover wiretapped senators and congressmen.[6] It is known that he wiretapped various Supreme Court justices.[7] The existence of extensive records on political figures was well known, and these files ensured that Hoover got much of what he wanted. Congress exercised little oversight of the FBI's affairs, wiretapping included.

One example is particularly illustrative of the way in which Hoover was able to use the FBI's investigative powers to protect its interests. In 1965, a Senate subcommittee undertook an investigation of electronic surveillance and mail covers. The FBI, a major focus of this review, was concerned. One FBI memo noted: "Senator Long ... has been taking testimony in connection with mail covers, wiretapping, and various snooping devices on the part of federal agencies. He cannot be trusted." (Jones 1965) Two high-ranking Bureau officials met with Edward Long (the chairman of the subcommittee) and a committee counsel. There is no indication that there were any briefings of other subcommittee members, nor is there any reason to believe that during the 90-minute meeting Long was told any details of FBI electronic surveillance, such as the bugging of a congressman's hotel room during the sugar lobby investigations (see below), the bugging and wiretapping of Martin Luther King, or the wiretapping of a congressional staffer, two newspaper reporters, and an editor of an anti-Communist newsletter (USSR 94 *Intelligence Activities: Staff Reports*, p. 309). The FBI men suggested that the senator issue a statement saying that he had held lengthy conferences with FBI officials and was now completely satisfied "that the FBI had never participated in uncontrolled usage of wiretaps or microphones and that FBI usage of such devices has been completely justified in all cases" (DeLoach 1966a). When Long said that he did not know how to write such a press release, the FBI officials said they would be happy to do so—and they did (ibid.).

The Long Subcommittee chose not to hold hearings on FBI electronic-surveillance practices. An internal FBI memo noted: "While we have neutralized the threat of being embarrassed by the Long Subcommittee, we have not eliminated certain dangers which might be created as a result of newspaper pressures on Long. We therefore must keep on top of this situation at all times." (DeLoach 1966b) The FBI's determination to control such investigations can be inferred from the fact that it maintained files on all members of the subcommittee (USSH 94 *Intelligence Activities: FBI*, p. 477).

A year later Senator Long again took up the fight, introducing a bill to limit FBI electronic surveillance to national-security cases. Hoover was not pleased. Shortly afterwards *Life* broke a story that the senator had received $48,000 from Morris Shenker, Jimmy Hoffa's counsel (Lambert 1967). The article intimated that the money was a bribe to prevent or reverse charges against the Teamsters' leader. The senator's career ended, and his electronic-surveillance bill died.

Hoover remained firmly in power through eight presidencies and 48 years. His long tenure as director is now recognized as a period during which the FBI routinely engaged in the sort of widespread political surveillance usually associated with totalitarian regimes.

In part Hoover's control came about because of abdication of responsibility by other members of the government. Truman's 1946 authorization of wiretaps in investigating subversive activities required the attorney general to authorize their use. There were, however, many occasions on which an attorney general was informed of the wiretaps only after they had been installed (USSR 94 *Intelligence Activities: Rights of Americans*, p. 63). In 1965, under a directive from President Johnson, Attorney General Katzenbach tightened requirements for electronic surveillance. By then Johnson opposed wiretapping except in cases of national security, and in a directive that went out to heads of agencies he wrote: "In my view, the invasion of privacy of communications is a highly offensive practice which should be engaged in only where the national security is at stake." (Johnson 1965) Installation of bugs now needed written approval from the attorney general, and both bugs and wiretaps were subject to 6-month time limits, after which new authorization from the attorney general was required (USSR 94 *Intelligence Activities: Rights of Americans*, p. 105).

Even with the tightened restrictions, the Department of Justice did not feel that it controlled the FBI's use of wiretapping. In 1972, former Attorney General Ramsey Clark told a judge: "Reports by FBI agents on electronic surveillance had caused the Department [of Justice] 'deep embarrassment' many times. Often we would go to court and say that there had been no electronic surveillance and then we would find out we had been wrong. Often you could not find out what was going on ... frequently agents lost the facts." (Theoharis 1974, p. 342)

The FBI was not the only law-enforcement agency to engage in wiretapping without regard for the views of legislators and judges. State police also wiretapped, sometimes legally, sometimes not.

Wiretapping was made illegal in Illinois in 1927, but Chicago's police intelligence unit ignored the law. The police saw no reason to try and change the legislation; the effort would only raise controversy. They merely continued wiretapping, aware that their actions were illegal (Dash et al. 1959, p. 222).

In California, wiretaps were similarly banned, but the police there took a different tack. When law-enforcement officers wanted to do a wiretap investigation, they would hire a private investigator, who was told to deny any official connections if caught in the act of tapping (ibid., pp. 164–165).

Even in states that permitted police wiretapping, numbers could be misleading. Official records in the State of New York showed fewer than 3,000 wiretap orders for the period between 1950 and 1955. A careful study by Samuel Dash and his two co-authors led to different conclusions. They observed that the official figures omitted wiretaps installed by plainclothesmen. Those numbers were surprisingly large. Dash et al. (ibid., p. 68) concluded that for every 10 wiretaps installed by plainclothesmen who had obtained court orders, another 90 taps were installed without court authorization. This interpretation would indicate that the New York police performed between 16,000 and 29,000 wiretaps a year.

Oddly enough, in Massachusetts, a state with a particularly liberal wiretapping law that required only written permission from the attorney general or the district attorney, the Boston Police Department did not use wiretaps in criminal investigations. The public was opposed to it, and the police believed wiretapping was a "dirty business" to which "only a lazy police department" would resort (Dash et al. 1959, p. 147).

Wiretaps versus Bugs

Microphone surveillance and wiretapping play similar investigative roles. But the Federal Communications Act said nothing about microphone surveillances, and the *Nardone* decisions therefore left such bugging legal. Although ultimately the Supreme Court whittled away at the warrantless use of electronic bugs, its early decisions condoned the practice.

In 1942 the Supreme Court ruled that it was legally permissible for law-enforcement officers to plant a bugging device on a wall adjoining that of a suspect's office (*Goldman v. United States*, 316 US 129). In 1954 it upheld a state court conviction based on evidence obtained by microphones concealed in walls of the defendants' homes during warrantless break-ins by the police (*Irvine v. California*, 347 US 128).

In *Irvine*, in sharp contrast to later decisions, the Supreme Court ruled that, because the case in question had been a state prosecution within a state court, the conviction could stand. Members of the Court were disturbed, however, by violations of the Fourth and Fifth Amendments. The bug had been placed in a bedroom, and the justices were deeply offended by the invasion of privacy. Attorney General Herbert Brownell "clarified" policy for J. Edgar Hoover, warning that the language of the Court "indicates certain uses of microphones which it would be well to avoid" because "the Justices of the Supreme Court were outraged by what they regarded as the indecency of installing a microphone in the bedroom" (Brownell 1954b).

In 1961, in *Julius Silverman et al. v. United States* (365 US 505), the Supreme Court changed its direction, holding that a search occurred whenever a bug was used, even if the walls of the target's apartment had not been breached. The case arose when District of Columbia police, suspecting that gambling was taking place in a row house, pushed a foot-long "spike mike" into a space under the suspect's apartment from the vacant house next door. The spike hit a solid object (most likely a heating duct), which "became in effect a giant microphone" (ibid., p. 509).

Telephone conversations were picked up from the room above, amplified, and taped. The Court held that, in the absence of a search warrant, the evidence was inadmissible. In the *Nardone* cases, the Court had ruled on the narrow basis of the FCA. In *Silverman*, it brushed away these technicalities ("In these circumstances we need not pause to consider whether or not there was a technical trespass under the local property law relating to party walls" (ibid., p. 511)) and ruled on the basis of the Fourth Amendment. With this decision, the Court began moving in the direction of constitutional protection against warrantless electronic surveillance.

In 1967, the Supreme Court established the doctrine of "legitimate expectation of privacy." In *Charles Katz v. United States*, (389 US 347) the Court determined that evidence obtained from a warrantless electronic bug placed in a public phone booth was inadmissible:

> ... [W]hat [Katz] sought to exclude when he entered the [telephone] booth was not the intruding eye—it was the uninvited ear.... No less than an individual in a business office, in a friend's apartment, or in a taxicab, a person in a telephone booth may rely upon the protection of the Fourth Amendment. One who occupies it, shuts the door behind him, and pays the toll that permits him to place a call is surely entitled to assume that the words he utters into the mouthpiece will not be broadcast to the world. To read the Constitution more narrowly is to ignore the vital role that the public telephone has come to play in private communication. (ibid., pp. 511–512)
>
> ... Wherever a man may be, he is entitled to know that he will remain free from unreasonable searches and seizures. (ibid., p. 515)

With this decision, the Supreme Court changed the doctrine underlying US wiretap law. Unlike the *Nardone* decisions, which relied on statutory interpretation, *Katz* is based on underlying principles of the Constitution. In ruling that electronic bugging was illegal without a search warrant, the Supreme Court arrived at the current view of bugs and wiretaps as a form of search: that they are permissible, but subject to the limitations and protections laid down in the Fourth Amendment.

Wiretapping and Organized Crime

When asked to explain the need for wiretapping, police often cite organized crime as the principal application. In order to attack organized crime, police must find a way of penetrating the tightly knit organizations. This is not easy. The difficulty is compounded by the fact that "the core of organized crime activity is the supplying of illegal goods and services ... to countless numbers of citizen customers" (President's Commission on Law Enforcement 1967, p. 187), and the consensual nature of these crimes means that "law enforcement lacks its staunchest ally, the victim" (Duke and Gross 1993, p. 107). Organized crime is particularly effective in using the code of *omerta*—silence—to keep its participants and its victims from speaking to law-enforcement agents (ibid., p. 198). A common police response has been the use of electronic surveillance. The 1967 *Katz* decision took this tool from the police just as the visibility of organized crime was increasing.

Organized crime had been a force in American society for years. With the repeal of Prohibition, it had shifted its activities from bootlegging into gambling, loansharking, and control of legitimate businesses, including garbage disposal, garment manufacturing, real estate, restaurants, vending machines, and waterfront activities (ibid., p. 195). Yet between the 1930s and the 1950s the FBI pursued bank robbers, kidnappers, auto thieves, and the Communist Party, but not organized crime (although from time to time individual members of crime "families" made their way into the FBI's net).

There are various theories as to why J. Edgar Hoover ignored organized crime's very existence for some 30 years. It may be that any concerted federal effort would have involved the FBI in interagency cooperation, something that the turf-conscious Hoover abhorred. It may be that an investigation of organized crime would have been a drawn-out affair that would not have yielded the statistics on criminals caught and property recovered that were the "heart and soul" of Hoover's annual speech before the Senate Appropriations Committee (Sullivan 1979, pp. 117–118). It may be that investigating organized crime would have risked corruption of the investigators to a much greater extent than other types of criminal investigation.

Organized Crime Becomes Visible

Sergeant Edgar Croswell, a New York State trooper, Robert Kennedy, Attorney General under his brother John, and Joseph Valachi, a 60-year-old "soldato" in the Genovese crime family, brought organized crime to the nation's attention in the late 1950s and the early 1960s.

On Saturday November 15, 1957, Sergeant Croswell was doing his morning rounds in Apalachin, a small village in the Southern Tier of New York State, near Pennsylvania, when he discovered limousine after black limousine turning into the country estate of Joseph Barbara, a bottler and distributor of soft drinks in nearby Endicott. His suspicion aroused, Croswell set up a roadblock near the estate to check for vehicle identification—a legal search under the state laws. Many of the crime bosses fled, but 67 of them were identified—some through the roadblock, others through various forms of carelessness, including registering under their own names at hotels in the area. The next day, the nation awoke to headlines like "Royal Clambake for Underworld Cooled by Police" and "Police Ponder NY Mob Meeting; All Claim They Were Visiting Sick Friend."

In 1961 Robert Kennedy became US attorney general. Five years earlier, while counsel to the Senate Select Committee on Improper Activities in the Labor or Management Field, Kennedy had investigated labor racketeering and had uncovered ties between the unions and organized crime. When he became attorney general, Kennedy made organized crime a priority (Lewis 1961). Congress soon passed legislation that Kennedy had requested to fight organized crime (*New York Times* 1961).

In 1963, Joseph Valachi broke the code of silence that had made investigations of organized crime so unrewarding. Imprisoned for heroin trafficking, Valachi killed a man whom he believed had been sent to assassinate him in jail. Soured by the belief that the "family" had tried to have him "hit" and facing murder charges for defending himself, Valachi turned government witness and began to talk. In the Senate's staid hearing rooms, Valachi laid out the complex system of bosses, soldiers, and international links that characterized organized crime. Charts describing the succession of gang control in the New York area graced the walls. The information was shocking to those who had been repeatedly told by the FBI's director that organized crime did not exist.

The combined effect was to transform organized crime from a myth to a priority in federal eyes. The government concluded that its involvement was crucial in working against organized crime because criminal networks flourished largely as a result of corruption in local police forces. In the words of Ramsey Clark, Kennedy's successor as attorney general: "The presence of any significant organized crime necessarily means that local criminal justice has been corrupted to some degree." (Clark 1970) Furthermore, because the criminal organizations spanned state borders, nationwide coordination of investigations was viewed as crucial.

By 1966, even J. Edgar Hoover had caught up on organized crime. Testifying before the House Subcommittee on Appropriations, Hoover said: "La Cosa Nostra is the largest organization of the criminal underworld in this country, very closely organized and strictly disciplined." (Carroll 1967)

Title III: Wiretaps Made Legal

In their report on the Valachi hearings, members of the Senate Governmental Operations Committee called for legislation authorizing wiretapping. Their view was echoed by a presidential commission on law enforcement: "A majority of the members of the Commission believe that legislation should be enacted granting carefully circumscribed authority for electronic surveillance to law enforcement officers...." (President's Commission on Law Enforcement 1967, p. 203)

Not all experts agreed with the commission's conclusions. Attorney General Clark prohibited all use of wiretaps by federal law-enforcement officers. He told Congress: "I know of no Federal conviction based upon any wiretapping or electronic surveillance, and there have been a lot of big ones.... I also think that we make cases effectively without wiretapping or electronic surveillance. I think it may well be that with the commitment of the same manpower to other techniques, even more convictions could be secured, because in terms of manpower, wiretapping, and electronic surveillance is very expensive." (Clark 1967, p. 320) Clark pointed out that in 1967, without using wiretaps, federal strike forces had obtained indictments against organized crime figures in nine states, and that "each strike force has obtained more indictments in its target city than all federal indictments in the nation against organized crime in as recent a year as 1960" (ibid., pp. 79–80).

In 1965, the Chief Judge of the US District Court in Northern Illinois,

William Campbell, told Congress of his strong disapproval of wiretaps: "My experiences have produced in me a complete repugnance, opposition, and disapproval of wiretapping, regardless of circumstances. This invasion of privacy, too often an invasion of the privacy of innocent individuals, is not justified. In every case I know of where wiretapping has been used, the case could have been made without the use of the wiretap. Wiretapping in my opinion is mainly a crutch or shortcut used by inefficient or lazy investigators." (USSR 90-1097 *Omnibus Safe Streets and Crime Control*, p. 1495)

Detroit's police commissioner, Ray Girardin, was also opposed to wiretapping: "I feel that [wiretapping] is an outrageous tactic and that it is not necessary and has no place in law enforcement.... [T]he only exception to this that I would entertain at this time would be in a situation where the security of the nation has to be protected against an outside power." (Girardin 1967)

At other times, in other places, other officials had denounced wiretapping. A 1961 congressional survey revealed that the attorneys general of California, Delaware, Missouri, and New Mexico opposed federal laws permitting wiretapping (USSH 87 *Wiretapping and Eavesdropping Legislation*, pp. 545, 547, 554, and 560). Daniel Ward, State's Attorney for Cook County, Illinois, testified in 1961: "I do not think that one can honestly say that wiretapping is a *sine qua non* of effective law enforcement." (Ward 1961)

Ramsey Clark's opposition was sustained by President Johnson, who proposed to Congress in 1967 that wiretapping be limited to national-security cases and that it be performed only by federal officials (*Congressional Quarterly Weekly* 1967, p. 222). But in the aftermath of the Crime Commission's report, and during a time of domestic unrest (riots in the ghettos, the Vietnam War protests, and several political assassinations), Congress saw the issues differently. Despite the lack of unanimity, even among police, Congress chose to legalize wiretapping as a tool for law-enforcement investigations in criminal cases. Title III of the Omnibus Crime Control and Safe Streets Act of 1968 (18 USC §2510–2521) established the basic law for interceptions performed in criminal investigations: wiretaps are limited to the crimes specified in the act—a list including murder, kidnapping, extortion, gambling, counterfeiting, and sale of marijuana. The Judiciary Committee made clear that organized crime was a central motivation for Title III.[8]

In order to conform to the standards of the Fourth Amendment, Congress required law enforcement to obtain a warrant before initiating a wiretap.[9] To receive a court order, an investigator draws up an affidavit showing that there is probable cause to believe that the targeted communication device—whether a phone, a fax, or a computer[10]—is being used to facilitate a crime. A government attorney prepares an application for a court order, and approval must be granted by a member of the Department of Justice at least at the level of Deputy Assistant Attorney General.

Observing that "wiretaps and eavesdrops are potentially more penetrating, less discriminating, and less visible than ordinary searches," Congress decided that review of the application by a federal district court judge was in order (National Commission 1976, p. 12). The judge must determine that (i) there is probable cause to believe that an individual is committing or is about to commit an indictable offense; (ii) there is probable cause to believe that communications about the offense will be obtained through the interception; (iii) normal investigative procedures have been tried and either have failed, appear unlikely to succeed, or are too dangerous; and (iv) there is probable cause to believe that the facilities subject to surveillance are being used or will be used in the commission of the crime. Only if all these criteria are satisfied will a judge approve a wiretap order.[11]

After a court order for electronic surveillance is approved, it is taken to a service provider (e.g., a telephone company) for execution. The provider is required to assist in placing the wiretap,[12] and is compensated for all expenses. Surveillances are approved for at most 30 days; any extension requires a new court order.

Almost all states have also passed statutes permitting wiretaps by state and local law enforcement officers for criminal investigations. Under Title III, state acts were required to be at least as restrictive in their requirements as the federal code, and many are more so.

In an effort to prevent repetition of the concealment practiced by J. Edgar Hoover, Congress required that records on electronic surveillance be available to the public. Each order, whether filed under Title III or under a state statute, must be reported, and every year the Administrative Office of the United States Courts issues a report detailing the duration of each order, the number of persons intercepted, the type of surveillance used, the outcome of the case, and other information.

A New Wrinkle: "Domestic National Security"

There are different codes of conduct in wartime than in peacetime. Soldiers are permitted, under appropriate circumstances, to kill without the elaborate procedures required for a civil execution. The most common "appropriate circumstance" is, of course, war with a foreign power. Less commonly and more controversially, a country's military may be deployed against its own citizens in times of insurrection. Under these conditions, which constitute a threat to the state itself, it is generally felt that peacetime restraint would place the state at too great a disadvantage and permit the country to be conquered or the government to be overthrown.

Similar reasoning applies, even in peacetime, to the state's actions with respect to spies, subversives, and revolutionaries—people who are not, or do not consider themselves, bound by the country's laws. The state will, where possible and appropriate, treat such people as lawbreakers and prosecute, but it may also take other actions that are not taken against citizens in the normal course of events.

It is not surprising, therefore, that in matters affecting national security the legal requirements for the placement of wiretaps should be relaxed. This is the reasoning behind the 1940 authorization granted to J. Edgar Hoover by Franklin Roosevelt. In a more general sense, it is the reasoning under which all intelligence agencies conduct communications intercept operations against foreign targets.

The relaxed requirements for national-security wiretapping are an attractive target for abuse. A number of presidents, including Kennedy, Johnson, and Nixon, used "national security" as a pretext for employing wiretaps in domestic political intelligence.

Kennedy abused electronic surveillance only a month into his presidency. The Dominican Republic was pressing Congress to pass a bill that would allow more Dominican sugar to be imported. Kennedy opposed this legislation, and his administration suspected that congressmen were being bribed to support it. Such bribery would be a legitimate national-security concern. Attorney General Robert Kennedy requested that the FBI initiate an investigation to see what pressures the Dominican Republic was putting on Congress (Wannall 1966). The inquiry lasted 9 weeks. A dozen wiretaps and three microphone surveillances were involved. Three members of the Department of Agriculture and a Congressional staffer had their home phones tapped (Hoover 1961a,b). One lobbyist was wiretapped both at home and at the office (Hoover 1961b). No bribes were discovered, but the wiretaps provided the president with important political information. One FBI summary sent to the attorney general said that a lobbyist "mentioned he is working on the Senate and has the Republicans all lined up" (FBI 1962). The administration bill won, and the FBI concluded that its surveillance work had been a major factor (Wannall 1966). Most of the wiretaps were taken off in April 1961, but two remained even after the administration's bill passed (USSR 94 *Intelligence Activities: Staff Reports*, p. 330).

President Lyndon Johnson was even more blatant in requesting national-security wiretaps. During the 1968 election, Johnson (who was not a candidate for reelection) asked the FBI to conduct surveillance of Anna Chennault, a prominent Republican who, Johnson claimed, was attempting to undermine the US-Vietnam peace talks in Paris. The White House asked the FBI to institute physical coverage of Chennault and both physical and electronic coverage of South Vietnam's Embassy (USSR 94 *Intelligence Activities: Staff Reports*, p. 314). Summaries of the information obtained from the physical surveillance were later given to the White House (DeLoach 1968b; FBI 1968). Apparently the FBI was concerned about being involved in such a political case and eschewed electronic surveillance of Chennault. The electronic surveillance of the embassy was an indirect way of accomplishing the same goal (DeLoach 1968a; USSR 94 *Intelligence Activities: Staff Reports*, p. 315).

President Nixon went yet further, and ultimately the illegal electronic surveillance that started early in his administration played a pivotal role in toppling his presidency.

On May 9, 1969, a front-page story appeared in the *New York Times* reporting that the United States was bombing Cambodia and had been doing so for some time. By recording false coordinates for the sorties (placing them in South Vietnam), the Air Force had hidden the location of the attacks. News of the cover-up had been leaked to *Times* reporter William Beecher by someone in the government. Outraged, Henry Kissinger, Nixon's national security advisor, told J. Edgar Hoover to find the leaker(s). Within hours the FBI had focused on Morton Halperin, a Kissinger aide who "knew Beecher" and whom Hoover "considered part of the Harvard clique" (Hoover 1969a). Without written approval from the attorney general, a wiretap was installed on Halperin's home line that same afternoon. Seventeen people, some in the government and some in the news media, had their phones tapped during the investigation,[13] but no leakers were identified.

Some of the wiretaps were on for a matter of weeks, others for more than a year. The longest was the one on Morton Halperin, which remained active for 21 months, even though by September 1969 Halperin had left the White House. In February 1971, when the tap was finally removed, Halperin had been working for months for Edmund Muskie, a Democratic senator from Maine and a candidate for his party's presidential nomination (Halperin 1974, p. 296). Hoover was forwarding various tidbits of political information, such as the fact that former president Johnson "would not back Muskie" for the White House (Theoharis and Cox 1988, p. 415). By that time, information from the wiretaps was being sent directly to H. R. Haldeman, Nixon's political advisor (USSR 93 *Electronic Surveillance for National Security*, pp. 296 and 351).

Daniel Ellsberg's turn came next. Ellsberg had made copies of classified histories of US involvement in Vietnam—the 47-volume "Pentagon Papers"—available to *New York Times* reporter Neil Sheehan. On Sunday, June 13, 1971, portions of the papers, and articles based on them, began appearing in the *Times*. The government sued to block publication, but lost. Excerpts were printed in the *Times* and in the *Washington Post*.[14]

Although the Pentagon Papers were highly critical of the Democrats who had preceded him in the presidency, Nixon was infuriated by their publication. Ellsberg was indicted on charges of violating the Espionage Act, but the Nixon White House went further.

John Ehrlichman, Nixon's domestic affairs advisor, authorized a burglary of Ellsberg's psychiatrist, then covered it up in an attempt to keep

Ellsberg's trial from ending in a dismissal of charges. Instead, as a result of the inquiries into the Watergate break-in, the burglary of the psychiatrist was discovered. Further inquiry revealed that Ellsberg's phone had been tapped by the government,[15] but the wiretap authorizations, logs, and other records could not be found. The judge declared a mistrial, and charges against Ellsberg were dismissed (*United States v. Susso*, CD Cal. 9373-WMB, 1973).

Charles Radford II, a Navy yeoman who was serving the Joint Chiefs of Staff, was wiretapped after a syndicated newspaper column by Jack Anderson described a conversation between Nixon and Kissinger concerning the administration's decision to "tilt" toward Pakistan in the India-Pakistan conflict—a column which was to contribute to Anderson's winning a Pulitzer Prize a few months later. Radford was a suspect in part because he and Anderson attended the same church (Smith 1973; Hersh 1983, p. 470). The FBI was asked to wiretap Radford, although the government did not plan any prosecution of the leak (Smith 1973). Shortly afterward, Radford was transferred to the Naval Reserve Training Center near Portland, Oregon. He moved in with his stepfather until he could find permanent housing. A wiretap was put on the stepfather's phone. After Radford found a place of his own, a wiretap was placed on his new phone. The wiretap on the stepfather's phone remained for another 2 months. There were also wiretaps on the phones of two of Radford's friends, one a State Department employee and one a former Defense Attaché. None of the wiretaps were authorized by the attorney general, and at no time was prosecution planned (USSR 94 *Intelligence Activities: Staff Reports*, pp. 326–327).

Wiretapping Requires a Court Order

Even after the Supreme Court's ruling in the *Katz* case and the passage of Title III, national security continued to serve as a cover for warrantless electronic surveillance in domestic political intelligence. In 1972 the Supreme Court ordered an end to warrantless wiretapping even for "national-security" purposes.

John Sinclair, Lawrence Plamondon, and John Waterhouse Forrest were charged with bombing a CIA office in Ann Arbor, Michigan. It turned out that the federal government had tapped Plamondon without

prior judicial approval. Plamondon requested copies of the tapes to determine if the government's case had been tainted by the wiretaps. The
District Court ordered the government to give Plamondon copies of the
tapes.

The government refused and appealed. It argued before the Supreme
Court that this was a matter of national security and that Title III requirements for a search warrant were not an attempt to limit surveillance in
such cases. While the Supreme Court agreed that Title III was limited
to criminal investigations, it did not buy the argument that domestic-
security surveillance could justify departure from Fourth Amendment
protections:

> ... these Fourth Amendment freedoms cannot be properly guaranteed if do-
> mestic security surveillances may be conducted solely within the discretion
> of the Executive Branch.... (*United States v. United States District Court for
> the Eastern District of Michigan et al.*, 407 US 297, 1972, p. 316–317)

> Official surveillance, whether its purpose be criminal investigation or on-
> going intelligence gathering, risks infringement of constitutionally protected
> privacy of speech. Security surveillances are especially sensitive because of
> the inherent vagueness of the domestic security concept.... We recognize, as
> we have before, the constitutional basis of the President's domestic security
> role, but we think it must be exercised in a manner compatible with the
> Fourth Amendment....
>
> We cannot accept the Government's argument that internal security mat-
> ters are too subtle and complex for judicial evaluation. Courts regularly deal
> with the most difficult issues of our society.... If the threat is too subtle or
> complex for our senior law enforcement officers to convey its significance
> to a court, one may question whether there is probable cause for surveil-
> lance.... (ibid., p. 320)

Surveillance of domestic organizations now required a court order
whether the national security was involved or not. The Court pointed
to the absence of law on wiretapping for national-security purposes and
invited Congress to fill the gap. Instead the country spent almost 2 years
mesmerized by the president's illegal activities.

Watergate

The Watergate affair began early on the morning of June 17, 1972, with
a break-in at the Democratic National Committee's headquarters in the

Watergate Office Building in Washington. The intruders were a group paid by the Committee to Re-Elect the President, Nixon's campaign committee. The burglary was the second attempt to install a bug on the phone of Lawrence O'Brien, the Democratic Party's chairman. A first attempt had only succeeded in placing a working bug in his secretary's phone. This provided information on the secretary's social life, but not on O'Brien's political plans.

The Watergate affair reached its climax on August 9, 1974, when Nixon, facing impeachment, became the first president to resign from office. Had he not done so, he could have expected to be impeached on counts that included the following:

> [Nixon] misused the Federal Bureau of Investigation, the Secret Service, and other Executive Personnel ... by directing or authorizing such agencies or personnel to conduct or continue electronic surveillance or other investigations for purposes unrelated to national security, the enforcement of laws, or any other lawful function of his office; ... and he did direct the concealment of certain records made by the Federal Bureau of Investigation of electronic surveillance. (USHH 93 *Impeachment Inquiry*, Book III, pp. 2255–2256, Article II, §2)

> [Nixon] failed to take care that the laws were faithfully executed by failing to act when he knew or had reason to know that his close subordinates endeavored to impede and frustrate lawful inquiries ... concerning the electronic surveillance of private citizens.... (USHH 93 *Impeachment Inquiry*, Book III, pp. 2256–2258, Article II, §4)

The Senate Investigates

The revelations of Watergate engendered a general distrust of government as high official after high official was implicated in the illegal actions of the Nixon presidency. There was deep concern about the involvement of intelligence agencies in many of the questionable proceedings of the Nixon administration. In January 1975 the Senate appointed an eleven-member special committee to investigate government intelligence operations to determine the extent to which "illegal, improper, or unethical activities" were engaged in by government agencies (USS 94 *Resolution*).

The Senate Select Committee to Study Governmental Operations with respect to Intelligence Activities (commonly known as the Church Committee, after its chairman, Senator Frank Church), began its study with the year 1936, which marked the reestablishment of domestic intelligence programs in the United States after a hiatus of about a decade. The committee uncovered a long history of presidential wiretapping, including Truman's wiretapping of the lobbyist Thomas Corcoran, Kennedy's "Sugar Lobby" taps, Johnson's surveillance of Anna Chennault, and Kennedy's and Johnson's wiretapping and bugging of Martin Luther King. It also examined the wiretap abuses of the Nixon era, including those enumerated above. The hearings revealed many illegal covert operations by the intelligence agencies, and the Church Committee concluded:

> Too many people have been spied upon by too many Government agencies and [too] much information has been collected. The Government has often undertaken the secret surveillance of citizens on the basis of their political beliefs, even when those beliefs posed no threat of violence or illegal acts on the behalf of a foreign power. The Government, operating primarily through secret informants, but also using other intrusive techniques such as wiretaps, microphone "bugs," surreptitious mail opening, and break-ins, has swept in vast amounts of information about the personal lives, views and associations of American citizens.... (USSR 94 *Intelligence Activities: Rights of Americans*, p. 5)

Among these were "the interest of the wife of a US Senator in peace causes; a correspondent's report from Southeast Asia to his magazine in New York; an anti-war activist's request for a speaker in New York" (ibid., p. 108). The committee observed: "The surveillance which we investigated was not only vastly excessive in breadth ... but was also often conducted by illegal or improper means." (ibid., p. 12)

Although much of this surveillance had occurred before Title III, the illegal activities of the Nixon administration had not. When Congress had regulated the use of electronic surveillance for criminal investigations under Title III, it had believed that the legislation would put tight restrictions on the use of wiretaps and electronic surveillance. Nonetheless, Title III had been circumvented by the Nixon administration, which had used the "national-security" justification for many investigations.

The Church Committee's inquiries emphasized not "Who did it?" but "How did it happen and what can be done to keep it from happening again?" (ibid., p. viii) The committee felt that Title III was appropriate as written, but set forth a list of recommendations for explicit laws on electronic surveillance in national-security investigations.[16]

The underlying premise of the Church Committee's recommendations was that Americans should be free of the type of surveillance the hearings had exposed. The CIA should refrain from electronic surveillances, unauthorized searches, or mail openings within the United States,[17] and NSA should not monitor communications from Americans, except in cases where the person is involved in terrorist activities or intelligence work, and then only when a search warrant has been obtained.[18] The committee also outlined the form a law permitting electronic surveillance for foreign intelligence might take. [19]

The Foreign Intelligence Surveillance Act

In 1978 the Foreign Intelligence Surveillance Act (FISA), much of it based on the Church Committee's recommendations, became law. FISA (50 USC 1801 et seq.) governs wire and electronic communications with "United States persons"[20] who are within the country. It does not apply to those of US persons overseas, excepting communications with a US person resident in the United States. Under FISA, US persons in the US may be subject to surveillance if they are suspected of aiding and abetting international terrorism.

A court order is required for a FISA wiretap.[21] Such an order may be granted by the Foreign Intelligence Surveillance Court, which is made up of eleven District Court judges specially appointed by the Chief Justice of the United States.[22] The order must be applied for by a federal officer, and approved by the attorney general, who is required to inform the House and Senate Committees on Intelligence of all FISA wiretap activity twice a year.

The attorney general is required to furnish an annual report to the Administrative Office of the US Courts on the number of FISA applications and orders. All other information on FISA wiretaps is classified. In that sense, FISA represented a move away from establishing public safeguards —notice to targets, oversight, and minimization (in FISA minimization is limited to minimization of information about US persons)—that both provide accountability and limit the ability to misuse surveillance authority.

For a long time after 1979, there were an average of slightly over 500 FISA wiretap orders per year[23]; by 1995, more than 8000 requests had been made by the government for surveillance under FISA. None had been turned down. The reason for this is a matter of dispute. Proponents of FISA say it is because surveillance applications are carefully prepared and reviewed before being presented to the Foreign Intelligence Surveillance Court. Opponents argue that it is because the court is only a rubber stamp (Cinquegrana 1989, p. 815). A critical report by the Foreign Intelligence Surveillance Court in 2002 shed a bit of light on these issues (see chapter 11).

The Electronic Communications Privacy Act

The next major federal wiretapping statute was the 1986 Electronic Communications Privacy Act (Public Law 99-508), which updated Title III to include digital and electronic communications. In recognition of new telecommunications switching technologies, ECPA allowed for approvals of *roving wiretaps*—wiretaps without specified locations—if there was demonstration of probable cause that the subject was attempting to evade surveillance by switching telephones. It required a warrant for wiretapping the non-radio portion of a cellular communication. The widespread availability of radio scanners made the latter stronger in legal terms than it was in practice.

Under ECPA, pen registers and trap-and-trace devices require court orders. Any government attorney can file for such an order, and there is no requirement of probable cause for a search warrant to be issued.

The late 1980s witnessed two major changes in telecommunications. The breakup of the Bell System into a long-distance carrier and seven regional companies encouraged a proliferation of competitors in the industry; this made seeking a wiretap more complicated for law-enforcement agents, who now had to contend with a plethora of new companies instead of a monolithic "Ma Bell." The other change came when call forwarding, call waiting, cellular phones, and fax machines appeared on the market.

8

Communications in the 1990s

Briefings on the dangers of encryption by Raymond Kammer (the acting director of the National Institute of Standards and Technology) and Clinton Brooks (assistant to the director of the National Security Agency) gave rise to fear in the FBI that wiretaps would soon be rendered useless. In 1992 the FBI's Advanced Telephony Unit predicted that, because of encryption, by 1995 the FBI would have access to only 60% of intercepted communications.[1] Facing a technology it understood poorly, the FBI wanted a fast remedy. As a first step, it sought congressional action to protect wiretapping from technical encroachments. The idea was that, because wiretapping was a tool with accepted legal standing, new legislation could be presented as a matter of maintaining the status quo.

Digital Telephony

Unless continued access to traffic is maintained, neither targeting nor analysis of electronic communications will be possible. The post-1982 breakup of AT&T presaged a period of tremendous growth in telecommunications products and businesses. For the FBI's wiretapping activities, it was a period of growing complexity. In executing wiretaps, agents could no longer deal with a single telephone company; they now had to deal with equipment supplied by a variety of companies and with service provided by numerous carriers.

In 1992 the FBI put forth what it called a Digital Telephony Proposal, which mandated the inclusion of provisions for authorized wiretapping in the design of telephone switching equipment. When the FBI approached Congress with that proposal, it was far from clear that the FBI was experiencing anything other than a crisis of confidence. The proposed bill required that all telecommunications providers, both public carriers and private branch exchanges (PBXs—the private switching centers typically used within large companies), design their systems to accommodate government interceptions. Common carriers had 18 months to comply; the PBXs had twice as long. All costs of redesign were to be borne by the companies (FBI 1992b). The FBI claimed that new switching technology and such improvements as cellular telephones and call forwarding had made it difficult to install court-authorized wiretaps.

Evidence to substantiate that claim was hard to find. On April 30, 1992, the *Washington Post* reported: "FBI officials said they have not yet fumbled a criminal probe due to the inability to tap a phone. ..." (Mintz 1992) To this suggestion that it was not actually having any trouble, the FBI countered that, because of anticipated technological problems, court orders had not been sought, executed, or completely carried out (Denning et al. 1993, p. 26). Meanwhile, Freedom of Information Act litigation initiated by Computer Professionals for Social Responsibility found not a single example of wiretaps' being stymied by the new telecommunications technology.

Industry objected to the Digital Telephony Proposal and major players in computers and communications protested. Among industry's concerns were the cost (estimated to be in the hundreds of millions of dollars[2]) and the effect on privacy. A built-in "back door" for government wiretapping could easily become a back door to illicit surreptitious surveillance.

The General Accounting Office briefed Congress, expressing concern that alternatives to the proposal had not been fully explored (USGAO 1992). The General Services Administration characterized the proposal as unnecessary and potentially harmful to the nation's competitiveness (USGSA 1992). In an internal government memo, the National Telecommunications and Information Agency observed that facilitating lawful government interception also facilitated unlawful interception by others, and described the bill as "highly regulatory and broad" (NTIA 1992).[3] There were no congressional sponsors for the proposal.

As internal memoranda show, the FBI had given the Digital Telephony Proposal only a 30% chance of passing (McNulty 1992, p. C-14). But the proposal established an FBI beachhead.

The Value of Wiretapping

The Digital Telephony Proposal took as given the proposition that wire-taps were essential to law enforcement. Wiretapping for that purpose had been legal for a quarter of a century and, despite the fact that most Americans disapproved of it,[4] had come to be an accepted practice.

Decades of court-authorized wiretapping have provided a powerful analytic tool with which to assess its value: the reporting provision of Title III provides that the Administrative Office of the Federal Courts should annually publish a list of all wiretap orders issued under Title III and associated state statutes. This annual account is commonly known as the *Wiretap Report*, although, like Title III itself, it covers all forms of electronic surveillance, including microphone surveillance. Despite its value as an analytic tool, the *Wiretap Report* has severe limitations. In its statistical summaries it does not distinguish wiretaps from bugs. Another difficulty is establishing precise numbers. For example, in its statistical information on intercepts, the report uses the term "intercepts installed" to mean installed intercepts for which reports have been received. This probably results in a slight underestimate of those actually installed, since some are not subsequently reported upon. The 1988 report shows that there were 754 orders for electronic surveillance but indicates that 11 of these were never executed. Of the remaining 743 court orders, an additional 67 did not have an after-the-fact prosecutor's report.[5] Thus the report lists 676 "intercepts installed," although the actual number lies between 676 and 743. In our analysis, we have used the *Wiretap Report's* "intercepts installed" figure, as there is no way to ascertain the higher number.

Limitations of the Data

From the *Wiretap Report* we can discern who the prosecutor was on a wiretap order, who the judge was, how long the order was for, and for what crime the order was authorized.[6] We can see how many arrests were made and how many convictions occurred.[7] What we cannot discern from the *Wiretap Report* is what was said in the court hearings.

Court transcripts often reveal information that dry numbers hide. In 1972, law professor Herman Schwartz studied the 4 years' worth of court-ordered wiretaps that had resulted from Title III. He observed:

[T]here is an interesting item in the late J. Edgar Hoover's 1971 [FBI annual] report: right after a reference to the vital importance of electronic surveillance to the fight against organized crime, four major convictions are listed, including one of a Nicholas Ratteni. A check with counsel in the case disclosed that there was indeed a wiretap—against a co-defendant who was *acquitted*. (Schwartz 1974, pp. 186–187)

So the FBI dissembled. Schwartz uncovered a number of other results that undercut claims of wiretapping's usefulness:

[I]n *United States versus Poeta*, the US Court of Appeals opened its opinion by observing that the tap-derived evidence was unnecessary to the conviction; in another case, *Uniformed Sanitation Men versus Commission of Sanitation*, the Court made the same observation. In a 1971 report, a Nevada prosecutor reported two indictments in a kidnapping case in which wiretapping was used ... but candidly added that the indictments were "not as a result of the interception." (Schwartz 1974, pp. 185–186)

The law requires that the number of interceptions, both incriminating and non-incriminating, be reported. Schwartz found that law-enforcement personnel were prone to exaggerate the number of incriminating intercepts:

In *United States v. King*, the Government claimed that 80–85% of the conversations overheard in a drug case were incriminating and so it reported in the 1971 report, Order #35. The Court, however, found that the contemporaneous reports showed that the percentages were really between 5 and 25%.[8]

Naturally, such anomalies continue, of course. One enterprising New York lawyer has used wiretaps made by police to argue that his client had been entrapped.[9] In a 1995 terrorist conspiracy case, the FBI worked hard to establish a connection between Sheik Omar Abdel Rahman (a blind Egyptian cleric living in Brooklyn) and Razmi Yousef (an alleged terrorist and bomb expert). The FBI emphasized the wiretap evidence. In fact the wiretap transcripts only revealed that there had been several calls between Rahman's phone and Yousef's (Fried 1995; McKinley 1995b)—something that could have been discovered by less intrusive investigative methods. Both Rahman and Yousef were involved in terrorist activities but they never spoke to each other on the tapped telephone. Wiretaps of Rahman's conversations were described in the above-cited *New York*

Times stories as "not incriminat[ing]" and containing "no references to violence." In a situation reminiscent of the Ratteni case described by Schwartz, lawyers for the defense in the 1997 trial of Timothy McVeigh used wiretaps made by federal agents to demonstrate that a prosecution witness was unreliable (Brooke 1997b).[10] Such examples make clear that the data provided by the *Wiretap Report* give only a partial picture of wiretapping cases.

Where Wiretaps Are Used

Here is how things stood in the mid 1990s, as the Communications Assistance for Law Enforcement Act and encryption were being debated.

In 1968, wiretapping was seen as a tool for targeting gambling—the main source of income for organized crime. In the first 5 years after the "Title III" legislation, 64% of the 2362 reported intercepts were for investigations of gambling cases, 27% of the wiretaps were for narcotics cases, less than 5% for homicide and assault investigations, 2.5% for investigations into bribery, and under 1% for cases of "arson and explosives" (AO 1978, p. xvi). (The *Wiretap Report* denotes as "narcotics" any case in which the most serious offense includes drugs of any sort, including marijuana.) With at least 38 states running lotteries, and gambling legalized from Connecticut to Nevada, it is difficult to recall that gambling was once targeted as a serious crime.

In 1977 the number of narcotics investigations employing electronic surveillance began increasing. In 1982 President Reagan declared a national "War on Drugs," and gambling investigations using electronic surveillance dropped while narcotics investigations rose. In 1994 narcotics investigations made up 77% of the cases using electronic surveillance, gambling investigations less than 8%. Cost is a limiting factor. The average cost of a wiretap has risen, from $1358 in 1968 to $49,478 in 1994; part of this increase was due to a doubling in the length of the average electronic surveillance, from approximately 20 days in 1969 to just under 40 in 1994. Drug investigations, which frequently involve wiretaps of several months' duration, are particularly costly.

Drug investigations are the major focus of wiretaps and the seizure of major shipments of drugs is sometimes accomplished through their use. In a style similar to the Vietnam War's body counts, law-enforcement agencies set great store by the tonnage of drugs seized. However, not even the law-enforcement community is unanimous in believing that seizure is the best solution to the drug problem. Eliminating drug sources has had only limited success.[11] And money spent on drug busts may not be money well spent. A report prepared by a special committee of the American Bar Association concluded: "While law enforcement has had little effect on drug use, drug prosecutions have had a profound effect on the criminal justice system. In the cities the Committee visited, drug cases have overwhelmed the courts.... In light of the Committee's findings in the area of drugs, a simple but significant truth should be faced: conventional law enforcement methods are not controlling the drug problem." (American Bar Association 1988, pp. 46–47)

Sample Data from the *Wiretap Report*

Court-Authorized Electronic Surveillance, Title III, 1988–1994								(1994 Only)	
Year	1988	1989	1990	1991	1992	1993	1994	Federal	State
Orders Authorized	738	763	872	856	919	976	1154	554	600
Orders Denied	2	0	0	0	0	0	0	0	0
Orders Installed[1]	678	720	812	802	846	938	1100	549	551
Main Offense[2]									
Arson, Explosives, and Weapons	3	0	0	0	0	0	0	0	0
Gambling	126	111	116	98	66	96	86	8	78
Kidnapping	1	3	2	5	9	1	11	7	4
Narcotics	435	471	520	536	634	679	876	435	441
Racketeering	80	89	90	114	90	101	88	68	20
Telephone Wiretaps[3]	549	621	591	591	632	679	768	397	371
Electronic Bugs[4]	61	65	62	62	38	55	52	42	10
Average Number/Order									
Persons	129	178	131	121	117	100	84	112	58
Intercepts	1251	1656	1487	1584	1861	1801	2139	2257	2030
Incriminating Intercepts	316	337	321	290	347	364	373	374	372

1. *The Wiretap Report* uses the term 'Intercepts Installed' to mean intercepts installed *and* reported upon. This is likely to be a slight underestimate of those actually installed, since some surveillances are not actually reported upon. We use the Administrative Office's terminology here.
2. As determined by judge issuing surveillance order.
3. This number does not include telephone wiretaps that were part of a "combination" tap involving more than one type of surveillance.
4. This number does not include electronic bugs that were part of a "combination" tap involving more than one type of surveillance.

While wiretaps may contribute to large drug busts (although we are not aware of any study that has compared employing wiretapping with spending comparable funds on other types of law-enforcement activities focusing on drugs), it is not clear that this effort makes any real difference in the underlying problem. Indeed, a 1994 RAND study observes that $34 million invested in drug treatment programs achieves the same consumption reduction as does $246 million spent for domestic law enforcement, $366 million spent for interdiction, or $783 million spent for source-country control.[12]

Where Wiretaps Are Not Used

Although kidnappings are frequently touted as a reason for the need for electronic surveillance,[13] wiretapping does not seem to be useful in such cases. Between 1969 and 1994, wiretaps and microphone bugs were reported to have been used in 80 kidnapping cases; thus, on average, wiretapping played a role in two to three of those cases each year.[14] Though there is no reason to doubt that the court orders for surveillance were correctly made, there is some reason to doubt the necessity. Meanwhile, in recent years there have been approximately 500 kidnapping cases per year.[15]

It is not surprising that wiretaps rarely figure in kidnapping cases, since investigators are typically unaware of the kidnappers' locations. But wiretapping's role is small for another reason: the interception of ransom calls from kidnappers does not require any court authorization if the recipient of the call consents—a form of interception called a "consensual overhear."[16]

Domestic terrorism is sometimes given as a reason for wiretap surveillance. There were 59 wiretapping cases between 1968 and 1994 involving arson, explosives (the most frequent form of domestic terrorism), and weapons, or about two a year. The period 1988–1994 saw four Title III wiretap investigations of firearms and none of arson or explosives.[17] It is possible that no wiretaps were used in cases involving explosives. However, since intelligence wiretaps may also be used for investigating domestic terrorism if there is alleged foreign involvement, it seems more likely that federal authorities found it more expedient to employ the Foreign Intelligence Surveillance Act, under which nothing need be reported but the total annual number of surveillances.

The States and Wiretapping

Forty-four states have their own wiretapping statutes, but not all of those states strongly support wiretapping. It took California more than 20 years to pass a wiretapping statute, and when the law finally did pass it was loaded with restrictions, including a limitation to drug investigations. At first the state of California performed few wiretaps a year—8 in 1994 compared to 71 federal wiretaps (AO 1995, pp. A-2–A-7 and A-58). "If it's anything big, you should let the Feds do it," explained San Francisco Chief Assistant District Attorney Dick Iglehart, who was California's Chief Assistant Attorney General, Head of the Criminal Division, at the time of the passage of the California statute.[18] By 2004 California had caught up with New York and the other big players, and had performed 180 state investigations using wiretaps. Over half of those were in Los Angeles, where there was reason to believe that there had previously been substational underreporting of state wiretaps.[19] The Public Defender's office filed suit over the practice, and a Los Angeles police officer testified that the hand-off procedure was standard practice and had been used "hundreds of times" since the mid 1980s with "10 percent to 15 percent of the cases involving wiretaps concealed from the defense" (*New York Times* 1998b). Despite strong objections from the Public Defender's office (Quant 2006), the judge ruled that the hand-off was permissible. But then something very interesting happened regarding the number of state wiretaps reported for Los Angeles. Despite the Superior Court ruling and the LAPD claim that the practice was "standard procedure," the number of legally-authorized state wiretaps in Los Angeles jumped from 37 in 1998 (AO 1999, p. 14) to 62 in 1999 (AO 2000, p. 15) and then 78 in 2001 (AO 2002, p. 15). It is, or course, impossible to know what fueled this increase: better reporting, or a real increase in the number of state wiretaps. The *Wiretap Report* reports 48 jurisdictions that permit wiretapping: the Federal, Puerto Rico, the Virgin Islands, the District of Columbia, and 44 states.

Convictions from Wiretaps

How many convictions result from wiretaps?[20] In 1988 wiretaps were used in 609 investigations, including 45 in which both wiretaps and clectronic bugs were employed. There were 279 court cases and 1808 convictions in these cases.[21] These convictions had cost $33.5 million for wiretapping alone.

Of course, it is impossible to know in how many of these cases wiretap evidence was crucial in obtaining a conviction or a guilty plea (whether directly or through evidence obtained as a result of the information gleaned from a wiretap). The numbers above provide only an upper bound on the potential effectiveness of wiretapping in law enforcement. It is essential to note that a wiretap is ordered only if there is already probable cause that the person being investigated is involved in a serious and indictable crime.

Who Else Is Being Tapped?

Since Title III has been in force, the number of conversations intercepted has increased, the number of nonincriminating conversations intercepted has increased, and the number of incriminating conversations intercepted has remained the same. More specifically, according to data on the period 1968–1994 released by the Administrative Office of the US Courts, the average annual number of incriminating conversations intercepted remained between 200,000 and 400,000 per year, while the number of intercepted conversations increased steadily from about 400,000 in 1968 to over 2 million.[22] In 1994, for example, 1137 court orders for electronic surveillance resulted in the interception of 2.35 million conversations. Only 15% of the intercepted conversations were incriminating; the remainder of the wiretapped conversations were not related to illegal activities.[23]

Gambling accounted for 29% of the incriminating wiretap intercepts in 1994. Gambling skews the statistics, for it is an activity with a high number of incriminating intercepts. It is also low on convictions. If we look to the 1988 numbers now that there has been enough time for trials to have taken place, we see that gambling accounted for 27% of the incriminating wiretap intercepts but for less than 14% of the convictions (244 of 1808). It is not surprising that wiretaps authorized for a gambling investigation should yield incriminating calls, since there has to be probable cause for the authorization. Neither is the low level of convictions a surprise; if a bookmaking operation is tapped, the incriminating calls are likely to be people placing bets. Few, if any, of these low-level bettors are prosecuted.

Thus, from the raw data of the *Wiretap Report* we can observe that fewer than one-sixth of the intercepted calls resulted in anything of use to law enforcement.

How Things Stand Now

The numbers above describe the situation in 1998, when this book first appeared. With only minor exceptions, things have not changed substantially in the past decade. Except for a dip in 2000, the number of Title III wiretaps continues to rise, with 1773 Title III wiretaps reported by the Administrative Office of the US Courts for 2005 (AO 2006, p. 7). There has been a shift, with state and local wiretaps constituting an ever-increasing percentage of Title III wiretaps. Undoubtedly one reason for this shift is the number of wiretaps from the state of California, which was 235 in 2005; New York, with 391, New Jersey with 218, and Florida with 72 make up 80% of the state applications for wiretap orders (ibid., p. 7). In 2005, fully 81% of all Title III wiretaps were for drug cases (ibid.), up slightly from the 72%–77% range of previous few years. The average wiretap in 2005 cost $55 thousand, lower than the cost in 2004, and wiretaps were used for an average of 43 days (AO 2006, p. 9). As had been the case previously, contrary to public testimony, wiretaps are not particularly efficacious in kidnapping cases. Over the last nine years, wiretaps have been sought in an average of only five kidnapping cases a year and installed in an average of fewer than four kidnappings a year (AO 1998, 1999, 2000, 2001, 2002, 2003, 2004, 2005, 2006).

Two changes do stand out, however. The first is the move to portable devices. In 2001, the most common devices being wiretapped were portable devices (AO 2002, p. 8). To no one's surprise, that trend has continued. The second change from the mid 1990s is not unrelated to the first. Portable devices have driven an increase in the number of daily communications. So it should also be no surprise that the government is conducting surveillance on an increasing number of communications. In 2004, the number of intercepted communications under Title III was a record 4.9 million, slightly over a million of which were deemed incriminating (AO 2005, p. 21). In 2005, the government's success ratio improved slightly: the number of interceptions was down to 4.8 million, while the number of incriminating ones was slightly higher that it had been in 2004, so that 22% of communications tapped under Title III warrants were incriminating (and 78% were not).

In 2000 Congress extended the reporting requirement for the Administrative Office of the US Courts, which was due to expire and added a twist: reporting on encryption problems encountered during wiretapping cases.[24]

Reports were to be aggregated, so that there would no break-out of the cases in which law enforcement encountered encryption. The result is quite different from what the FBI had anticipated in 1992, when the Advanced Telephony Unit predicted that it expected fully 60% of wiretapped calls would be encrypted (Advanced Telephony Unit). Law enforcement has encountered encrypted communications. There were 22 state and local cases in 2000, 34 state and local cases in 2001, 17 state and local cases in 2002, one state case in 2003, 41 state cases and one federal case in 2004, and 13 in 2005. Of these exactly one caused difficulty and the ciphertext could not be decrypted.

What has given rise to these curious figures is hard to determine.[25] Another oddity is that federal investigators, seemingly more likely to encounter encryption than state and local investigators, reported encountering only a single case of encryption. Whatever the mechanisms, the Wiretap Report indicates that, in contradiction to the predictions of a decade ago, encryption in wiretapped communications is simply not a problem for law enforcement.

What Sways the Courts?

Citizens and legislators alike have generally accepted police claims of wiretapping's indispensability as a crime-fighting tool but have felt a sense of disquiet about the ease with which it lends itself to invasion of privacy and political spying. As a result, Congress gave with one hand and took away with the other, allowing wiretapping but hedging it about with stiff warrant requirements and a strict reporting regime.

It is impossible to be sure what determines a guilty verdict. It is possible to ask jurors why they voted as they did but it is not clear that the jurors really know. Indeed, getting inside the mind of a jury is notoriously difficult. In trials with high-priced attorneys, a lot of time and energy is spent developing jury profiles. Much of this effort is guesswork.

The exacting standards for issuing wiretap warrants make it difficult to judge whether wiretaps were actually needed in the cases in which they have been used. The fact that we know of no definitive evidence for the value of wiretaps may mean that we are ignorant, that no one has ever tried to develop such evidence, or that there is none.

Transactional Information

The Digital Telephony Proposal came along when electronic surveillance was on the rise. Title III phone wiretaps were up from approximately 620 per year in the late 1980s to an average of 870 five years later.[26] In 1987 some 1600 pen registers had been installed; by 1993 the annual number was 3400. In 1987 the FBI, the Drug Enforcement Agency, the Immigration and Naturalization Service, and the US Marshals Service requested a total of 91 trap-and-trace devices; in 1993, the number was over 2000.[27] Part of the reason for this expansion in obtaining transactional information undoubtedly lies in improvements in telephone signaling that have made the information easier to obtain.[28]

As we noted earlier, transactional information cuts both ways. As in the case of the World Trade Center bombing, it can be useful in determining the structure of a conspiracy. It can also be used to forage into people's private affairs. Probes of Hillary Clinton's involvement in the Whitewater affair included reconstructions of how the first lady spent the days immediately after Vincent Foster's death.[29] Long-distance phone records were also used to expand the FBI's investigation of CISPES.

Constraints on Wiretapping

Title III requires that there be a court order before a wiretap can be installed.[30] In contrast with what is done when examining bank records or phone logs, a law-enforcement investigator who wants to wiretap has to draw up an affidavit showing that there is probable cause to believe that the targeted communication device—whether a phone, a fax machine, or a computer—is being used to facilitate a crime. The crime must be serious and indictable. At the federal level, a wiretap request must be signed by a member of the Department of Justice at least at the level of Deputy Assistant Attorney General. Applications are decided upon by a federal District Court judge. Taps are approved for at most 30 days; extensions require a new court order.

The law obliges agents to tap only what is relevant to the investigation. This is called *minimization* and it requires an investigator to stop listening if the suspects are discussing issues not related to a potential crime, then turn it on several minutes later to check if the conversation has returned to indictable actions. The expense of having investigators do this provides a practical limit on the use of wiretaps. For example, in the Illwind investigation of Pentagon fraud, which ultimately tapped 26 telephone lines, several hundred agents were needed just for monitoring (Pasztor 1995, p. 190).

Criminals, whether they be defense consultants, drug dealers, or anyone else involved in a complex business, speak cryptically, and wiretapped lines reveal only part of the puzzle. In a recent FBI case, a wiretap picked up a conversation in which a murder plot was being discussed. The agents could not understand the street slang and jargon employed by the criminals and were unable to prevent the crime (Dam and Lin 1996, p. 89).

If the criminal evidence being sought is obtained before the end of the 30-day period, the law requires that the interception be terminated immediately. The raw numerical data provided by the *Wiretap Report* are insufficient to establish how carefully this rule is followed. In 1994, 22% of the wiretaps ran for the full authorization period but some wiretaps are terminated very quickly because there is a "pervasive pattern of innocence." In one case police discovered that one of their suspects had a very active telephone and moved to get a wiretap warrant. A brief period of monitoring revealed, however, that the calls were not evidence of an abundance of drug dealing but only of a popular teenage daughter.[31]

These restrictions, like others on wiretapping, are artificial. Though built into the law, they can be changed.[32] Indeed, one important restriction, minimization, has exceptions. For example, if the suspects are speaking a language not understood by the agent who is listening, tapes can be made for later use when an agent who comprehends that language is available (Dam and Lin 1996, p. 400, note 17).

The FBI has consistently maintained that wiretapping is an expensive technique (Pasztor 1995, p. 186) and thus never likely to be put to broader use. An Australian-cabinet investigation of telecommunications interception, however, reached quite different conclusions, estimating the cost of a day's wiretapping as $Aus570, as compared to $Aus1376 for a listening device and $Aus2772 for vehicle tracking (Barrett 1994). Most of the cost of American wiretaps appears to due to the requirement that, in most cases, someone must monitor the tap in real time, turning a recorder on and off as the conversation drifts from innocent to incriminating and back.

The essential question about the future costs of wiretapping under US law is whether the courts will hold that minimization can be done by machines in a legally satisfactory way. Since electronic evidence gathered using search warrants is already handled in this way,[33] it seems possible that the answer will be yes. Should this occur, it would pave the way for much broader surveillance by law-enforcement agencies in the fashion currently practiced by intelligence agencies.

The FBI Makes a Case for Wiretapping

Beginning with the 1992 Digital Telephony Proposal, the FBI began a massive lobbying effort for passage of a wiretapping bill, presenting facts and figures that made a case for the importance of electronic communications to law-enforcement investigations. In the period 1985–1991, court-ordered taps conducted by the FBI reportedly figured in 7324 convictions, almost $300 million in fines levied, and over $750 million in recoveries, restitutions, and court-ordered forfeitures (FBI 1992a). Since the FBI conducts fewer than one-third of the non-FISA wiretap investigations, it can be assumed that the numbers above would be substantially higher if all such surveillance were taken into account. In 1992, during the presidency of George H. W. Bush, some White House staffers objected to the way the FBI calculated the losses due to organized crime, and disputed the FBI's claim that all 7324 convictions were due to electronic surveillance: "[S]ome [of these] convictions could probably still be obtained absent surveillance." (Anderson and Buchholz 1992) The Treasury Department observed: "It is difficult to do a critical analysis of DOJ's cost benefit package without a full explanation of how DOJ arrived at its cost/benefit figures, and what costs and benefits were included in those figures. It is not clear that DOJ knows, or could know, all the costs and benefits involved but this should be clearly stated." (Levy 1992) The vice-president's office also had trouble with the calculations, noting: "In several places in the analysis, figures are cited without reference to their sources or to how they were derived. For example, on p. 4 a figure of $1.8 billion is cited for potential economic loss...." (McIntosh 1992) Despite the doubts cast upon these numbers, they appeared and reappeared in various briefings—most notably in 1993. Less than a month after President Bill Clinton took office, his senior director for intelligence programs received an FBI briefing paper on encryption, in which the FBI's questionable data were quoted (Sessions 1993b, p. 6).

The Digital Telephony Proposal Reappears

In 1994 the FBI prepared a revised Digital Telephony Proposal that limited wiretapping to common carriers and allocated $500 million to cover their costs. Carriers would have 3 years "after the publication by the Attorney General of a notice of capacity requirements"[34] to comply; after that, failure to fulfill a wiretap order could result in a fine of up to $10,000 a day. The revised proposal, the "Digital Telephony and Communications Privacy Improvements Act of 1994," was submitted to Congress in March 1994.

FBI Director Louis Freeh pressed for the passage of the new bill. Again the FBI claimed that the new technology was impeding its wiretapping ability. In a February 17, 1994, speech to the Executives' Club of Chicago, Freeh said: "Development of technology is moving so rapidly that several hundred court-authorized surveillances already have been prevented by new technological impediments with advanced communications equipment." In March, testifying before Congress, Freeh presented a lower estimate, citing a 1993 informal survey of federal, state, and local law-enforcement agencies, which revealed 91 instances of recent court orders for electronic surveillance that could not be fully implemented (Freeh 1994b, p. 33). Even those numbers were not well substantiated.

Freeh's testimony had some curious gaps, the most serious of these was that, although Freeh was speaking in support of the Digital Telephony bill, his examples of electronic surveillance included electronic bugs.[35] Freeh himself seems confused about the distinction. When Senator Arlen Specter congratulated Freeh on the timeliness of his appearance (the FBI had just arrested Philadelphia mobster John Stanfa), Freeh readily agreed. The tape used in the case had come from an electronic bug.[36]

In April 1994, under an agreement that the details not be publicly released, Freeh supplied to the House and Senate Judiciary Subcommittees details of 183 instances in which the FBI had encountered difficulties in conducting court-authorized interceptions (USHR 103-827 *Telecommunications Carrier Assistance*, p. 14). The General Accounting Office, which earlier had complained about the FBI's lack of specificity in its electronic surveillance requirements, confirmed that the FBI did face technical problems in wiretapping as a result of the use of new digital technologies, including call forwarding, optical fiber, and ISDN (ibid., pp. 14–15). Despite Freeh's efforts, by late fall the Digital Telephony bill was in trouble. Freeh returned to Congress in a "last-ditch lobbying effort," pushing hard for passage of the bill that he had made his agency's highest priority (Chartrand 1994). His work paid off and the Digital Telephony bill became law under a new name: Communications Assistance for Law Enforcement Act (CALEA).[37]

CALEA put the government right in the middle of the process of designing telephone switches. It provided that, subject to federal appropriations to cover the costs of modification, telecommunications networks deployed after January 1, 1995 had to be configured to meet law-enforcement interception requirements,[38] whereas systems installed earlier did not have to be in compliance until the "equipment, facility, or service" was replaced or substantially upgraded (Communications Assistance for Law Enforcement Act, Public Law 103–414, §109). A cryptography provision was included in CALEA: "a telecommunications carrier shall not be responsible for decrypting, or ensuring the government's ability to decrypt ... unless the encryption was provided by the carrier and the carrier has the information necessary to decrypt the communication" (ibid., §103(b3)). The law authorized the expenditure of $500 million to cover costs of the modifications. It also empowered the attorney general to determine the appropriate level of surveillance standards the telephone companies would have to meet. Attorney General Janet Reno decided that the FBI, the agency that had written and lobbied for CALEA, would be in charge of determining those very standards.

Within a year, the attorney general was to publish notice of the maximum capacity required by law enforcement. In October 1995 the FBI announced its analysis, for purposes of which the United States was divided into three parts on the basis of previous rates of telecommunications surveillance. In Category I (the area with the highest-density of communications interceptions, which presumably included the New York metropolitan area and Dade County, Florida[39]) the phone companies were to "expeditiously" increase the capacity for monitoring until 1% of the "engineered capacity" could be intercepted. (Interceptions might mean pen registers, trap-and-trace devices, or actual wiretaps.) In Category II areas these numbers were halved and in Category III areas the requirements for maximum surveillance were halved again. According to the FBI, "engineered capacity" is about 10% of the number of telephone lines, or some 15 million lines over the whole country.

These requirements translated to an extremely large number of simultaneous intercepts. There were approximately 160 million phone lines in the United States (FBI 1997b). Category I included about 12.5% of these, or a bit under 2 million lines; Category II was another 12.5% of the lines. Category III covered the remaining 75% of telephone lines. Thus the FBI requirements would result in capacity to wiretap approximately 30,000 lines simultaneously (EPIC 1995b). That is over 4 times the *annual* number of phone surveillances (total number of trap-and-trace devices, pen registers, and FISA and Title III wiretaps) in 1993 and 20 times the *annual* number of FISA and Title III wiretaps. Were the telephone carriers' engineered capacity to increase, so would law enforcement's ability to wiretap and track transactional activity.

A few months later there were complaints about other FBI interpretations of CALEA. The FBI proposed that the cellular telecommunications group adopt a standard enabling law-enforcement agencies to determine the precise location of a wireless user within half a second (Markoff 1996). "In 1968 when they passed the original wiretap legislation, phones didn't move," said James Kallstrom.[40] "The notion that we in law enforcement should not be able to take advantage of the technology is a crazy notion (Markoff 1996)."

The industry objected. "The FBI is asking us to go beyond the legislation and ... turn all wireless phones into location beacons," fumed Ronald Nessen of the Cellular Telecommunications Industry Association (McGee 1996a). The legislators had been explicit on this very point: "... such call-identifying information shall not include any information that may disclose the physical location of the subscriber (except as can be determined from the telephone number)" (CALEA, §103 a2B). In his congressional testimony, Louis Freeh had pledged that CALEA would not expand wiretapping powers.[41] In response to the industry objections, the FBI agreed to redraft its proposed cellular standards (McGee 1996a).

The FBI also reexamined the capacity issue. Scrutinizing surveillance activity county by county across the United States, in early 1997 the FBI proposed capacity numbers based on the maximum simultaneous surveillance that had occurred during the period from January 1993 to March 1995, defining "simultaneous surveillance" as surveillance that had occurred on the same day. Then the FBI added together all forms of telephone surveillance—wiretaps, penregisters, and trap-and-trace devices—to arrive at a baseline number. This was then multiplied by a growth factor to reflect the fact that the numbers were to apply in 1998.[42] The new regulations asked for the ability to conduct 39,767 "actual" simultaneous surveillances by 1998 and 57,749 "maximal" ones.[43] The latter number is 8 times the total electronic surveillances (wiretap, electronic, combination (USAO 1993), pen register, and trap-and-trace devices) conducted in 1993. Some of the increase arises as a simple consequence of the county-by-county approach—if a county had no surveillance activity during the time period, the FBI gave it a baseline of one (FBI 1997b)—but this inflation of the total figure may not affect Americans' privacy as much as other aspects of the FBI's capacity requirements.

The lines between wiretaps, pen registers, and trap-and-trace devices were blurred under the new proposals. The FBI's technique of combining the numbers for wiretaps, pen registers and trap-and-trace devices meant that it was requesting the capability to perform 57,000 simultaneous surveillances, which could mean 57,000 uses of trap-and-trace devices or 57,000 wiretaps.

Freeh made several other efforts to increase the wiretapping capabilities of law enforcement. In 1995, immediately after the bombing of the Murrah Federal Office Building in Oklahoma City, Freeh proposed new legislation that would permit law-enforcement agents to obtain roving wiretaps (taps on a suspect who moves from phone to phone) without having to get individual court orders for each tap (Purdom 1995). Before the Oklahoma bombing, the FBI had paid little attention to right-wing militia groups and it is hard to imagine how expanded wiretapping capabilities could have prevented the act. Indeed, it was old-fashioned police work—including catching a speeder on a highway—that netted the suspects only a few days after the crime.

Responding to Oklahoma City, the White House sought expanded capabilities for electronic surveillance, including an expansion of Title III to cover *any* federal felony, the ability to use illegally obtained electronic surveillance information in court so long as the evidence had not been obtained in "bad faith," and the ability of the FBI to obtain long-distance telephone billing information without a court order. The White House also sought full funding for CALEA. Congress turned the president down on most of these requests; however, it approved funding for CALEA,[44] and it approved the use of subpoenas (as opposed to search warrants) to obtain local phone records.[45]

After the mysterious explosion of TWA Flight 800, in August 1996, Louis Freeh and James Kallstrom (in 1995 the latter became an FBI Assistant Director in Charge, New York Division) again urged an expansion of law enforcement's wiretap authority. President Clinton proposed that terrorist actions be included among the crimes for which electronic-surveillance orders could be obtained under Title III (terrorist actions were already included under FISA).[46] Clinton also recommended more liberal provisions for roving wiretaps, 48-hour emergency warrantless wiretapping, and profiling of airline passengers through electronic records (billing information and the like) to determine whose baggage should be the subject of careful searches for explosives.

The roving wiretaps were to be roughly akin to electronic writs of assistance. Whereas the Title III required judicial approval of each wiretap placed on a suspect, the government now sought the ability to wiretap any telephone the suspect might be using—at a bar, a coffee shop, a gym, or a pizza parlor—without specific prior judicial approval of the wiretap's location. These proposals did not pass. The National Transportation Safety Board inquiry into the Flight 800 disaster, concluded that the explosion was the result of a spark in the fuel tank rather than terrorist action (NTSB).

The International Connection

In a behind-the-scenes effort, simultaneous with its domestic lobbying efforts for CALEA, the FBI worked the international front. Certain countries, including Britain, could be expected to be sympathetic to the FBI's viewpoint. In a worldwide context the British legal system appears similar to the American but there are sharp differences in the area of communications interception. There have, for example, been numerous charges of wiretaps on British labor unions and on the British "green movement"[47] and, in one instance, the private phone line of an Assistant Chief Constable who had initiated proceedings in the 'Industrial Tribunal' after failing to receive a promotion for which she had applied eight times (Donohue 2006, pp. 1166–1167).[48] The lack of a constitution and the British experience with terrorism (Irish terrorism has been a constant in England for at least a century) has led to a wiretap law which is less restrictive than the American one.[49]

The FBI briefed the international community on problems in communications interception at its research facility in Quantico, Virginia. Shortly afterward, the European Union (EU) opened discussions on interception.[50] The influence of the FBI was clear, although point 2 of the EU resolution lamely attempted to put some European control on the matter ("requirements of the member states will be conveyed . . . in order to avoid a discussion based solely on the requirements of the FBI").

A little over a year later, and without any public debate, the European Council passed a resolution on "realtime" monitoring capabilities. Like CALEA, the EU resolution required telecommunications providers to give law-enforcement bodies access to transactional data and call content. But the EU resolution went farther; it required providers of mobile services to give the locations of their subscribers.

The origins of the European initiative were clarified by a Memorandum of Understanding (MOU)[51] which extended the agreement to non-EU nations that chose to sign. Nations interested in participating were told to contact the General Secretary of the EU Council or the Director of the FBI for further information.

The EU's resolution and memorandum were not publicized. Although the resolution had the force of law on EU members, it was not brought before various parliamentary bodies. When, in the British House of Lords, the chairman of the Select Committee on European Affairs sought information on the resolution and the accompanying MOU, he was told that it was simply a "set of practical guidelines" and that it was not of parliamentary "significance."[52]

Standards bodies *were* kept informed, however. Service providers and manufacturers of telecommunications equipment were told that they would have to adhere to the standards of the resolution if they were to provide service or sell equipment to EU members or to signers of the MOU. In late 1996 the European Council began inviting non-EU members to participate in the MOU.[53] By early 1997 the FBI seemed to have made significant headway internationally in its attempt to develop law-enforcement access to telecommunications systems.

The FBI pressed for these surveillance technologies in a world where human rights guarantees are quite different from those of the United States. In the context of satellite-based communications systems, the European Union continued to look at surveillance issues. The EU Police Cooperation Working Group considered the possibility of "tagging" each subscriber "in view of a possibly necessary surveillance activity."[54] The United States may limit its use of wiretap surveillance to serious crimes and require a court order for such surveillance; however, as these excerpts from the US Department of State's Country Reports on Human Rights Practices for 1996 demonstrate, other countries do not.

- El Salvador: Wiretapping of telephone communications by the government is illegal, but occurs.

- Colombia: Despite a law, various state authorities monitor telephone conversations without obtaining authorization.

- Spain: Investigation continues into allegations of wiretapping by the National Intelligence Agency of private conversations between the king, various ministers, and other prominent figures.

- Greece: On occasion the government placed international and domestic human rights activists under surveillance.

- Angola: The government maintained a sophisticated security apparatus dedicated to surveillance, monitoring, and wiretapping of certain groups, including journalists, opposition leaders, and diplomats.

- Nigeria: Human rights and prodemocracy leaders reported that security agents regularly cut off or monitored their organizations' phones.

- Singapore: The authorities have the capability to monitor telephone and other private conversations. While there were no proven allegations that they did so in 1996, it is widely believed that the authorities routinely conducted surveillance on some opposition politicians and other critics of the government.

- China: All public dissent against party and government was effectively silenced. The 1982 constitution states that "freedom and privacy of correspondence of citizens are protected by law." In practice, however, authorities frequently record telephone conversations of foreign visitors, businessmen, diplomats, residents and journalists as well as Chinese dissidents and activists and others. Authorities also open and censor international mail.

- Indonesia: Security forces engaged in selective monitoring of local and international telephone calls without legal restraint. (USDoS 1997)

Hong Kong was an invited participant in these meetings from the beginning, despite the fact that in 1997 sovereignty of the British colony reverted to China, which has an abysmal record on human rights.

Encryption and Wiretapping

The ability to wiretap is substantially less useful if the conversations under surveillance are encrypted but the FBI sought to downplay CALEA's connections with encryption and, in particular, with escrowed encryption. After the introduction of the key-escrow program in 1993, many people raised questions about such connections. The government denied any ties between encryption and surveillance.[55]

Then in September of 1994, weeks before CALEA passed, FBI Director Freeh said otherwise. Asked how the FBI would respond should it encounter non-escrowed encrypted conversations in wiretapped communications, Freeh replied that he would go to Congress and ask for laws banning non-escrowed encryption.[56]

The White House disavowed the remarks, saying that Freeh tended to go his own way. In the following months, Freeh repeated his position, often quite strongly. (See e.g. Freeh 1996.) This did not surprise many of those who had opposed the original Digital Telephony Proposal. They had always expected such a response and that expectation was borne out by documents obtained by the Electronic Privacy Information Center through Freedom of Information Act litigation.

A 1991 NIST Public Key Status Report mentioned that the FBI was "working on draft legislation to control and license all cryptography" (USDoC 1991c). A memo written by National Security Advisor Brent Scowcroft on January 17, 1992 said that two days earlier the president had approved that the Department of Justice "should go ahead now to seek a legislative fix to the digital telephony problem, and all parties should prepare to follow through on the encryption problem in about a year." The Scowcroft memo continued: "Success with digital telephony will lock in one major objective; we will have a beachhead we can exploit for the encryption fix, and the encryption access options can be developed more thoroughly in the meantime."[57] (Scowcroft 1992)

CALEA passed in 1994. One half of the fix was now in, ready to be exploited for the encryption "problem."

9

Cryptography in the 1990s

Pretty Good Privacy

In 1990, a programmer from Boulder, Colorado, Philip Zimmermann, wrote Pretty Good Privacy (PGP), a program for protecting the privacy of email, and made it available over the Internet. Under the State Department's interpretation of the Arms Export Control Act, this constituted an illegal export.

Zimmermann might not have had any trouble had he not offended another vested interest. The PGP program was in blatant infringement of the Rivest-Shamir-Adleman patent and it bore a remarkable resemblance to a program called Mailsafe (written by Ron Rivest) marketed in the mid 1980s by RSA Data Security. Zimmermann recalls receiving a visit from puzzled customs investigators, who told him they had received a complaint from RSA Data Security alleging the theft and international shipment of stolen intellectual property. The customs inspectors did not really understand what was at issue. Patent infringement wasn't their responsibility. Disks stolen out of warehouses and smuggled out of the country were, however, and this is how Zimmermann believed they had interpreted the complaint. A federal prosecutor in San Jose, California, began an investigation, and a grand jury in that city heard testimony on the subject for over a year. The experience was disquieting for all involved —not least the prosecutor and the grand jury, who were not used to investigating in a fish bowl. Many witnesses reported their experiences on the Internet and the cryptography community followed the events attentively.

Meanwhile, PGP spread out of anyone's control. Because the RSA patent held only in the United States, foreign users were not at risk of being sued for contributory infringement. A worldwide group of programmers began further development on the program and later versions were said to have been developed abroad and imported to the US Midway through the course of criminal investigation, the patent-infringement aspect of the case became moot when RSA Data Security changed the license for its reference implementation of the RSA cryptosystem in a way that permitted a "legal" US version of PGP (PGP 2.6).

The investigation, of which the grand jury was only the most visible part, ended when the Department of Justice decided not to prosecute. The government's reasoning is not known. Quite independent of the central legal issue (whether posting code on the Internet, where foreigners can get at it, constitutes export under American law or is merely the exercise of a free-speech right to publish), the case was an evidential nightmare. Zimmermann had not actually posted the code himself; someone else had done it with his permission. More important than this, however, was an unquestioned act of publication. The MIT Press, with its thumb firmly on its nose, published the code of PGP as a 600-page hardbound book (Zimmermann 1995) printed in an OCR font, and sold it though its usual worldwide distribution channels. Had the government prosecuted Zimmermann and not gone after MIT, it would have invited scorn. But MIT was three times as old as NSA, just as well funded, and even more influential in the military-industrial complex. The Department of Justice let the case drop.

Free of the threat of prosecution, Zimmermann founded a company and began to expand his product line. Today, PGP has a worldwide following, and it has entered the mainstream by means of an easy-to-use interface to the popular Eudora email program. In writing PGP, Phil Zimmermann did something for cryptography that no technical paper could do: he gave people who were concerned with privacy but were not cryptographers (and not necessarily even programmers) a tool they could use to protect their communications.

A National Encryption Policy

In the period immediately following the 1989 NIST/NSA Memorandum of Understanding, from a public vantage point encryption policy seemed to be lurching along without direction. At the FBI's request, the chairman of the Senate Judiciary Committee, Joseph Biden, introduced a nonbinding sense-of-the-Congress resolution recommending that, under appropriate legal authorization, telephone companies provide the plaintext of encrypted messages they encountered while wiretapping. Biden later withdrew the resolution, which had been part of an anti-terrorism measure (Markoff 1991). Industry complaints about restrictive export controls on cryptography resulted in agreement on a slightly loosened export policy: seven-day approval for software employing RC2 and RC4, RSA Data Security algorithms that used 40-bit keys. Meanwhile DES continued to be restricted for export. The lack of clear direction complicated the situation for industry and thus vastly slowed the development of secure systems.

Various groups sought a clarification of federal encryption policy. The Computer System Security and Privacy Advisory Board, a NIST review committee created through the Computer Security Act,[1] requested a national review of cryptography (Brooks 1992, p. C-13). A bill in Congress included a requirement for presidential analysis of aspects of encryption policy (ibid.).

The Brooks-Kammer briefings[2] of the FBI had created a confluence of interest in law-enforcement and national-security circles. NSA urged discussion and adoption of a "national encryption policy." What NSA had in mind was a "national policy" decreeing that "because of legitimate law enforcement needs in the US the US government will have to have a carefully controlled means of being able to decrypt information when legally authorized to do so" (ibid., p. C-12).

The FBI was pursuing passage of the Digital Telephony bill, and NSA was working on an algorithm to satisfy the FBI's need for strong but accessible cryptography. The Digital Telephony effort was known to the public; the encryption work was not. NSA opposed any public debate on cryptography.[3]

The US government's technique for attacking the spread of strong cryptography was also changing. The Clipper program attempted to use standardization and federal buying power to influence civilian use of cryptography. After the effective failure of this program, the government turned to the only other tool available without new legislation: export control. The most notable reason for this shift was that, as cryptography entered the mainstream market, exportability became essential for successful mass-market products.

NSA's work with NIST had been directed toward cryptography used in computers, so it was with some surprise that in 1992 the federal government faced the threat of deployment of strong, relatively inexpensive cryptography in telephones.[4]

Cryptography and Telephony

In the past decade, secure telephones using advanced key management have become widespread in the national-security community. During the 1980s, approximately 10,000 second-generation (STU-II) secure telephones used by the US government were replaced with third-generation STU-IIIs. By the mid 90s more than 300,000 STU-IIIs had been produced, and the unit price had dropped from several thousand dollars to about $1500. Each of the three producers of STU-IIIs—AT&T, Motorola, and Lockheed Martin[5]—also made commercial derivatives using DES and exportable versions using trade-secret algorithms. These versions are generally presumed to be less secure than STU-IIIs, and because of smaller production volumes they are more expensive. At least one, however, has a flexible key-management system that makes it more suitable to the commercial environment than a STU-III.[6]

The secure-phone market is plagued by the existence of too many different kinds of secure telephones, most of which will not interoperate. (It has been jokingly said that the number of types of secure phones exceeds the number of instruments.) The US government now has several and is in the process of introducing more. The centerpiece of the new efforts has been the ISDN-based Secure Terminal Equipment (STE) designed to interoperate with and ultimately replace the STU-III. Unlike the STU-III, the STE is not a controlled cryptographic item. All of the secret components are contained in a PCMCIA card (the Type I cousin of Fortezza) along with the keying material. There are also a number of other Type I voice security devices—some working over traditional phone lines, some using Voice over IP, and some wireless—with various sorts of interoperability.[7]

Voice-encryption systems for the commercial market have also been a staple of companies such as Gretag and Crypto AG in Switzerland and Datotek, Cylink, and TCC in the United States. It was only in 1992, however, that an attempt at selling a modern piece of equipment for secure telephony to a mass market occurred. AT&T announced the Telephone Security Device Model 3600 (TSD 3600) for an initial price of $1295.

In the fall of 1991, David Maher, an AT&T engineer who had been the chief architect of the AT&T STU-III, realized that it had become possible to design a secure phone using a single digital signal processing chip.[8] This permitted a piece of equipment for secure telephony to be built at a total cost of between $100 and $200. Like all modern secure telephones, it works by digitizing the voice signal and then encrypting the bitstream, using keys negotiated by public-key techniques. The beauty of the TSD 3600 is its size and simplicity: it is a 1-pound box smaller than this book. After installing it in the cord between the handset and the body of a standard phone, the user has only to push a "go secure" button to operate it.

As the development of the TSD 3600 proceeded, the head of Maher's division, who had been hired in part for his excellent connections in the intelligence community, discussed AT&T's new security venture with NSA.

NSA was interested in using the TSD 3600 in government applications, but also expressed concern over the problems it might pose for law enforcement. The agency suggested a key-escrow scheme for the phones, promising to deliver the appropriate chips to AT&T by the fall of 1992 so as not to delay the project. AT&T agreed to incorporate the escrow algorithm in some models of the TSD.

The promised chips did not arrive on schedule, and sample TSD 3600s using DES were lent to prospective customers in the fall of 1992. At the time AT&T promised that the DES version would shortly be joined by another model containing a yet-to-be-announced federal encryption

Figure 9.1
The AT&T TSD 3600. (Photograph by Eric Neilsen.)

standard. The model with the new "Type IIE" encryption algorithm would enjoy the benefit of easy exportability and certification for use in government applications.

Bill Clinton became president on January 20, 1993. Six days after the inauguration, Clinton's Senior Advisor for Intelligence was briefed by the FBI on encryption and "the AT&T problem" (Sessions 1993a). The new administration agreed with the current plans. On April 16, 1993, the White House announced the Escrowed Encryption Initiative, a Federal Information Processing Standard intended to "improve security and privacy of telephone communications" (White House 1993).

The Escrowed Encryption Standard

The Escrowed Encryption Standard (EES) was designed to fit a set of seemingly contradictory requirements: strong cryptography, yet readily exportable, with messages accessible to law enforcement under proper legal authorization. The trick was key escrow.

EES consisted of a classified algorithm (Skipjack) that was to be implemented on tamper-resistant chips (Clipper) with escrowed keys. The chips were to be fabricated in a secure facility (the original facility was run by Mykotronx), and escrow agents would be present during the process. Keys would be split into two components, with each piece stored at a secure facility under the control of a federal executive-branch agency. Each half of the key would be worthless without the other. Only under "proper legal authorization" would keys be released to law-enforcement agents. According to Senate testimony, the escrow agents would cost $14 million to set up and $16 million per year to run (USS 103a, p. 95).

When a Clipper chip prepares to encrypt a message, it generates a short preliminary signal called the Law Enforcement Access Field (LEAF).[9] Before another Clipper chip will decrypt the message, this signal must be fed into it. The LEAF is tied to the key in use, and the two must match for decryption to be successful. The LEAF, when decrypted by a government-held key that is unique to the chip, will reveal the key used to encrypt the message.

The proposed standard was limited to encryption of voice, fax, and computer information transmitted over a telephone system (USDoC 1994b, p. 6003). At the initial Clipper announcement, the administration stated that it was neither prohibiting encryption outright, nor acknowledging Americans' right to unbreakable commercial encryption (White House 1993). In later briefings, the administration gave assurances that it would not seek legislation limiting the use of encryption products (USDoC 1994b, p. 5998; McConnell 1994, p. 102).

The key-escrow program provided a widely available form of cryptography of sufficient strength to satisfy the "Type II" requirement for protection of sensitive but unclassified government communications.[10]

This program had two essential elements: the algorithm was secret and was available to approved manufacturers in the form of tamper-resistant integrated circuits and the cryptosystem contained a trap door that permitted US authorities to exploit intercepted traffic when required.

Packaging cryptography in hardware provides the best security and has always been standard practice in the Type I systems used to protect classified information. In such environments, the restriction to isolated (separate chip) hardware implementations represents less additional cost, since the isolated implementation would be necessary for security reasons anyway.

The Clipper Controversy

As required by law, NIST provided a period for public comments on the newly proposed Escrowed Encryption Standard.[11] The response was vociferous and loud. Supporters outside the government were few, while opponents were many and varied, ranging from the American Civil Liberties Union to Citicorp bankers to a large segment of the computer industry. During the public comment period NIST received 320 letters on the proposed standard. With the exception of letters from Motorola (a major manufacturer of secure telephones that may have been contemplating developing devices to meet the new standards), a professor of computer science at Georgetown University, and "no comment" statements from a number of government agencies, the remainder of the letters were negative—including several from government agencies.[12]

The major objection to key escrow was that the mechanism compromises an individual's privacy *even if the escrowed keys are never accessed*. The knowledge that the government has the technical ability to read all communications creates a perception that no communication is private, even if the vast majority of communications are never intercepted or read.

Concern with privacy was not, however, the only ground for objection. Escrowed keys represented a major step back from the encryption techniques that had been developed in the mid 1970s. One purpose of public-key cryptography is to facilitate secure communication in a diverse community by reducing the trust that must be placed in centralized resources. Another is to limit the lifetimes of keys; by extending these, escrow creates vulnerabilities both for society and for the individual.

The decision to escrow keys as part of the standard and to include the LEAF led naturally to the implementation of the algorithm in a tamper-resistant chip. But such a contrivance was most unusual for a Federal Information Processing Standard, and the implicit inclusion of classified portions in a Federal Information Processing Standard effectively changed it from a mechanism for promoting interoperability among communication products to one for exercising control over those products and the industry that produces them. Rather than being able to read the standard, implement conforming products, and submit samples for certification, companies would be required to purchase tamper-resistant chips from authorized suppliers. Both the diversity of sources and the availability lifetime of parts would be outside the company's control.

Formal government secrecy of a technology amounts to the most extreme form of regulation and to a great extent removes both the government and a segment of industry from accountability to the public. The EES stated that the government would regulate which companies would be allowed to include the new encryption product.[13] Companies would not only be beholden to the authorized suppliers of Clipper chips; they would be beholden to the government for permission to purchase them. The computer industry has been characterized by rapid and nimble developments; to many observers, this federal standard seemed to bode steep bureaucratic hurdles for any product that included security.

If the introduction of key-escrow technology were successful, a vast body of traffic would be transmitted under its "protection." Much of this would have been sent by radio or satellite, and there would be no way of estimating how much of it was recorded and by whom. Under these circumstances, escrow agents become an intelligence target of unprece dented proportions. Compromise them and all that has been recorded can be read.[14]

There is also a vulnerability that does not depend on even the continued existence of the escrow agents. Although the standard contains no statement as to the length of either the device-unique key or the family key, it has been stated elsewhere that, like the session keys, these will both be 80 bits. Under these circumstances, it appears that an opponent who knows the Skipjack algorithm, the LEAF creation method, and the escrow authenticator can recover the device-unique key in at most a small multiple of 2^{80} operations. A message so valuable that someone would attempt to perform 2^{80} operations to read it strains the imagination today. It is less strain to imagine a cipher chip whose history is such that after it has been in service a decade or more someone might perform a similar number of operations to acquire easy access to its lifetime traffic.[15]

Despite the strong protests, on February 9, 1994, NIST adopted the Escrowed Encryption Standard as a Federal Information Processing Standard (USDoC 1994b). To objections that the standard was a first step toward prohibition of non-escrowed encryption, NIST responded that the standard was voluntary. To concerns that the system might infringe on individual rights, NIST responded that decryption would occur only when legally authorized. To protests over the secrecy of the algorithm, NIST responded that there are no known trap doors or weaknesses in it. To objections that the standard would be ignored by people engaged in criminal activity, NIST responded that EES would make strong encryption widely available and that, to the degree that it was successful, non-escrowed encryption would become harder to obtain. Escrow agents remained undetermined, and NIST acknowledged that the standard lacked sufficient detail to function as an interoperability standard.

The standard was limited to voice, fax, and computer information communicated over a telephone system. But at the very last minute, NSA had attempted to scuttle that limitation. In memos between NIST and NSA days before EES was approved, NSA modified the standard to cover "telecommunications systems" instead of "telephone communications." NSA also expanded the coverage of the standard to include PCMCIA (Personal Computer Memory Card International Association) cards (USDoD 1994). Apprised of the changes, NIST scientists objected, and the modifications disappeared. Had they remained, EES would have become a standard for both voice and data communications. In addition, EES would have given Fortezza—a PCMCIA card that performs key exchange, computes digital signatures, and encrypts using Skipjack—a free pass around the laborious exception-approval process.[16]

AT&T ultimately developed half a dozen models of the TSD 3600, only some of which could interoperate. These included the D model, which used DES with a 768-bit modulus for Diffie-Hellman key exchange. D models were able to interoperate only with other D models. There were exportable F models that used a Datotek algorithm with a 512-bit modulus, and non-exportable P models running an algorithm developed by the Swiss company Gretag AG. The S models had Clipper, P, and F algorithms, so they could interoperate with the F model, the P model, and the government G model (equipped with Clipper).

AT&T anticipated a large market for these devices, expecting them to appeal to executives in businesses facing aggressive international competition. The original TSD 3600 with DES encryption might have achieved its market objectives. The "improved" Clipper model saw disappointing sales.

By the fall of 1995, total sales of all TSD 3600s were about 17,000. The largest single block were the 9000 Clipper models bought by the FBI in an attempt to seed the market. Most of the remainder were an exportable version exported to buyers in Venezuela and several Middle Eastern countries.[17] According to the government, the Escrowed Encryption Standard was developed to "make strong encryption broadly available and affordable" (USDoC 1994b, p. 6000). The immediate effect of EES, however, was to kill off the first secure phone device targeted at a mass market. By 1997 no secure phone product had come along to take the TSD's place, and telephone conversations remained unencrypted and unprotected. A National Research Council panel cited the lack of encryption between cellular telephones and base stations as a serious problem[18] and recommended it be fixed forthwith (Dam and Lin 1996, p. 327).

The Larger Plan: Capstone et al.

Far from being the whole of the key-escrow plan, Clipper was only the beginning. Paralleling its development, and perhaps started earlier,[19] was the data-oriented Capstone program.

Like the Clipper chip, the Capstone chip implemented the Skipjack algorithm and key escrow. It also provided key management via the Key Exchange Algorithm (KEA), a name it has been claimed was merely NSA's way of concealing use of Diffie-Hellman key exchange. The claim was made plausible by Capstone's third major capability: performing the NIST Digital Signature Algorithm, which uses the same arithmetic mechanism as Diffie-Hellman.

The major use for the Capstone chip was as the heart of a PCMCIA card originally called Tessera[20] and later renamed Fortezza. The initial use of the Fortezza card to provide security for the Military Messaging System (a form of email used by the Department of Defense) was expected to "bootstrap" the use of Fortezza cards for a wide range of computer applications.

Clipper II, III, IV

By the fall of 1995, it was clear that the Clipper chip was not popular. Only the AT&T product was using it, and only a few thousand of these had been sold. During the previous year, joint work between NIST, Georgetown University, and Trusted Information Systems (a small security company with headquarters in Maryland) had produced a software mechanism remarkably similar in function to the Clipper chip, using public-key cryptography where the Clipper chip had used physical tamper resistance. NIST issued a set of ten principles for software key escrow and scheduled two meetings to discuss the idea with industry representatives.

The project had a certain oddity to it. The promise was that systems complying with the ten principles would be exportable, but the meetings were hosted by an organization (NIST) without any role in the export process, and even its parent, the Department of Commerce, plays a role secondary to the Department of State in the issue of cryptography export. (Everyone in the game knows that if the Department of State agrees to "Commerce jurisdiction" for a product, export permission follows.) There was talk of a Federal Information Processing Standard for key escrow, but this too was odd. Each of the two extant cryptographic FIPS says, in effect, "This system is good enough for some category of government traffic." The proposal for software key escrow said nothing about cryptographic quality; indeed, it only specified a particular type of weakness. In the end no FIPS was ever proposed.

The essence of the "Ten Commandments," as they came to be known, was to limit the keys of exportable cryptosystems to 64 bits. Such systems must allow recovery of the key from traffic in either direction. They must not interoperate with unescrowed versions of the same systems.[21] Most important, the escrow agents would have to be in the United States[22] or in countries having bilateral agreements with the United States that guaranteed the US government access to the keys.

In 1996, derivatives of the software key escrow proposal evolved and eventually became part of the export regime. Technical developments included dropping the key-length restriction and relaxing the non-interoperability requirements, but the real developments were in marketing.

Intentionally blurring the distinction between communication and storage, proponents of key escrow have pushed the notion that key escrow is something that users need in order to be able to recover their data if they lose their keys. Along with this notion goes a new name, "key recovery," and a claim that key recovery is substantively different from key escrow. In respect to stored data, there is much to be said for this view. If you have encrypted all the copies of a file, then the keys are as valuable as the information the file contained. If you lose the keys, you lose the information. Under these circumstances, spare keys are more than a good idea; they are essential. On the other hand, the same is not true of communication. There is no reason to want to decrypt the ciphertext of a secure phone call after the call has ended. If either of the callers wanted a recording of the call, the right thing would be to record the plain text at one end of the line; that does not require escrowing any keys. Some forms of communication, such as email, do blur the distinction between key escrow and key recovery. Encrypted email is sometimes decrypted and reencrypted in a local storage key and sometimes left encrypted in the transit key (which is retained).

The other marketing angle was to present key recovery as an essential capability of the key-management infrastructure.[23] The message here is that users won't trust cryptographic systems unless they are sure that they can always get their data back.

These notions were set forth in the late spring of 1996 in the report of an interagency committee assembled to study cryptographic policy (White House 1996). In the fall, a proposed set of regulations containing a new sort of incentive followed. For two years, beginning on January 1, 1997, the government would allow export of unescrowed systems with 56-bit keys (mostly DES systems, presumably) in return for promises from the exporters that they would implement key-recovery systems in their products. Essentially simultaneously, IBM formed a coalition with other companies to implement key-recovery technology and announced what it claimed were fundamentally new and secure techniques for satisfying everybody. For nearly a year, IBM treated its new techniques as trade secrets, but in September 1997 they were made public in a technical report (Gennaro et al. 1997).

The Multi-level Information Systems Security Initiative

After the success of the STU-III project, NSA broadened its objectives and began a project that was originally called the Future Secure Data System (paralleling Future Secure Voice System, the developmental name of STU-III) and later the Secure Data Network System (SDNS). The SDNS project developed protocols for security at several levels of network architecture, addressing such issues as network layer encryption and key management.

The Secure Data Network System evolved into a substantial program called the Multi-level Information System Security Initiative (MISSI), the main goal of which is to solve a much broader range of computer security problems using encryption embodied in individually carried PCMCIA cards. A user sitting down at a workstation on the Defense Message System inserts a PCMCIA card that encrypts and decrypts email, for example. The Type II portion of the program uses the Fortezza card from the Capstone program and is entirely tied to key escrow.[24] After a brief flirtation with a Fortezza+ card, the Type I portion evolved a new PCMCIA card (called Krypton) with much higher performance.

The Computer Security Act of 1987 appeared to have put NSA out of the mass-market cryptography business in the late 1980s, but MISSI certainly looked like an attempt to get back in.

The National Research Council Report

The Clipper controversy convinced Congress that an independent study was needed. In 1994 the National Research Council (NRC) was asked to conduct a "comprehensive independent review of national encryption policy" (PL 103-160, Sec. 267). Everything was to be considered, including the effect of cryptography on the national-security, the law-enforcement, commercial, and privacy interests of the United States, and the effect of export controls on US commercial interests.

The NRC put together a panel of 16 experts from government, industry, and science, 13 of whom had received security clearances.[25] The chairman, Kenneth Dam, had been Deputy Secretary of State under President Reagan; other panelists included General William Smith (former Deputy Commander in Chief of the European Command, and President

Emeritus of the Institute for Defense Analyses), Ann Caracristi (former Deputy Director of NSA), and Benjamin Civiletti (Attorney General under President Carter).[26] Many opponents of the government's policies anticipated that such a group would support the Clinton administration's conservative directions in cryptography policy, but in its 1996 report it did not. Arguably its most important finding was that "the debate over national cryptography policy can be carried out in a reasonable manner on an unclassified basis" (Dam and Lin 1996, p. 298). The NRC panelists declared that, although classified information was often important in operational decisions, it was not essential to deciding how cryptography policy should evolve. This ran counter to the long-standing position of the intelligence community, and it was a striking conclusion to have come from a panel that included so many members of the national-security establishment.

The panel argued for broader use of cryptography ("on balance, the advantages of more widespread use of cryptography outweigh the disadvantages") and emphasized that there should be "broad availability of cryptography to all legitimate elements of US society." Current US policy, they said, was inadequate for the security requirements of an information society (ibid., p. 300–301), and current export policy hampered the domestic use of strong cryptosystems.[27] The panel urged that the market be allowed to decide the development and use of commercial cryptography.

Panelists urged an immediate loosening of export-control regulations. They recommended that products using DES for confidentiality purposes immediately be made easily exportable (ibid., p. 312). Observing that escrowed encryption was a new technology, and that new technologies come with potential flaws, the panel urged the US government to go slow with escrowed encryption—to experiment with the technique, but not to aggressively promote the concept until it had experimented with it on a small scale and knew how to adapt it for large-scale practice (ibid., pp. 328–329). Echoing the First Amendment and contradicting FBI Director Louis Freeh, the panel said that "no law should bar the manufacture, sale, or use of any form of encryption within the United States" (ibid., p. 303).

The panelists recognized that some of their recommendations would complicate law enforcement and they urged that the government take steps to assist those responsible for law enforcement and national security in adjusting to the new technical realities (ibid., p. 322). In an analogy to the statute that criminalizes the use of the mails in commission of a crime, they suggested that the government consider legislation that would criminalize the use of encryption in interstate commerce with criminal intent. They also urged that law-enforcement agencies be given resources to help them handle the challenges posed by new technologies.

The short message of the report was that the United States would be better off with widespread use of cryptography than without it (ibid., p. 299). This was not a message the Clinton administration wanted to hear.

Soon thereafter, the insecurity of DES was shown decisively. Using custom-designed chips and a personal computer, the Electronic Frontier Foundation created "DES Cracker," a $250,000 dollar machine built in less than a year. In July 1998 DES Cracker broke a DES-encoded message in 56 hours. There was some luck involved; the key was found after only a quarter of the key space was searched (rather than the expected half). There was nothing particularly novel about the decryption machine except that it was actually built rather than merely designed. DES Cracker was scalable: with an additional $250,000 dollars and a link between the resulting machines, there would be a DES "Double-Cracker" capable of decoding DES-encrypted messages twice as fast.

International Lobbying

When the Clipper effort ran into problems at home, US government officials began lobbying for it—quietly—in other countries. In 1994, under the influence of such lobbying, the Australian government reported that the biggest *current* threats to telecommunications interception were digital telephony and encryption (Barrett 1994, p. 4). This was at a time when the only mass-market telephone encryption device available was the TSD 3600, most examples of which were either Clipper models bought by the FBI or export models with weak encryption.

The US lobbying had more profound success in Great Britain.[28] Beginning shortly after the announcement of the Clipper program in the United States, the Department of Trade and Industry began to sponsor research on public-key-based escrow schemes at the Cryptologic Research Unit of the University of London. At the same time, development was going on behind the scenes on a draconian legal framework that would effectively outlaw the use of non-escrowed cryptography.[29]

Bilateral agreements on key escrow did not materialize, and the White House took a more public route through the Organization for Economic Cooperation and Development. The OECD is an association of industrialized democracies[30] that seeks to foster—not impede—international trade.

Cryptography was a natural topic for the OECD, which had a distinguished history in privacy policy.[31] Having developed policy guidelines for transborder data flows in 1980 and for information security in 1992, the OECD tackled encryption in early 1996.

The Clinton administration saw the OECD's efforts as a chance to get an international stamp of approval on its key-escrow plans and sent a delegation glaringly different from those usually seen at meetings of international economic-development organizations. Most often, Scott Charney, head of the Department of Justice's Computer Crime Unit, acted as chairman of this delegation. Also included were current and former members of the security establishment, such as Stewart Baker, former general counsel of NSA (who at one point took minutes for the OECD Secretariat), and Edward Appel of the National Security Council staff. With members representing the White House viewpoint, the US delegation pressed for adoption of key escrow. Initial reactions by the other delegates ranged from skepticism (the Japanese delegation wanted to know what would prevent criminals from using their own cryptography systems—see Baker 1997) to mild support for the US position (most notably from the British delegation).

In the economic-development setting of the OECD, key escrow was difficult to sell. Unlike law enforcement, business has little need for real-time access to communications, encrypted or otherwise. Other nations did not see the issues as the United States did. The Danish government's Information Technology panel recommended that no limits be placed on a citizen's right to use encryption (ITSC 1996). The Dutch delegate spoke in opposition (Rotenberg 1996, p. 7). The Nordic countries argued for strong cryptography without trap doors.[32] Meanwhile, German companies, taking advantage of the restrictions on their US competitors, were selling strong cryptography, and the German government had little interest in restricting such sales.[33] Behind the scenes, and kept very much in the background, was Phil Reitinger, a member of the US Department of Justice Computer Crime Division, who was seconded to the OECD to write a draft policy. Yet even this influence was insufficient to convince OECD member nations to support the US policy.

In late March of 1997 the OECD issued its cryptography guidelines, which sidestepped key escrow and emphasized the importance of trust in cryptographic products ("Principle 1: Market forces should serve to build trust in reliable systems"). The OECD recommended that cryptography be developed in response to the needs of "individuals, businesses, and [lastly] governments," and urged that "the development and provision of cryptographic methods should be determined by the market in an open and competitive environment, and that the development of international technical standards, criteria and protocols for cryptographic methods should also be market driven" (OECD 1997). Despite the intense lobbying efforts by the Clinton administration, mandatory key escrow did not make it into the OECD's cryptography guidelines.

Seven months later the European Commission dealt a further blow to the US position. In a policy paper on a European framework for digital signatures and encryption, the commission was cool to key escrow. It observed that such schemes are easily circumvented and that the involvement of a third party increases the likelihood of message exposure (European Commission 1997, pp. 16–17). The Commission expressed concern about the difficulty of key escrow across national borders. The report said that any such scheme should be limited to what is "absolutely necessary" (ibid., p. 18)—hardly the ringing endorsement the US was seeking.

The US Congress' Response

Congress entered the fray in March of 1996 when Senator Patrick Leahy introduced the Encrypted Communications Privacy Act of 1996 (S 1587), a compromise bill that allowed for a relaxation of export controls, affirmed the right to use any form of encryption domestically, created a legal framework for escrow agents, and criminalized the use of encryption in the furtherance of a crime. Less than a month later, Senator Conrad Burns proposed the more strongly pro-cryptography Promotion of Commerce On-Line in the Digital Era (PRO-CODE) Act (S 1726). Burns's bill prohibited mandatory key escrow, enshrined the freedom to sell and use any type of encryption domestically, and liberalized export rules. But 1996 was a presidential-election year, and the complex legislation did not go forward.

Burns reintroduced PRO-CODE in 1997 (S 377). In the House, Representative Bob Goodlatte proposed the Security and Freedom through Encryption Act (SAFE) Act (HR 695). Under both bills, the freedom to sell and use any type of encryption would be unconstrained, and mandatory key escrow would be prohibited. Export of cryptography would be under the control of the Department of Commerce, and export of strong encryption would be permitted if similar products were available overseas. The SAFE bill would criminalize the use of encryption in the furtherance of a crime; the PRO-CODE bill did not address that issue.

In his trademark cowboy hat, Montana Senator Burns seemed like an unusual legislator to be pressing for liberalization of laws on high technology. Burns saw PRO-CODE as having a significant impact on rural areas, where distances preclude face-to-face communication, and where substantial economic growth in recent years has occurred exactly in activities that would greatly benefit from secure electronic communications (Carney 1997).

When Congress reconvened at the end of the summer, the tables turned again. At a Senate Commerce Committee markup, the PRO-CODE bill was sidetracked and replaced by one introduced by Senators Bob Kerrey and John McCain. The Secure Public Networks Act (S. 909), tightened rather than loosened control over the export of encryption products and created incentives for many organizations to introduce key escrow.

Cryptography was also in trouble in the House of Representatives. Despite repeated assurances from the Clinton administration that it would not move for domestic regulation of cryptography, FBI Director Louis Freeh pressed Congress for restrictive laws. The House International Relations and Judiciary Committees had reported the SAFE bill out positively, but the House National Security Committee listened closely to

Freeh's requests, and accepted an "amendment in the nature of a sub-stitute," introduced by Representatives Porter Goss and Norman Dicks, which turned Goodlatte's measure around completely. It not only tight-ened controls on export, but proposed legal controls on the use of cryp-tography. With various versions of the SAFE bill in the House, and differ-ent measure pending in the Senate, it was far from clear what direction Congress would take.

10

And Then It All Changed

The Advanced Encryption Standard

When did the Third Millennium begin? On January 1, 2000? A year later? On September 11, 2001? In cryptography, it began on January 2, 1997 with an inconspicuous notice in the *Federal Register* that marked the beginning of a project to replace the aging US Data Encryption Standard (DES).

The contrast with the events that led to the adoption of DES two decades earlier could hardly have been greater. Although the bones of the formal process were the same—a call for proposals in the *Federal Register* and ultimate selection by the Department of Commerce with the advice of the National Security Agency—everything else was different. In 1997, the notice called, not for algorithms, but for comments on proposed algorithm specifications. Whereas the previous standard appeared to have been designed to be just strong enough for non-national-security applications, the new proposal aimed for the highest grade security: a 128-bit block size and keys of 128, 192, or 256 bits.

Two rounds of comments on criteria were followed by a call for algorithms, due June 15, 1998. Twenty-one algorithms were submitted. Of these, fifteen met NIST's complex set of requirements for documentation, implementation, tests, and rationale intended partly to facilitate evaluation and partly to discourage frivolous submissions. At first glance the response was international—ten algorithms submitted by groups outside the US and five by groups within. At second glance it was even more international than that. All but one of the US candidates included non-US nationals on their design teams.[1]

For nearly a year and a half, all fifteen were under study, a process highlighted by two public conferences, one in California and one in Rome. In the late summer of 1999, the number was reduced to five: three American (MARS, from IBM; RC6, designed by Ron Rivest and colleagues from RSA Data Security; and Twofish, designed by Bruce Schneier and his colleagues) and two European (Serpent, designed by cryptographers from the United Kingdom, Israel, and Norway; and Rijndael, designed by two Belgian cryptographers).

Differences between the Data Encryption Standard and its advanced descendant were more than programmatic. The description of DES in FIPS-46 was entirely in engineering terms, speaking of lookup tables and bits and shifts. Over time these structures came to be viewed in more abstract mathematical terms and the mathematical descriptions gave rise to cryptanalytic techniques. The description of Rijndael was given in mathematical terms (Landau 2004). Certainly it had lookup tables, and bits, and shifts, but these followed from rather than preceded the mathematical notions. In a quarter-century, the field had matured. The Advanced Encryption Standard was truly a second-generation cipher.

Why the difference? In the early 1990s two powerful cryptanalytic techniques had been developed in the public research community: differential cryptanalysis, which infers key bits by comparing input and output differences of pairs of encrypted texts, and linear cryptanalysis, which infers them from linear relationships between the input and output bits.[2] At the same time, other researchers were using ideas about algebraic structure to develop block-structured cryptosystems resistant to mathematical attacks. Polynomials proved key to this. Building on approaches developed for error-correcting codes, researchers used algebraic structure, particularly the theory of finite fields, to create a methodology for constructing cryptosystems provably secure against differential and linear cryptanalysis.

Study of the five finalists continued through another year, and another conference before the winner was announced on October 2, 2000. It was the Belgian submission, Rijndael. Bureaucratic processes dragged on for more than a year before the standard received the required signature of the Secretary of Commerce, but on November 26, 2001, the United States adopted a cryptographic system designed outside its own borders as the "Advanced Encryption Standard."

The adoption of the new standard created an odd situation. AES seemed to be as strong a cryptosystem as anyone could ask for, yet its standing was the same as that of its never-too-strong predecessor. Why couldn't AES be approved as a Type II algorithm, allowing equipment whose functioning was public to be applied to a wide range of government applications? The question hung fire for a year and a half. In June 2003, it was answered in a way that even AES's strongest proponents hadn't dared expect. AES was declared a Type I algorithm, approved for the protection of all levels of classified traffic.

The instrument of this approval was Policy Number 15 of the Committee on National Security Systems (CNSS 15). Curiously, the memorandum was For Official Use Only. It was announced, however, in a virtually identical fact sheet (CNSS15) which appeared in August. The memo is written in a tedious bureaucratic style largely devoted to warning its readers that having an approved algorithm whose workings they know does not entitle them to use anything other than approved implementations for protecting classified information. The important paragraph, however, is perfectly clear:

> The design and strength of all key lengths of the AES algorithm (i.e., 128, 192 and 256) are sufficient to protect classified information up to the SECRET level. TOP SECRET information will require use of either the 192 or 256 key lengths.

Although the Advanced Encryption Standard was now approved for protecting all levels of classified information, this declaration was very much a matter of principle and would remain so until actual equipment was designed, built, approved, and fielded. Nonetheless, the memo had real significance. It showed COMSEC equipment manufacturers a new route to serving their classified markets. Equipment would inevitably follow.

Elliptic Curves, Secure Hash Algorithms, and Suite B

The adoption of AES was not the only radical change working its way through cryptography or through government cryptography in particular. Advances in computing and discrete mathematics had begun to make the Diffie-Hellman and RSA public-key cryptosystems uncomfortably expensive to operate securely. The computer scientists Arjen Lenstra and Eric Verheul compiled tables of the equivalences of the workfactors of a variety of cryptosystems (Lenstra and Verheul 2000). In order for either of the traditional public-key systems to match the security of AES, they would need to employ keys thousands of bits long and do millions of instructions in each operation.

Fortunately, a solution had been coming to hand since the mid-1980s when two mathematicians, Neal Koblitz from the University of Washington and Victor Miller from IBM,[3] developed a new approach, nearly simultaneously. Put briefly, by using more complicated arithmetic than RSA or Diffie-Hellman, you can make the numbers smaller and get the same level of security. The arithmetic in question grew out of the solutions of algebraic equations of mixed degree. The equations give rise to pretty objects called *elliptic curves* and the new approach was called *elliptic-curve cryptography*. Rather than requiring thousands of bits to implement a secure Diffie-Hellman key negotiation or an Elgamal-type signature,[4] elliptic-curve cryptography requires about twice as many bits as AES to achieve comparable security.

To all appearances, elliptic-curve cryptography has been developed at least as much in the public world as in the secret. Although research and development in the area have been done at many places, one company, Certicom of Mississauga, Ontario, has been completely focused on the field and very vocal in claiming the subject as its own. The company has a large patent portfolio—and alleges a larger portfolio of patent applications—which it has been brandishing at other would-be practitioners. Certicom's claims received a substantial boost in October 2003 when the US National Security Agency paid Certicom $25 million for a very broad license to its technology. The license allowed NSA to sublicense its rights broadly for the development of products to support US national security (Certicom 2006).

Practical application of digital signatures, public-key cryptography's less surprising but perhaps more important facet, requires a new "conventional" cryptographic component, called a *message digest function* or *secure hash function*. Of the two terms, "message digest" gives a better picture of what is going on, but the name "secure hash" is used for some of the most important standards, and we will use the two phrases interchangeably.

A message digest function begins with an arbitrarily large data object and produces a small (at most a few hundred bits) one that is inextricably tied to the larger, much as the digest of a book or article is tied to the larger work. A message digest is an example of what is called a *one-way* function, a function that is easy to compute forward but difficult to invert, so that the original message cannot be recovered from the digest. One-wayness, however, is not sufficient for a message digest; it also needs to resist all attempts to produce two messages with the same digest.[5] The latter property is difficult to achieve and secure hash functions have had a troubled history. For many years, an algorithm called MD5, designed by RSA inventor Ron Rivest, has been a mainstay of commercial computing. MD5 was designed to have a workfactor only a little greater than that of DES, or 2^{56}. In the early 1990s, NIST acting as the public face of NSA, put forth a standard (FIPS-180) intended to have a workfactor of 2^{80}. Within a few years, NSA had broken its own algorithm[6] and replaced it with a variation called SHA-1. This algorithm stood untroubled until the summer of 2005, when Xianyun Wang and Hongbo Yu (professors at Shandong University in Beijing) and Yiqun Lisa Yin (an independent security consultant) showed that it could be broken with significantly less effort.

The attack by Wang et al. on SHA-1 will probably not become a practical threat for several years, sufficient time to move to new algorithms. Although a movement is underway to intensify the study of hash algorithms with a view to designing some with an entirely different architecture, the government had earlier put forth standards corresponding to the various key sizes of AES.[7] The very size of the new algorithms would seem to mean that the attacks seen so far will have no practical impact on their functioning.

Cryptographic security depends not only on the quality of the individual algorithms used but on careful coordination of the strengths of algorithms used in combination. A set of cryptographic algorithms all selected to support the same workfactor is called a *suite*. In 2005, NSA extended the CNSS15 approach and announced a full suite of public algorithms that, like AES, were approved for the protection of all levels of classified information. The new construct was called Suite B, apparently by contrast with a previous Suite A, a collection of secret algorithms with colorful names like Juniper and Mayfly. Suite B is made up primarily, though not entirely, of Federal Information Processing Standards. In addition to AES, it contains two secure hash algorithms, SHA-256 and SHA-384, an elliptic-curve version of the Digital Signature Standard, and two key negotiation algorithms elliptic-curve Diffie-Hellman, and MQV, an elliptic-curve algorithm named for Menezes, Qu, and Vanstone from the University of Waterloo in Ontario (NSA 2005). MQV is preferred because it provides authentication at little additional cost.

Suite B is intended to serve a number of objectives. By employing unclassified algorithms, NSA began a convergence between commercial encryption equipment and that used to protect classified information—perhaps eventually making them identical. In so doing, the NSA hopes to draw the major computer and communications manufacturers into a market now dominated by more specialized producers, thereby lowering the cost of acquiring security equipment. These financial objectives of Suite B are bolstered by two interoperability goals.

One is national. A consequence of the increasing interconnectedness of the world and particularly the ever declining distinction between internal and external is the need for greater interoperability among communication systems: those of the military and the intelligence community, those of the police, and those of other "first-responders" like fire departments and ambulance services. Because these systems are intended for responding to terrorist attacks as well as natural disasters, they need to be secured. The two requirements suggest and may actually demand a uniform cryptographic methodology throughout.

The other interoperability requirement is international. Wars are increasingly being fought by ad-hoc coalitions assembled for the occasion. Long standing coalitions like NATO have achieved some degree of cryptographic interoperability among their members.[8] A coalition assembled in weeks can have interoperability only if it plans for it in advance. The central element in this planning is to adopt common cryptosystems, a course of action that makes the traditional secrecy about cryptography meaningless. The only practical solution is to move toward the use of civilian systems and standards by the military wherever possible. Ultimately, this will lead to secure, open, standards accessible throughout the world, another advantage of the Suite B approach.

Export Control

Throughout the late 1990s, the government's rhetoric in regard to cryptography was completely intransigent. Nonetheless, the same period saw a diverse sequence of events coming together to force a change. Every year, almost like clockwork, the government was confronted with a new problem.

In 1996, Daniel Bernstein, a graduate student at the University of California in Berkeley, decided that rather than ignore the export-control regulations as most researchers had, he would assert a free-speech right to publish the code of a new cryptographic algorithm electronically. Bernstein did not apply for an export license, maintaining that export control was a constitutionally impermissible infringement of his First Amendment rights. Instead, he sought injunctive relief from the federal courts. Bernstein won in both the district court[9] and the Appeals Court for the Ninth Circuit.[10] Unfortunately for the free-speech viewpoint the opinion of the appeals court was withdrawn in preparation for an *en banc* review—a review by a larger panel of Ninth Circuit judges—that never took place. With the appearance of new regulations, the government was able to ask the court to declare the case moot, which it did. This indefinitely postponed what the government perceived as the danger that the Supreme Court would strike down export controls on cryptographic source code as an illegal prior restraint of speech.

In 1998, a US intercept network called ECHELON became embarrassingly public (Campbell 1999). The ECHELON system is a product of the UK-USA agreement, an intelligence association among the US, the UK, Canada, Australia, and New Zealand. published earlier (Hager 1996), a 1999 report prepared for the European Parliament (EP) stated that ECHELON was targeting major commercial communication channels, particularly satellite systems. This caught Europe's attention. The implication was that the system's purpose was commercial espionage, a view confirmed, at least in part, by former CIA Director James Woolsey's article "Why We Spy on Our Allies" (2000).[11]

The natural European reaction to this evidence that they were being spied on was an interest in improving the security of European communications but strict regulations on cryptography were an impediment to any such program. The European governments responded by relaxing their rules on the use, manufacture, sale, and export of cryptography, thereby putting the US under pressure to relax its own export rules.

Export control regulations also created direct problems for the government in its role as a major software customer. The military was trying to stretch its budget by using more *commercial off-the-shelf* hardware and software. Economic realities meant that restrictions on cryptography often forced companies to omit security features altogether rather than supporting distinct domestic and foreign versions of the same product. As long as export regulations discouraged the computer industry from producing products that met the government's security needs, the government would have to continue to buy more expensive *government off-the-shelf* equipment for its own use. This was becoming uneconomical to the point of infeasibility. The only way to induce the manufacturers to include sufficiently-strong encryption in domestic products was to allow them to put it in the products they exported as well.

From 1997 to 1999, the US government attempted to bribe the computer industry by allowing the export of products containing 56-bit DES by companies that had made plans to implement key escrow in their products and were making satisfactory progress on their plans. Because development by the companies involved was a commercial secret held in confidence between the companies and the government, the success or lack of success of this program cannot be determined but in the fall of 1999, as noted in the previous chapter, it was abruptly abandoned.

Earlier in 1999, a bill called SAFE (Security And Freedom through Encryption), which would have forced the administration to change the export regulations, passed the five committees with jurisdiction and was headed to the floor of the House of Representatives, when the administration announced that the regulations would be revised to similar effect. By capitulating, the White House avoided the loss of control that would have resulted from a change in the law.

On September 16, 1999, US Vice President and presidential candidate Albert Gore Jr.[12] announced that the government was changing its policies. Beginning with regulations announced for December—and actually promulgated on January 14, 2000—keylength would no longer be a major factor in determining the exportability of cryptographic products.

The new regulations shifted away from the strength of cryptography as the principal determiner of its exportability. In its place, the new regulations considered two issues: whether the hardware or software was for a commercial or government customer and whether the product was off-the-shelf (the term used in the regulations was "retail") or whether it was adapted to the needs of individual customers. The object was to provide the cryptography needed by industry and commerce while making it difficult for military organizations with special requirements and installed bases of cryptographic equipment to make use of the newly available products. For most purposes, the export control portion of the "crypto wars" had been won by industry.

The key-escrow battle ended more quietly, with most (perhaps all) of the government's programs being canceled. In June of 1998, the Skipjack algorithm that underlay Clipper was declassified so that the military could use it in software for secure email. Despite being an elegant algorithm and strong enough for many purposes, Skipjack was tainted with its key escrow past and has not been widely employed.

The notion of key escrow has not died, however. It has been built into products for commercial use where employers are in a better position to require their employees to submit to potential spying than the government was to impose the same requirement on the citizenry in general. Key escrow, better called key recovery for this purpose, is also essential when cryptography is used to protect stored information rather than communications.[13]

Cryptography after Deregulation

Contrary to the expectation of many of its fans, the deregulation of cryptography did not produce any immediate explosion in either the number of available cryptographic products or the frequency of their use. Anyone who expected most email and phone calls to be encrypted overnight was surely disappointed.

The reasons for slow growth in the cryptographic business, however, seem fairly clear. Foremost, cryptography, at least communications cryptography, is a phenomenon of the interaction among people. To engage in encrypted communication you must make an investment in hardware or software. The value of your investment is proportional to the number of other people similarly equipped and so the market has natural exponential growth.

The benefits of exponential growth are well known in successful fields and the term is applied willy-nilly to anything that is growing quickly. The downside (more precisely the slow upside) of exponential growth is exhibited by public-key infrastructure. Keys, certificates, directory servers, and revocation lists are of very limited use until most people have PKI-enabled products. Consequently, their up-front costs are not offset by a robust revenue stream and must be supported by an investment that is slow to produce returns. This burden has been borne by the US military who have built themselves an electronic key-management system at a cost of tens of millions of dollars but a similar commitment is difficult for the commercial sector.[14]

Other phenomena act to advance or retard the basic exponential progress. Most conspicuous of technical problems is the lack of uniform standards. Several non-interoperable suites of cryptographic algorithms and formats for keys, certificates, and cryptograms have significant shares of the commercial market. Exponential growth is thereby fragmented and must occur independently within each sector.

Regulation, which in the nineties acted primarily to decrease the use of cryptography, may come to play a supportive role as the value of cryptography for protecting private data against the compromises that are so frequently in the news becomes more widely recognized. Similarly, the popularity of laptops and the ease of laptop theft has created a market for cryptographic protection of their file systems. Many companies have imposed requirements—similar in effect to regulations—that the laptops carried by employees be so protected. In June 2006 the US Office of Management and Budget put into place a recommendation that all sensitive data on laptops be encrypted unless exempted by the department's Deputy Secretary (or his designee) (Johnson 2006).[15] Cryptography may get a big boost from regulations giving *safe haven* in the event of compromises of personal data when the data are properly encrypted.

What promotes the use of cryptography more than anything else is its default inclusion in products and its automatic operation without the need for user action. The Secure Socket Layer protocol that comes built-in to all browsers may be the most widely deployed cryptosecurity system of all time. A plausible competitor for this title is the A5 algorithm used in GSM telephony. Another is the use of cryptography in smart cards. A comer in this category is the automatic encryption of phone calls by Skype, a popular Voice over IP system. All of these are commercial products, which have far outrun their military ancestors in deployment.[16]

DRM, DMCA, and TCG

If cryptography has exhibited a growth area, it is one rather different from what cryptography's early pundits anticipated: protecting the interests of purveyors of intellectual property.

The incredible advantage of information in digital form—it can be readily and inexpensively moved and copied—is a disadvantage from the conventional marketing viewpoint, which, roughly speaking, knows how to charge for what is scarce. Digital products, unlike antibiotics for example,[17] can readily be copied. If you have one copy of a program or a digital copy of a picture, you can readily and inexpensively have another but one antibiotic tablet is of little help in producing more.

Attempts to prevent ready copying of digital products have come to be called *Digital Rights Management*. In essence, DRM is a regime in which the goods are kept in encrypted form everywhere except in controlled environments in which the digital products can be viewed, listened to, or otherwise used. An approach of this sort is proactive and imposes a prior restraint on would-be users of digital products. Another approach, more in line with conventional enforcement of copyright is to label each copy of a product in a way that cannot readily be altered, an approach called *watermarking*. When an unauthorized copy of a watermarked work is discovered, it is possible to examine the label and trace the copy back to its source. The tamper resistant, and often covert, labeling of digital objects uses *steganography*, a cryptographic technology that hides messages rather than merely making them unreadable.

One widespread cryptography-based copy protection system is the Content Scrambling System (CSS) used to encrypt the contents of DVDs.

Although intended to prevent the copying of DVDs, particularly onto other media like hard drives, CSS had the side effect of preventing the implementation of DVD players on computers running open-source operating systems, particularly Linux. CSS is administered by the DVD Content Control Association,[18] which is unwilling to license its technology for use in to open-source software. In October 1999, however, *deCSS*, an independent implementation of CSS developed in Norway by Jon Lech Johansen and unknown associates, became available over the Internet. As it turned out, cracking copy-protection systems is not illegal in Norway, and attempts to prosecute Johansen failed.

When NSA director Bobby Ray Inman tried to acquire the legal power to control cryptographic publication in the early 1980s, the idea was widely condemned. The American Council on Education panel that was created in hopes that it would recommend the idea came nowhere close. Ironically, a legal system of censorship of cryptographic research has since grown up to serve commercial ends with no comparable condemnation.

The Digital Millennium Copyright Act shepherded through Congress by the entertainment industry gives legal protection to technical systems used in protecting copyrighted material. Its functioning is comparable to laws against breaking and entering: if you lock your door, even with a very poor lock, you acquire a measure of legal protection that is lacking if you leave your door unlocked. Anyone who breaks your lock and enters your home is guilty of breaking and entering, a crime more serious than mere trespass. In a similar way, the DMCA created both a tort and a crime of defeating copyright protection measures. The law made it a crime to defeat such measures, even when the objective was to use the material in a manner permitted under copyright law. The objective was to criminalize attempts to defeat copyright protection mechanisms independent of any issue of the protection of particular copyrights. An exception was created for research but it was painfully narrow, requiring the researchers to give notice in advance to the owners of the system under study.

This time the crypto community was remarkably docile. Two cases set the tone.

In 2000 the Secure Digital Music Initiative (SDMI), an industry group consisting of about 150 companies and organizations, put together a challenge to researchers to break the audio watermark they had developed. The contest rules were stringent: three weeks to remove the watermark without badly degrading audio quality (the latter was not an announced contest rule). Felten and his colleagues decided to participate without actually officially joining the contest, thus preventing them from competing for the prize but also allowing them to avoid signing the required confidentiality agreement.

Instead of competing for the cash award, Felten, his students, and fellow researchers at Rice University wrote a technical paper showing how to defeat the SDMI technology. Felten et al. intended to present the research at the Fourth Annual Information Hiding Workshop, held in Pittsburgh on April 25–27, 2001. Through a particularly foolish action on the part of the SDMI, the Recording Industry Association of America (RIAA), and Verance Corporation, Felten and his colleagues were threatened with legal action if they presented their work. The argument was that the Princeton and Rice University computer scientists had violated the anti-circumvention aspects of the DMCA. No matter that DMCA has an escape clause for research—§1201 (g)(2) (B), which permits circumvention if the "act is necessary to conduct such encryption research"—or that the ensuing publicity was likely to cost SDMI and RIAA far more than permitting Felten and his colleagues to go public with their research.

Princeton University declined to provide counsel to defend the scientists and so the researchers withdrew their paper from the meeting. Instead the Electronic Frontier Foundation (EFF), a civil-liberties group focused on citizens' rights in the digital world, stepped in. The researchers filed suit in federal court, seeking a "declaratory judgement" that publication of the research paper would fall within the plaintiffs' First Amendment rights (*Felten v. RIAA*, US DC NJ Case #CV-01-2669). The recording industry backed down, the scientists presented their paper at a different—and more widely attended—venue, the USENIX Security Symposium (Craver et al. 2001), and the case was dismissed for lack of standing.

With Niels Ferguson, a case of interest to cryptographers, the situation worked out differently. Ferguson, an established cryptography researcher and consultant, claimed to have broken the High Bandwidth Digital Content Protection (HDCP) system, an Intel cryptographic system for encrypting digital video communications between cameras and players (HDTV, etc.). Licensing for the system was available through Digital Content Protection LLC, a subsidiary of Intel.

Ferguson, a Dutch citizen living in Holland, believed that were he to publish his results, he would be subject to arrest for violation of the DMCA, anytime he was in the United States.[19] Ferguson did not make his work public; instead he submitted a letter to the chair of the 2001 ACM Workshop on Security and Privacy in Digital Rights Management describing the chilling effect of the DMCA. He also submitted an affidavit in the *Felten v. RIAA* case.

Other researchers studying HDCP did not react to the chill in quite the same way. In particular, a group of researchers from the University of California at Berkeley, Carnegie Mellon University, and the Canadian company Zero-Knowledge Systems also found an attack on the HDCP system (Crosby et al. 2001). Wanting to publish but realizing the possible conflict with DMCA, the researchers proceeded with caution. One of them, Berkeley professor David Wagner, consulted with University of California lawyers, who made clear they would be defended in any civil suit.[20] Next, in accordance with the requirement of the DMCA, the researchers "made a good faith effort to obtain authorization before the circumvention"[21] and met with engineers from Digital Content Protection LLC, who appeared to appreciate the advance notice they received about problems with their technology.[22] Then the security researchers published their work at the same ACM workshop that Ferguson had avoided.

Trusted Computing Technology

In the late 1990s five major computer companies, AMD, HP, IBM, Intel, and Microsoft, formed a consortium called the Trusted Computing Platform Alliance to develop standards for a broad new approach to copy protection and a variety of other computer security problems. The idea was to add dedicated security hardware called *Trusted Platform Modules* (TPMs), hardware capable of monitoring and controlling all activity, to computers. The work of the TCPA was widely perceived as an attempt to reduce personal computers to the status of such consumer-electronic devices as television sets, devices on which the owner's control over what programs could be run would become comparable to the viewer's control over what TV programs were available to watch.

Partly in response to the criticism, the TCPA later reorganized and reincorporated itself as the Trusted Computing Group. It also expanded its core membership (the Promoters) from five to seven, adding Sun and Sony, and making provision for adding more as time went on.

The basic objective in adding the sort of security hardware for which TCG is developing standards is to achieve tighter control over the software running on computers. This technology has many possible applications. It can, for example, make it far more difficult for viruses and worms to infect a machine. It can enable system administrators to ensure that only approved programs or only licensed copies of programs or only the latest versions of programs can be run on a system.

Conceptually, trusted computing technology begins with control of what operating system can be run. The process is called *secure boot*: the microcode built into the computer checks a signature on the operating system it is loading and will let the system run only if it bears the correct signature. The most influential developers of secure boot technology were William Arbaugh, Dave Farber, and Jonathan Smith at the University of Pennsylvania.

This approach may be very secure, but it is also very inflexible. For many purposes, it is desirable to allow a computer to run a variety of operating systems. In these cases, it may still be valuable for one system to be able to determine with certainty what set of programs another system is running. This technique is called *attestation*.[23] A computer may be capable of running any of the popular operating systems—Linux, Solaris, Windows—and any applications that those operating systems support. In interaction with other computers, however, it may be asked to attest to its configuration, i.e., to get a signed message from the tamper-resistant TPM describing the configuration of its hardware and software. This allows the computer that has demanded the attestation to decide whether it will allow interaction to proceed.

In some networks, for example those running the electrical power grid, TCG technology is entirely appropriate. Computers are not connected to that network by rights but to serve the interests of the power companies and their customers by managing the country's electric power. A similar argument can be made for enterprise networks in which all the computers are owned by the enterprise.[24] On the other hand, individual computer owners take the reasonable attitude that they should be able to run whatever programs they wish and fear that if trusted-computing technology becomes widespread, this freedom will be denied them. Other critics, with a more entrepreneurial view fear that trusted-computing technology will stifle innovation by allowing ISPs to discriminate against programs —browsers, for example—that were not provided by their preferred commercial partners. This can be done entirely without malicious intent, for example, by an ISP that is trying to limit the burden of maintaining compatibility with an ever-growing number of versions of a popular program.[25]

The Bigger Picture

The 1990s will be remembered long after the roaring nineties a century earlier, the roaring twenties, or the 1960s have been forgotten. Not only were the 1990s a boom period dotted with great feats and great fortunes, the 1990s changed the foundations of society in a way that may not be appreciated for decades.

The technologists will remember the era for the World Wide Web. Invented in 1989 at CERN (the European laboratory for particle physics), the Web began to take hold about 1993 and was flourishing by three or four years later. The Web made delivery of information over the Internet easy and natural rather than tedious and geeky. Within a decade of its birth, the Web was being used by businesses and governments as the primary way that they should be getting information out to and acquiring information from their customers. Great fortunes were amassed by those who got on the Web bandwagon early and such names as Amazon, eBay, Yahoo, and Google became household words. 'Google' even became a verb.

Deeper, less flashy developments also drove the move toward a society based on digital communications. For its whole previous history, the cost of telephony had been driven by the cost of long distance transmission. In the 1990s, optical fiber technology came of age. By placing fiber through conduits previously occupied by copper, communications

companies multiplied their bandwidth by a factor of 1000 and sometimes made money on the deal by selling the copper they had replaced.
The result was the development of a vast overcapacity that still exerts a
profound effect on the communications business. Fiber is the bedrock on
which the current high-speed Internet and plans for future higher-speed
internets are built.

It is hard to think of a dot-com that better captured the positive energy of the 1990s Internet more readily than Google, with its philosophy
"Do no evil." But from a security perspective there is a dark side to
the enabling technology of search engines, which give the ability to discover targets with known security vulnerabilities. Such information was,
of course, public before the Web, but like county court records, it was
largely inaccessible, leaving these systems relatively safe. However, now
through the use of search engines, it has become a rather trivial job to
discover and access public systems with known vulnerabilities (Landau
2006, p. 433).

The decade was also one of growing internationalization. Falling communication costs, falling shipping and travel costs, and falling barriers
to both trade and travel increased the tendency of businesses to expand
worldwide. Internationalization merged naturally with the growing business trend toward *outsourcing*, using contractors rather than employees
for many tasks, ranging from sweeping the floor to programming, accounting, and public relations. Once international communication became adequate to support close business relationships across intercontinental distances, it became apparent that capable well-educated work-
forces from countries with low labor costs were readily available. This
dramatically increased international business communication and consequently the need for its security.

Internationalization and outsourcing have been eagerly embraced by
US businesses. Less engaging from a US perspective is a decline in the
preeminent position the US has held in world commerce since the end
of World War II. Although the United States is the world's third-most-
populous nation, it has only about 5 percent of the world's population.
As the gross national products and living standards of many parts of the
world increase, US influence over high-tech policy issues will decline.

It is worth noting that whereas in the late 1990s 80% of Web content was in English (Wallraff 2000, p. 61), by 2002 the percentage had declined to less than 50% (Crystal 2004, p. 87). Indeed, by 1998, over half of the newly created websites that year were *not* in English (ibid.).

Conspicuous on the international scene is the rise of China as a major cultural and economic power and a major creditor of the United States. Indeed, among Internet users the second-most-common native language is Chinese (English is first).[26] China has begun developing standards for digital products in many areas. These standards, some of which are cryptographic, are in potential competition with European and American standards.

For a long time, computer communication involving humans consisted primarily of single-fixed-width-font text. In non-text interaction between computers and humans, the images were usually generated locally. This was natural enough. Network bandwidths were low. Fifty-kilobit backbones and modems running at a few thousand bits per second were considered fast.

The Web brought a new paradigm: HTML, the hypertext markup language. HTML is a crude typesetting language, which, however, provides hyperlinking—the possibility of including one document, by reference, in another. HTML is a small and in many ways primitive subset of an ambitious standard called SGML (Standard Generalized Markup Language) a construct so rich that people are forever creating subsets. A subset that has taken on great significance is XML, the Extensible Markup Language.[27]

XML has proved to be the most popular approach, not for human-to-computer communications but for computer-to-computer communications, communications that are not simply about moving "customer data" but communications that involve negotiation between computers about services, inventories, communications, and security.

The explosions in both the raw power (bandwidth) of communications and in the computational capability to manage the communicated material have created a world in which every sort of information, from names and addresses, to historical documents, to satellite photographs, is more readily available. On one hand, the ease with which real estate speculators or would-be home buyers can appraise property remotely makes some homeowners indignant. On the other, details of some government buildings have been made indecipherable in the aerial photographs most easily found on the Web.

In a world where more and more material is being brought under tighter and tighter control, covered by non-disclosure agreements, and protected by digital rights management systems, there is one major move toward openness: the open-source software movement.

By lowering the cost of both creation and dissemination, modern computer technology has enabled the user to be a creator in unprecedented ways. During Hurricane Katrina, many users were able to put together information from publicly available sources to enable people to discover whether their homes had been damaged. This is one of many examples of technology enabling individuals to provide services for themselves that could once have been provided only by governments or large corporations.

If a communications network is to be flexible and encourage innovation, then it must perforce allow applications unanticipated at the time of the design of the network. The Internet does this is through the *end-to-end* principle, which is the idea that the communication endpoints —the applications—should implement the functions, rather than letting low-level function implementation be part of the underlying communication system. This principle been fundamental to Internet design since the beginning.[28] The Internet concentrates investment (and particularly "smarts") at the edges. The center is a computationally powerful but fundamentally dumb collection of routers and transmission channels. (In fact, the routers and channels are not so much dumb, as neutral to the application they are transmitting.)

This is fine in theory; in practice, as the Internet has evolved, it has departed somewhat from these principles. In particular, it has evolved into a fragmented network in which many nodes are walled off into private isolated networks that cannot communicate with each other.

The key elements of this "walling off" are *network address translators (NATs)* and Firewalls. The current Internet Protocol (version 4, or IPv4) has addresses that are only 32 bits long. There are about 4 billion possible addresses. This is a large space if addresses are chosen at random but a much smaller space if large sets are reserved to be in contiguous chunks. The fact is that address space is in too short a supply to be used as originally intended: every host computer has a unique address and therefore can be addressed by any other computer on the network.

Beginning in the early 1990s, organizations that could not get enough unused addresses began to allocate their internal addresses independent of the outside world.[29] Such common functions of the Internet as email do not actually make direct use of internet addresses. If you send email to president@whitehouse.gov, the fact that the IP address of the White House is 63.161.169.137 does not even come to your attention, and the individual address of the workstation on the President's desk is something you may not even be able to discover. The email servers will go from the email address president@whitehouse.gov to the IP address 63.161.169.137 and deliver the email there. At that point the "president" component of the address becomes important, and the White House mail server will take over and discover the correct internal address to which to forward the message. It makes no difference whether the internal address is unique; it is only accessible to computers inside the White House. This is network address translation. It is typical of institutional connections to the Internet.

Closely associated with NATs are *firewalls*, computers that manage and filter traffic between an internal network and the Internet. Firewalls serve several purposes but one of the main ones is explicitly to prevent unfettered access to the Internet. Rather than supporting (permitting) any form of access to the Internet, firewalls frequently allow only a limited range of services. The White House firewall described above might support no service other than email.

An attempt to expand the address space of the Internet has been underway for several years. Internet Protocol version 6, IPv6, has 128-bit addresses (256 billion billion billion billion possibilities) and can provide enough address space for the end-to-end connectivity originally envisioned. As noted earlier, however, the protocol that handles addressing is the protocol that defines the network and must be shared by all network users. Even when one protocol has been designed as an extension of another, transition is difficult.

Had IP originally been designed with larger addresses, firewalls might have been designed differently and NATs might never have established themselves. In the existing Internet, however, many organizations are delighted to have control of their user's communications, and a return to unfettered end-to-end communication seems unlikely.

Carnivore

On July 11, 2000, the *Wall Street Journal* disclosed that the FBI had been "wiretapping" the Internet (King 2000). The FBI had developed a program with the ill-chosen name of *Carnivore* to scan and record network traffic.[30] Despite a public inured to cookies that let web sites track user visits, reports of Carnivore hit a public nerve. What exactly was the FBI recording? Was the Internet versions of pen registers and trap and trace devices recording more information than they would for traditional land-line telephone systems? Was the mail of nontargeted individuals inadvertently being read? Why was an FBI device attached to an ISP, instead of, as had always been done by telephone taps, the Internet Service Provider (ISP) doing the sorting of relevant traffic for the bureau?

The truth turned out to be both more complex, and, on the surface, less threatening, than initial newsreports indicated. Carnivore was a *packet sniffer*, a program that analyzes network traffic; such a program can be configured to search for traffic to or from a particular user. ISPs serve as the local post offices of the internet world, sorting and delivering email and other services to their users. When Carnivore was placed at an ISP, it received all packets that traversed the Ethernet connection on which it was placed (Smith 2000, p. 13). Using filters, Carnivore recorded the traffic that fit pre-determined patterns, generally traffic to or from a particular user. Carnivore could be configured for full wiretap or pen register mode; in the latter, the data content was "X-ed" out (ibid., p. 56).[31] The FBI presented Carnivore (later renamed DCS, or Digital Collection Service 1000) as a natural application of wiretap law to Internet technology. Civil-liberties groups and computer experts disagreed. Because the Internet packet-routed architecture is fundamentally different from the circuit-switched telephone networks, the application of wiretap laws to the Internet is less straightforward than it would appear. Unlike the telephone system, transactional information (number called, number calling) is hard to separate from the content. Consider a targeted user reading a web page. The web page will be received from a website with an IP address. Carnivore will log the communication between the targeted user and the website. The web page, however, may contain hyperlinks to other websites. These will be resolved into IP addresses and Carnivore will log these as well. Even when the authorization is for a pen register and does not include recording the content of the web page itself, the content will be largely reconstructible from the pattern of IP communications.[32]

Carnivore also contains a mode for analyzing SMTP communications. (SMTP, the Simple Mail Transfer Protocol, is the standard for Internet email transmissions.) In this mode, email header information will be recorded. Email headers contain much more information than phone numbers. Some of this is analogous to information on physical envelopes and can be justified on the grounds that the same information would be obtained in a mail cover. Some, however, is not. If the entire mail header is captured, lists of addressees and copyees that would not be on an envelope will be included. A variety of version and status information about the mail clients and their configuration is also commonplace.

The other violation was not privacy, but engineering. CALEA had already created an odd situation in which the FBI was placed in the role of designing standards for the telephone network. Carnivore went a major step further, placing *the government's own search devices* directly onto the ISP's networks. In at least one case, this was done *over the ISP's objections.*[33]

Attorney General Janet Reno authorized an outside academic review of Carnivore. Several universities with strong research groups in computer security, including Dartmouth College, the Massachusetts Institute of Technology, Purdue University, and the University of California at San Diego, expressed interest in performing such a study but the restrictions the Department of Justice placed on the review—which included confining the study to narrow technical questions, pre-publication review of the study by DoJ, and DoJ approval of the final report—were such that these universities bowed out. Instead, the Illinois Institute of Technology (IIT), an institution not previously known for expertise in computer security, performed the study.

Like the Clipper chip before it, Carnivore suffered from design flaws. In the Clipper case, the design worked as advertised: it securely encrypted data under an 80-bit key. The flaw was that it was possible to spoof

Clipper so that one could use the Clipper chip to encrypt, but do so in such a way as to prevent law-enforcement access (Blaze 1994). In Carnivore's case, the potential problems affected the underlying security of the system. The IIT report found that the filter settings, which determined what traffic could be captured, could be changed by anyone with access to the system password, which was compiled into the Carnivore source code, and thus plausibly to anyone with access to the Carnivore installation, a poor security design.[34] IIT noted that the program had no audit trail, "Incorrectly configured, Carnivore can record any traffic it monitors"(Smith 2000, p. 17)—*any traffic* should not have been an available option—and, "except for FBI procedures and professionalism, there are no assurances against additional copies being made of an inadequately minimized intercept" (ibid., p. 60).[35] Such design runs contrary to the intent of wiretap law, which is very specific about data minimization, requiring that only content appropriate to the warrant be collected.[36]

When various members of Congress expressed concern about the limited nature of the IIT review, legislation restricting Carnivore use seemed possible. The Justice Department promised an internal review of the policy issues regarding Carnivore. But before it was completed, the events of 9/11 intervened, and the USA PATRIOT Act, section 216 of which legalized the application of Carnivore-like systems to packet-routed networks, was passed.

Meanwhile, use of Carnivore is down in favor of commercial systems to do the job (FBI 2003a,b).

Identity and Anonymity in the New World

To casual observation and in casual use, the World Wide Web appears to provide a sort of anonymity similar to that available to a shopper in a big city. You browse through sites and look at their contents. You are not asked for identifying information and you are not aware of providing any. Depending on your communications arrangements, however, you will be providing various sorts of information from which other information about you can be determined.

If you communicate from a fixed IP address, the web pages you visit will be able to recognize your visits as those of a single entity. As a practical matter, they or may not be able to convert your IP address into any of the usual customer data such as your name. If, as is more common, you communicate through a local ISP and get a different IP address from session to session, it will be far more difficult for the visited websites to track you but easy for the ISP. The latter may or may not be sharing your information with other commercial entities.

There are common and legitimate purposes for which casual anonymity is not sufficient. Investigators of many kinds from academics to reporters to police try to hide the patterns of their inquiries, even when they are consulting open sources. Many individuals value their privacy. For example, they wish to learn how to handle their illnesses without revealing their conditions to medical marketers or colleagues. Companies may be unable to do competitive analysis unless they can conceal their identities as they visit competition's web sites.

As AOL's indiscretion in releasing query logs for thousands of its customers revealed (Barbaro and Zeller 2006), a knowledge of the information a person seeks imparts a vast amount about the person's activities and plans and it is also likely to disclose the person's identity, even if that identity is not directly contained in the query record.

In response to the danger of Internet users being tracked and identified, various commercial entities have arisen to provide services that shield browsers' identities more effectively from the websites they visit. Such services, which include the late Zero-Knowledge Systems and Anonymizer.com of San Diego, find much of their customer base in commercial entities that want to survey their competitors websites in confidence that they are seeing the material that would be shown to typical browsers and not a show that has been put on especially for them.

In the 1980s David Chaum proposed several anonymity systems that wrap communications between the sender and receiver in layers of public-key cryptography. Central to these systems is the appropriately named *mix*. Each time the communication reaches a new mix on its journey from sender to receiver, the mix unwraps one layer of the cryptography, mixes up the order of the messages, and sends them all on. One descendent of Chaum's research is onion routing, the current version of which is called Tor (The onion routing) (Onion Routing 2006). Tor modifies mixnets by using anonymizing proxies (proxy servers sit between a client and a server fulfilling requests for the client if it is able—otherwise it ignores the request) (Reed et al. 1996).

When an application (perhaps a Web browser, perhaps an IM client), connects through the Tor network, a client proxy chooses a route for the traffic using server nodes called *onion routers*. The Tor proxy establishes a circuit using public keys belonging to onion routers in the selected path. Application data is passed using keys determined by the proxy and each onion router as the circuit was established. While each of the onion routers will know its predecessor and successor on the path, only the client proxy is aware of all nodes on the path which the communication traverses.

The Tor network is an overlay on the public Internet.[37] Widely used applications include anonymous web browsing (http and https), instant messaging (IM), and Internet Relay Chat (IRC). Tor has been running since October 2003. As of August 2006, the Tor network had about 750 Tor servers and the number was doubling approximately every 6–8 months. How many users does Tor have? Tor deliberately does not keep track.[38]

Initial work on onion routing began at the Naval Research Laboratory in 1995. Tor was begun in 2002 and was jointly designed by scientists at NRL and at the Free Haven project, working under contract to NRL. Most of the funding for onion routing, including Tor, has come from the Office of Naval Research and the Defense Advanced Research Projects Agency.[39] Given current law-enforcement concerns about tracking network users,[40] it might come as a surprise that the original funding sources for a public anonymizing network came from the Department of Defense. It should not. A group in the Navy, for example, who, during the last half decade, have been periodically stationed in the mid East have found Tor an excellent way to disguise their communication patterns. No one watching their ISP connection locally in country can learn their affiliation through tracking with which agency in the United States the Naval personnel contact and no one watching their communications in the United States can learn to which country—and *which house*—their communications are going.[41]

Naturally, this naval unit which had taken such pains to conceal itself using anonymizing technology wanted neither to be named nor to have its location identified. This illustrates a difficulty of explaining the value of anonymizing technologies: when an anonymity system is successfully used for a "good" task, its usage does not generally become public. Publicity only attends those cases when there is a problem with the anonymizing technology or where the anonymizing technology is used for some nefarious purpose. That the funding support for Tor came from a variety of agencies of the US Department of Defense makes it clear that the technology is beneficial in a wide variety of government situations.

Why might the Department of Defense fund an anonymizing system for the general public rather than build one just for military use? As the developers of Tor point out, "anonymity loves company" (Dingledine 2006). The more users an anonymity system has, the easier it is to hide the traffic. Thus a widely used anonymity system provides Department of Defense users the best protection from prying eyes.

At the same time that work was occurring on anonymizing technologies, there was also great effort underway to develop electronic identification systems of various types.

With the opening of the Internet to commercial traffic in the early 1990s, electronic commerce—ordering a book from Amazon, selling collectibles on eBay, making travel arrangements with Travelocity or borrowing money through eLoan—caught on faster with consumers than anyone but a few (now wealthy) backers expected. There was, however, a serious irritant: the constant need to reenter your name and password, not to mention credit-card information, billing and shipping addresses, and phone numbers each time you made a transaction over the Internet. Out of this difficulty was born the notion of single sign-on, a way for the user to sign on and authenticate herself to a single site and have that authentication carry over to multiple sites.

Microsoft developed the Passport system to which Hotmail users had immediate access (thus giving Passport a large installed base). Microsoft's approach centralized customer data and there were immediate objections from civil-liberties groups over the threat to privacy this entailed. Meanwhile, in 2001, in concert with American Airlines, Bank of America, Cisco, Nokia, Sony, and a number of other companies, Sun Microsystems proposed a federated system called the Liberty Alliance,[42] in which identity information (name and authentication) could reside with an "Identity

Provider," while site-specific information—the dates of car rental, the type of car desired—would be with the "Service Provider." The goal of the Liberty Alliance was a set of interoperable specifications for federated network identity allowing users to authenticate once—single sign-on—and link elements of their identities without centrally storing their data (Liberty Alliance). Liberty protocols provide pseudoanonymity in the communications between the Identity Provider and Service Provider, a featured designed to prevent data aggregation. The Liberty protocols have attracted wide interest, and various governments as well as large commercial enterprises are participating in the process.[43]

Anonymity and identity are among the many threads in human culture that have existed in uneasy harmony for millennia. The revolutionary changes of the 1990s—globalization, mobility, greater availability of information—brought many of these threads into open conflict and a new balance among them has yet to be found.

At a moment in human history, however, when reflection and tolerance might have served us best, the events pushed everyone in a direction that, by maximizing security, minimized privacy and individual liberty.

11

Après le Déluge

NSA in 2000

For NSA, the third millennium began in a confrontation with reality. After years of insularity, secrecy, and success with the same old methods, passive interception and an intense focus on cryptanalysis, they had run into difficulty, not with improved cryptography but with the complexity of modern communication (Hersh 1999). It is always difficult to judge intelligence agencies' claims about their problems. It is hardly in their interest to be saying "We are listening to your every word and you had better improve your security." Better to say, "You don't need to do anything special about security. The complexity of modern communication is so great that we can't find the traffic we want to read, let alone decipher it." Nonetheless, for reasons outlined in previous chapters NSA's story has a ring of some truth. Communications had changed dramatically in a very few years and it was not clear that NSA was putting its efforts into the right endeavors.

The agency's director, General Michael Hayden, set out to solve the problem with a change in style, placing a variety of outsiders in critical positions and, like businesses all over the world, outsourcing as much of its activity as possible in an attempt to gain efficiency and save money.

The agency gave its modernization projects heroic names like Trailblazer and Groundbreaker. Reports suggest that these have been less than entirely successful but, as with the initial reports of difficulty, such information is hard to judge. What seems likely is that some people and groups at the agency have a good understanding of the modern world of communications and are at the cutting edge of intercept technology and practice. NSA, however, is an organization with tens of thousands of employees and decades of established practice. Bringing the entire organization "up to speed" with its best pieces is a difficult task.

Censorship and Surveillance

Automation has allowed many organizations less well known and less formidable than NSA or other national SIGINT agencies to conduct surveillance and outright censorship on traffic.

Employer Web usage policies policed by firewalls

The police and intelligence agencies are supposed to tap your telephone calls only with a warrant. In many circumstances, however, you may be held to have given someone—typically your employer—blanket permission to listen to your communications. The federal government is allowed to monitor the communications into and out of its sensitive agencies to be sure that employees are not talking about classified subjects on unsecured lines. Commercial employees who deal with the public are generally subject to having their calls recorded for "quality-control purposes." To listen to the call that private-sector employees make from their desks at work, however, employers generally need a reason.

The same cannot be said of more modern forms of communication; email and Web browsing, which can be freely intercepted, inspected, and censored in the ordinary course of doing business. The justification offered has some appeal. Indiscreet telephone conversations can reveal critical tidbits but cannot convey programs or large org charts or chip masks. It is hard not to see some legitimacy in an employer's desire to insure that channels that are capable of carrying such information are only used for legitimate purposes. Employees may be unhappy if their employers treat them like children by limiting their web browsing but they are pleased when similar measures prevent their computers from being infected with viruses. Large organizations typically scan incoming mail for viruses and some also scan outgoing mail. In a sense, this is censorship but it certainly makes the online world more livable.

Cryptologie Après le Déluge

As is painfully well known, the terrorist attacks of 9/11/2001 reinvigorated the security and intelligence services of the whole world, and those
of the United States in particular. Curiously, they had no immediate effect
on the course of US cryptographic policy.

Adoption of AES and CNSS15

Proponents of cryptography worried that the government would renew
its attack on the field by claiming that it did not detect the hijacker's plot
because it had not been able to read the terrorists' encrypted messages.
Instead the response was surprisingly muted. On September 14, Senator
Judd Gregg proposed a ban on "unbreakable" cryptography (McAuliffe).
His proposal, however, found little support and was withdrawn in mid
October (McCullagh 2001b).

That the attacks might delay or derail final approval of the new Advanced Encryption Standard—whose adoption had informally been announced for September—still seemed possible. On November 26, 2001,
however, the Secretary of Commerce signed and AES was adopted as
Federal Information Processing Standard 197.

The Pentagon's approval of AES for protection of national security systems and NSA development of Suite B, both described in the last chapter,
followed. In the US, the struggle over privacy of communications versus
surveillance was moving in a new direction.

Things were not quite the same in Britain, which was adopting a course
intermediate between the old US policy of discouraging the free use of
cryptography and the current one of promoting it. As its name suggests,
the Regulation of Investigatory Powers Act[1] expands the snooping powers of British police and intelligence organizations. One of the more controversial provisions of RIPA is a requirement that users must disclose the
keys to encrypted data in response to investigators' demands. Although
RIPA was passed in 2000, implementation of regulations requiring key
disclosure did not begin until 2006. At the time of writing the consultation process has not been completed and no regulations have been
implemented. There is therefore no experience with how such regulations
work, whether it would be feasible to refute claims of forgotten keys,

how the law would apply to ephemeral keying, etc. It is equally unclear how the legal concept of key disclosure, even if it is successful in Britain, would fare under US law.

Expansion of Intelligence

After 9/11, as after Pearl Harbor, there was a universal feeling of "How could be have been caught by surprise this way?" The fall guy was the intelligence community. Why hadn't they detected the plot? The obvious question "Why should you expect to be able to detect every activity involving two dozen people and half a million dollars over two years?" was obscured by the fact that we seem almost to have discovered the plot in several places. If only intelligence and law enforcement had been more vigilant, perhaps the attacks could have been prevented.

The country's response was to retarget intelligence—which for decades had been shaped by the objective of watching the Soviet Union and its allies—toward attempting to watch the far more elusive target of "worldwide terrorism." In particular, signals intelligence expanded and shifted its focus: less effort listening to dedicated military communications; more effort intercepting commercial channels accessible to small less-well-funded organizations.

FISA taps, the foreign intelligence taps named after the legislation that authorized their use,[2] became a subject of public interest after the attack. How had the United States missed the plot? What had prevented the CIA from telling the FBI about Khaled al-Mihdhar and Nawaf al-Hazmi,[3] who were in the United States in the months before September 11, 2001? (They clearly weren't here to "visit Disneyland" (Wright 2006, p. 354).) Why hadn't the US government uncovered the plotters' plans? Should there be changes to wiretap law? In the aftermath of 9/11, the Department of Justice made a number of legislative proposals; an important one concerned the relationship between FISA and Title III wiretaps. Instead of foreign intelligence being the *primary* reason for a FISA wiretap, the USA PATRIOT Act,[4] the law hurriedly enacted after the 9/11 attacks—and which will be discussed in some detail shortly—modified FISA taps so that foreign intelligence had only to be a *significant* reason[5] for the taps.

The change was only a single word, but it was a very important single word. In issuing a wiretap warrant for a criminal investigation, the 1968 law, sometimes referred to as "Title III," requires probable cause that the individual named in the order is committing or is about to commit an indictable offense. FISA is much less stringent, stating simply that there is probable cause to believe the individual is an agent of a foreign power. Wiretapping, in which no notice is given until after the tapping is over (in the case of FISA, notice may never be given) and which may go on for months, is particularly insidious, and the less exacting requirements for FISA—establishing that the suspect is an agent of a foreign power rather than someone committing or about to commit an indictable offense— was deemed appropriate because foreign intelligence warrants are for information collection, not for criminal prosecution.

There was an escape clause: if during a foreign intelligence investigation, facts emerged indicating a federal crime had been, or was about to be committed, intelligence officers were to inform criminal investigators, who would begin an investigation. Such a policy had existed through several presidential administrations, and the policy was explicitly laid out in a memo from President Clinton's Attorney General, Janet Reno, a memo that was later endorsed by the incoming Bush administration in 2001 (Thompson 2001). The policy had explicit procedures for informing the FBI Criminal Division if a FISA investigation exhibited criminal aspects; it also made clear that the criminal division was not to "run" the FISA investigation (Reno 1995). Then 9/11 occurred and, with it, the Moussaoui case.[6] In such situations, it is often easier to act than to analyze. The change in FISA—along with the rest of the USA PATRIOT Act—was proposed and enacted within six weeks of the 9/11 attacks.

It was difficult to know if the modification was actually needed. In contrast with Title III wiretaps, of which there is a public account,[7] only the number of FISA orders (furnished annually to Congress) is disclosed, except in the rare case when the information gathered from FISA surveillance is used in a public trial.

In 2002, the Attorney General proposed a new set of procedures for FISA cases that would simplify the issue of the walls. The FISA Court was not pleased with these. Earlier the Justice Department had informed the court of mistakes in FISA applications. The court's opinion took these problems into account and was quite critical of the proposal (USFISC

2002a). Senate Judiciary Committee members Leahy, Specter, and Grassley requested that the FISA Court provide the Senate oversight committees with copies of this opinion; not only did the FISA Court decide to do so, but it went one step further and made the opinion public as well (Kollar-Kotelly 2002).

The FISA Court criticized FBI mishandling of the wall in foreign intelligence cases and criminal investigations:

- Information on FISA investigations had been shared with criminal investigators in the New York FBI field office without consultation of the FISA Court, as was required. (USFISC 2002a, p. 17)

- More than 75 FISA applications related to major terrorist attacks directed against the United States had misstatements or omissions of material facts including: an erroneous statement that a FISA target was not under criminal investigation; false statements concealing overlapping intelligence and criminal investigations, and unauthorized sharing of FISA information with FBI criminal investigators and assistant US attorneys; omission in FBI affidavits of a previous relationship between the FBI and a FISA target. (USFISC 2002a, p. 17)

- In another FISA case, where there was supposed to be a wall between the intelligence and criminal investigations, there was no separation. Instead the case was run by a single FBI squad and all screening was done by a single supervisor overseeing both investigations. (ibid., p. 17)

The Court went on to say that "in virtually every instance, the government's misstatements and omissions in FISA applications and violations of the Court's orders involved information sharing and unauthorized disseminations to criminal investigators and prosecutors. These incidents have been under investigation by the FBI's and the Justice Department's Offices of Professional Responsibility for more than one year to determine how the violations occurred in the field offices, and how the misinformation found its way into the FISA applications and remained uncorrected for more than one year despite procedures to verify the accuracy of FISA pleadings." (ibid., p. 18)

The FISA Court announced that it would no longer accept inaccurate FBI affidavits regardless of what caused the inaccuracy (ibid., p. 17). It barred an unnamed FBI agent from signing affidavits for the FISA Court (ibid., p. 17). And, for the period from March 2000, when the problem first came to light, until September 15, 2001, the court—in a manner more reminiscent of scolding an errant child than instructing federal prosecutors—required all Department of Justice personnel receiving certain FISA information to certify that they understood the wall procedures used to separate FISA investigations from criminal prosecutions (ibid., p. 18). Stewart Baker, former chief counsel of NSA and a man who might be expected to be on the side of federal investigators, deemed the FISA Court report a "a public rebuke" (Eggen and Schmidt 2002).

There was more. With respect to one particular FISA application, the court approved the wiretap order but ordered that

> law enforcement officials shall not make recommendations to intelligence officials concerning the initiation, operation, continuation or expansion of FISA searches or surveillances. Additionally, the FBI and the Criminal Division [of the Department of Justice] shall ensure that law enforcement officials do not direct or control the use of the FISA procedures to enhance criminal prosecution, and that advice intended to preserve the option of a criminal prosecution does not inadvertently result in the Criminal Division's directing or controlling the investigation using FISA searches and surveillances toward law enforcement objectives. (USFISC 2002b, p. 3)[8]

In order to handle the issue of the wall, the FISA Court used a chaperone as the protector of liberty: the Office of Intelligence Policy and Review. OIPR, an office of the Department of Justice responsible for advising the Attorney General on intelligence matters, was to 'be invited' to meetings between the FBI and the Criminal Division to prevent any further diversion of FISA searches and surveillances towards law enforcement objectives (ibid., pp. 12–13).

For the first time since the 1978 enactment of FISA, the US government appealed a FISA Court decision. The United States Foreign Intelligence Surveillance Court of Review convened, and decided in favor of the administration: the FISA Court had "misinterpreted and misapplied minimization procedures it was entitled to impose" (USFISCR) and had not properly interpreted the PATRIOT Act's change to FISA. The Review Court concluded that the government is not obligated to prove to the

FISA Court that the primary purpose in conducting electronic surveillance in a particular investigation is not criminal prosecution (USFISCR). The Review Court examined the reasonableness of FISA searches under the Fourth Amendment; it did not explicitly rule on the issue, but it did state "we think the procedures and government showings required under FISA, if they do not meet the minimum Fourth Amendment warrant standards, certainly come close."

Despite the Court of Review's failure to back the lower court's actions, there were other voices saying that bringing down the wall had gone too far. As Senators Leahy, Grassley, and Specter reported, in regard to FISA implementation failures (Leahy et al. 2003), there were a surprising number of serious problems: FBI headquarters did not properly support the FBI field offices in foreign intelligence issues and key agents were inadequately trained both in FISA *and* in aspects of criminal law. While secrecy about actual FISA cases is appropriate, secrecy within the FBI about FISA policies and procedures is not. As a result, the Bureau was hamstrung; the FBI's inability to analyze and disseminate intelligence information—"[T]he FBI did not know what it knew"—undermined the bureau's ability to do its job. The report also observed that these breakdowns resulted in a mishandling of the Moussaoui FISA application (ibid.). It looked as if the FBI, rather than the wall, was the real barrier hampering investigations.

The number of FISA taps continues to rise. From an average of about 500 in the 1990s, after 2001, the number of FISA wiretaps steadily increased. In the most-recently-reported year, 2005, there were 2072 FISA applications, all of which were approved by the FISA Court (which did make "substantive modifications" to 61 of these (Moschella 2006). The number of emergency FISA orders, which allow surveillance to be initiated *before* a court order has been approved and give the government seventy-two hours thereafter to obtain the order, was also substantially higher. Attorney General John Ashcroft testified that in 2002 he had signed 170 emergency FISA warrants. That number was more than three times the number of emergency FISA orders issued in the preceding 23 years (Ashcroft 2003).

The senators who had agreed to the PATRIOT Act change to FISA did not approve of the way this change had manifested itself. In hearings in 2002, Judiciary Committee Chairman Senator Leahy reminded the administration that "it was not the intent of [PATRIOT Act] amendments to fundamentally change FISA from a foreign intelligence tool into a criminal law enforcement tool." (USSH 107 *USA PATRIOT ACT in Practice*, p. 4) Senator Arlen Specter, the ranking Republican of the Judiciary Committee, complained: "When the purpose of the FISA Act was foreign intelligence and the court interpreted 'purpose' as 'primary purpose,' the change was made to 'significant purpose.' But then the Department of Justice came in with its regulation and said that since the PATRIOT Act said a significant purpose was foreign intelligence, then the primary purpose must be law enforcement—which is just, simply stated, ridiculous. The word 'significant' was added to make it a little easier for law enforcement to have access to FISA material, but not to make law enforcement the primary purpose." (ibid., p. 33) Leahy warned: "The Department is urging broader use of the FISA in criminal cases. And you are going to lose, ultimately lose public confidence both in the Department and in the courts, unless you can, by public reporting or otherwise show this is being used appropriately." (ibid., p. 37)

The USA PATRIOT Act

The Title III wiretap law was several years in the making and evolved through various studies and countless hearings. By contrast, the Anti-Terrorism Act of 2001, almost all of which became the Uniting and Strengthening America by Providing Appropriate Tools Required to Intercept and Obstruct Terrorism Act of 2001—the USA PATRIOT Act—was brought to Congress on September 19, 2001, just eight days after the terrorist attacks of 9/11 (O'Harrow 2005, p. 23). The PATRIOT Act was wideranging and represented a "shopping list" for the Department of Justice. As later studies showed, it was not the lack of tools that prevented US intelligence from finding the hijackers before September 11, but rather a problem of coordination between agencies (and within agencies). But in the fall of 2001 the atmosphere in the United States was tense and highly fearful and despite strong efforts by civil libertarians, the PATRIOT Act did not receive much public scrutiny. The act passed a little less than five weeks later, in the midst of the anthrax attacks.[9] In discussing the PATRIOT Act, our focus will be on those sections of the law related to wiretapping and electronic surveillance.

From a wiretapping perspective, the most significant aspect of the PATRIOT Act was the change we have already examined, namely the §220 shift in purpose in FISA wiretaps from foreign intelligence necessarily being a "primary" reason of the wiretap order to merely being a "significant" reason.[10] The PATRIOT Act wiretapping and electronic surveillance provisions include:

- Expansion of Title III list of serious crimes predicating a wiretap investigation:

  §201 added terrorism and production or dissemination of chemical weapons to the list of serious crimes under Title III. Since it was already possible to obtain a FISA warrant to investigate these crimes, the purpose of §201 was to extend the capability of investigating a US person suspected of domestic terrorism activities.

  §202 added "felony violation of section 1030 (relating to computer fraud and abuse)" to the list of serious crimes under Title III.[11]

- Ability of law enforcement to share electronic surveillance information with national-security officials:

  §203(b) permitted law-enforcement officials to share information obtained from a wiretap or other forms of electronic surveillance with "any other Federal law enforcement, intelligence, protective, immigration, national defense, or national security official."

- Emergency disclosure of electronic communications:

  §212 allowed ISPs to voluntarily release subscriber content and records to the government if there is reason to believe that there is an immediate danger of death or serious injury.

- Changes in laws governing pen register and trap-and-trace devices:

  §214 removed the requirement under FISA that the government prove the target is "an agent of a foreign power" before the court would approve the installation of a pen register or trap-and-trace device. This section included a provision prohibiting use of FISA pen register surveillance against a United States citizen where the investigation is conducted solely on the basis of protected First Amendment activities.

§216 modified the definition of these tools to make it clear that they applied to the Internet; amended the definition of the tools to include a prohibition on content collection; required that records of pen register and trap-and-trace devices must be provided under seal to the court within thirty days of installation.

- Single Application for Nationwide Wiretap and Surveillance Orders

§220 expanded the authority of a court-authorized wiretap order or pen register/trap-and-trace device order to apply nationally. Previously the orders had held only in the jurisdiction of the court, thus forcing law enforcement to apply in multiple jurisdictions for what was essentially a single wiretap or pen register/trap-and-trace device order.

- Roving Wiretaps:

§206 expanded FISA authority to include roving wiretaps (previously only Title III wiretaps could be roving).

- Other Issues:

§207 extended the duration of a FISA wiretap for non US persons who were agents of a foreign power to ninety days unless otherwise specified.

§210 expanded law enforcement's ability to gather information through subpoena. Previously law enforcement could obtain the "name, address, local and long distance telephone toll billing records, telephone number or other subscriber number or identity, and length of service or a subscriber to or customer of such service and the type of services the subscriber or customer utilized" from an ISP. §210 requires the service provider to disclose records of session times and duration; any temporarily assigned network address; and any means or source of payment.

§211 amended Title III to apply to cable operators providing telephone and Internet services.

§217 permitted warrantless government interception of the communications of a computer trespasser if the owner or operator of a "protected" computer authorizes the interception. "Protected" computers include any "used in interstate or foreign commerce or communication."

§223 permitted individuals to sue the government for unauthorized disclosure of surveillance information.

§225 eliminated civil liability for a carrier complying with a FISA wiretap or emergency order.

Certain clauses were due to "sunset" in 2005 unless explicitly extended by Congress. These included: §201 on adding terrorism to the list of serious crimes warranting a Title III wiretap search; §202 on similarly adding computer fraud to that list; §203 (b), on sharing criminal wiretap information with intelligence agencies; §206 FISA roving wiretaps; §207 extended the duration of FISA taps for non-US persons; §214 lowered standards for FISA pen registers and trap-and-trace devices; §217 warrantless interception of the communications of a computer trespasser; §218 the "significant purpose" provision, and §220 nationwide authorization for search warrants for electronic evidence. In the fall of 2005, because of civil-liberties concerns there was much dispute about extending these provisions and the PATRIOT Act was only temporarily extended.[12]

In the end, all the provisions were extended with modifications made to §206. The changes to §206 included a requirement that the wiretap order describe the *specific* target of the surveillance if the target's identity is not known,[13] that the FISA court must determine that the roving wiretap is needed based on specific facts about the target included in the application for the wiretap, and that whenever a new location is tapped under the order, the government must notify the FISA court within ten days[14] of[15] (i) the location of each new facility at which the surveillance is taking place, (ii) why these location changes occurred, (iii) a description of the minimization procedures being used if the minimization procedures are different from those in the original order, and (iv) the total number of surveillances being conducted (Yeh and Doyle 2006, pp. 17–18).

One noteworthy feature of the PATRIOT Act strikes at legal doctrines far older than electronic surveillance. The US took from British law the notion of *knock and announce*: police searching premises would at least attempt to inform the occupants at the time the warrant was being executed. The doctrine was gradually eroded by court tolerance of the need to make secret entries to install bugs. In the investigation of the spy Aldrich Ames Ames, however, the FISA court went a step farther and authorized a secret search. This might have derailed the Ames case, but Ames was persuaded to plead guilty by promises that if he did his wife would be treated more leniently. Because the Constitution does not address the issue of whether police must announce themselves for a search to be reasonable, this appears to be a matter of law rather than a constitutional issue. In 1994 FISA was amended to allow such "sneak and peek" searches in intelligence investigations, including cases of international terrorism (before that, the Attorney General had authorized such searches without judicial oversight). Section 213 of the PATRIOT Act authorizes the use of these delayed-notice searches, in which the target is told of the search but *after* (possibly quite some time after) the search has occurred, for any case in which providing notice might have an adverse effect on an investigation or unduly delay a trial. The Department of Justice has said that such searches have occurred in non-terrorism cases.

NIST's Computer Security Division

GISRA and FISMA

The White House Office of Management and Budget, which had been disturbed about the poor state of computer security in federal civilian agencies, thought the Congressional attention focused on Y2K[16] and encryption policy might lead to willingness for further action, such as on computer security in federal civilian agencies. This was the genesis of the Government Information Security Reform Act (GISRA),[17] which required agencies to do internal risk assessments and submit those results to the Office of Management and Budget, which followed up with reports to Congress.

GISRA applied for only a year; its provisions were reauthorized and strengthened by Title III of the E-Government Act of 2002,[18] the Federal Information Security Management Act (FISMA).[19]

FISMA included an enforcement provision; agency budgets were in danger if the agency did not comply with FISMA. Congress strengthened enforcement in another way. The Computer Security Act of 1987 permitted agencies to obtain waivers from NIST's cybersecurity recommendations; under FISMA, the Secretary of Commerce had authority to make NIST information system standards and guidelines mandatory. NIST's Computer Security Division now had an enforceable role in securing the government's non national-security systems.

CSD Stays at NIST

In January 2000 the White House announced the change in encryption export controls, and despite Senator Gregg's call, in the wake of 9/11, for encryption controls, the government's policy did not waver. Then an odd thing happened. In the summer of 2002, the White House announced its support for the creation of the Department of Homeland Security, a consolidation it had previously opposed. When the draft bill surfaced, all the usual suspects—the Coast Guard, the Federal Emergency Management Agency, the Transportation Security Administration—were in the new department. There was also a surprise: NIST's Computer Security Division, the group that provided cryptographic standards[20]

for US government civilian agencies, was also slated to be moved.

The rationale for the proposed move was that since protecting critical infrastructure properly belonged in the new department, so did the Computer Security Division (CSD). US industry did not like the idea. With the CSD part of the Department of Commerce, business and industry had input into the CSD standards development process. Indeed, the Advanced Encryption Standard development was a clear demonstration of how well government and industry were working together, to everyone's benefit. Moving CSD to Homeland Security would place the division in a department more focused with law-enforcement concerns than on commerce, and would likely reignite the crypto wars—or worse. There was opposition from industry and civil-liberties groups to the proposed move, and NIST's Computer Security Division ended up staying right where it had been. Standards for protecting critical infrastructure would be developed in the Department of Homeland Security, with help and coordination from CSD, as needed.

Funding, however, was a concern. CSD had been chronically under-funded from the beginning. One bright spot of the proposed move had been the potential for appropriate levels of funding. Various groups, from industry lobbyists to the Information Security and Privacy Advisory Board, a federal advisory committee, swung into action. They lobbied Congress, describing the broad value of the Computer Security Division's work to the federal government *and* to US industry,[21] and they cited the increased role of the Computer Security Division under FISMA. At a time when science agencies saw level or even reduced funding, the Computer Security Division's support doubled (from approximately $10 million to $20 million),[22] a more reasonable sum—though still small—given the enormity of the job.

Data Retention and Data Mining

In 1900 there were some 1.5 billion people in the world and no database of 1.5 billion items. A century later, there are 6 billion people, and a 60-gigabyte disk—a disk roughly big enough to store everyone's name —costs less than $100. Ten times that much storage—enough to store everyone's name and address and maybe a bit more—costs about $500. Within a few years, the storage needed to store a short biography (or dossier, if you prefer) of every person on Earth will be within the reach of many of those people.

There is probably as yet no list of everyone's name let alone a single collection containing everyone's biography but the implications for privacy are clear. The most obvious barrier to keeping records on everyone —the inability to manage the database—is falling, thereby preparing the way for compiling such a database.

The falling cost of storage has other implications. For over a decade now, video cameras have been proliferating in and beyond the industrialized world. Because videotape is cheap, there is no need to reuse; many of these cameras are always recording new tape and contributing to an ever-growing archive. Other sources of raw data are recordings of intercepted signals, billing records for communications and many other kinds of transactions, including such information as the movements of drivers who pay their tolls with "EZPass" devices.[23]

The information in these various existing repositories is not always very useful. The databases themselves are scattered and the information they contain is not always readily linked to human identities. Bit by bit, however, the technology to extract useful information form a welter of low-grade information is being developed and the extraction of information is following a path similar to that of gold, oil, and other valuable minerals. At one time only the richest gold mines and the richest oil fields were worth developing. As time passed, gold and oil grew more expensive and the means to extract them grew cheaper, so the gold diggers and oil drillers began to go after lower grade ore.

The analogy is lost on no one and the technology for extracting valuable information from low-value inputs is called *datamining*. The subject presents many difficult problems, most notably transcribing speech and recognizing people, but it is making rapid progress.

For information to be processed, of course, it must be available to the people who have the means and the desire to process it. Often valuable information is collected and soon disposed of either because it is no longer needed or as an explicit privacy protection measure. National police and intelligence organizations have sought to counter this problem by pushing for laws requiring *data retention*, the intentional storage of data beyond the time it would be needed for the purpose for which it was originally collected. The explicit intent of such laws is to make the information available for criminal investigations that might take years to develop but it also has another effect. The longer data are retained—particularly when they are retained under threat of serious penalties for their loss—the more likely they are to proliferate and to live beyond their intended lifetime.

The events of 9/11 derailed various efforts heading for increased citizen privacy and even turned some completely around. One such, previously described, was Carnivore. Another was data retention, the issue of to what extent communications carriers should routinely archive information about users' telephone calls, emails, and other communications. In this case, the first action was in Europe—or so it appeared.

In 2000 the European Commission issued a draft privacy proposal that included new protections for electronic communications. The European Council of Ministers, the EU's main decision-making body, did not oppose the effort but sought to include data-retention requirements in the proposal. In July 2001, the European Parliament's Civil Liberties Committee approved a draft directive stating support of "strict regulation of law enforcement authorities' access to personal data of citizens, such as communication traffic" (Lynch 2001). Then 9/11 occurred, and the UK and Dutch Members of the European Parliament strongly opposed the rules that had been drafted by the Civil Liberties Committee. So did someone else. In a letter from James Foster, deputy chief of the US mission to the European Union, the White House requested that the directive "permit the retention of critical data for a reasonable period" (Meller 2001). The United States had no such requirements.

Changes ensued in the European directive. Pressure from two Spanish MEPs and the European Council resulted in passage of a directive somewhat different from the original: "Member States, may ... adopt legislative measures providing for the retention of data for a limited period justified on the grounds [to justify national security (i.e., State security), defence, public security, and the prevention, investigation, and detection of criminal offences]" (European Union 2002, p. 34).

Implementation remained elusive, however. In the E.U., harmonization of such directives across the member states is critical, but in this case, the member states held sharply differing views on the privacy protections needed. Here is where European and American viewpoints sharply differ. European law requires *proportionality*: "proportionality of the measure in relation to costs, privacy (data protection), and efficacy" (Council of European Union 2005, p. 1). Thus a proposal for three-year data retention made by Denmark, Ireland, Sweden, and the United Kingdom was rejected (European Parliament 2005). The issue went back and forth and finally, in December 2005, the European Parliament passed a directive requiring telephone companies and ISPs to retain traffic data on *all* messages and phone calls for between 6 and 24 months (Best 2005). Data was to be kept *even* on unanswered calls. With this directive in place, member states were now free to implement the directive.

So were some non-member states. There had been public silence on this issue in the United States. But after the passage of the European directive, in the spring of 2006, Attorney General Alberto Gonzalez called on Congress to pass data-retention laws to combat child pornography. Several bills are in preparation.

CALEA Revisited

By 2003 the main issues, at least on the legal front, in applying CALEA to digital telephony had been resolved and the FBI turned its attention to a new issue: wiretapping VoIP (Bellovin 2006a). In a letter to the FCC in November 2003, the FBI gave notice about its growing concern (Milonovich 2003). Four months later the FBI submitted a formal petition asking the commission to clarify which VoIP services were subject to CALEA (FBI 2003). There was surprise in some circles: after all, during his 1994 testimony, FBI Director Freeh had made clear that the proposed law was limited to the telephone network.[24] As a result of complex negotiations, CALEA explicitly stated:

(8) The term 'telecommunications carrier'—

(C) does not include—

(i) persons or entities insofar as they are engaged in providing information services; and

(ii) any class or category of telecommunications carriers that the Commission exempts by rule after consultation with the Attorney General.[25]

and that the capability requirements for wiretapping did not apply to information services.[26]

CALEA applied to VoIP presents a number of complexities, complexities to which it appeared the FBI paid little attention. The real issue was not about wiretapping *per se*—there was no disagreement about the applicability of wiretap law to Internet communications—but about applying CALEA, and FBI design standards, to the Internet. The Public Switched Telephone Network and the Internet are two distinct communication networks. They may rely upon the same underlying transmission facilities and even share the same cables; both may use electronic routing and switching devices at central nodes to move bits efficiently through the network from one user to another; and both may use digital transmission and some form of time-division multiplexing (Bellovin 2006a, p. 9). But circuit-switched networks and packet-routed networks are very different and no amount of calling both "communications networks" can obliterate the differences: techniques that work for wiretapping in one are, in many cases, simply not feasible in the other. Although opponents of the FBI request believed that CALEA's lack of applicability to Internet applications was clear, not everyone agreed.

The FCC, in particular, sided with the FBI. In the summer of 2005, it announced that CALEA applied to two types of VoIP service: providers offering transmission or switching capabilities *on their own lines* between the end user and the Internet (facilities-based) and providers offering service that enabled the connection between an end user on the telephone network and VoIP.[27] The American Council on Education, whose members were concerned about the cost of applying CALEA to their internal networks, various civil-liberties groups, and the computer and telecommunications industry pressed the FCC for a stay; when that failed, they turned to the courts. In June 2006, somewhat to their surprise,[28] in a two-to-one decision, the US Court of Appeals agreed with the FCC and the FBI *American Council on Education, Petitioner v. Federal Communications Commission and United States of America*, No. 05-1404 et al. (D.C. App. June 9, 2006).[29]

These two particular types of VoIP services have architectures that fundamentally resemble the telephone network and thus their accommodation of CALEA is not particularly difficult.[30] That is not true for other VoIP services. VoIP, like much of Internet communication, is about mobility and mobility does not come for free. In some cases—intercept against a call from a fixed location with a fixed internet address connecting directly to a big Internet provider's access router—VoIP is equivalent to a normal phone call, and the interception does not present a technical challenge (Bellovin 2006a, p. 2). But if *any* of these conditions is not met—if the VoIP call is at all mobile—then the problem of assuring interception becomes enormously harder (ibid).

Before CALEA came into the picture, wiretapping was done somewhere along the *local loop*, the pair of wires running from the local telephone switch to the subscriber's phone. The local-loop wiretap receives all the information that travels down those wires, but it does *not* capture information, such as forwarded calls, that are diverted at the switch and do not travel on the local loop. The FBI sought CALEA to ensure that telephone standards would be designed to eliminate such problems, thus enabling full legally-authorized wiretapping. As discussed in chapter 5, the solution is to make the wiretap a silent participant in a conference call. Then all the information available to the local switch—call-forwarding, speed call lists, caller identities—is also available to the wiretap.

Internet users rarely know the IP addresses of the people with whom they communicate. In the old days computers were fixed and so were their IP addresses. Now a user may connect from an Internet cafe at 10, a conference room at noon, and airport lounge at 3, and each of these will have its own IP address—usually more than one. In the current scheme of things, Internet addresses are, more often than not, allocated dynamically, which means that the address the laptop had on Monday at 10 A.M. at the Cozy Corner Cafe is likely to be different from the one it acquires there on Tuesday. Thus the first step of a VoIP service is to take a familiar identifier—a user name, a telephone number, an email address—and transform it into a specific IP address where the user can currently be found. This is called a *rendezvous* service.

Once the association between name and current IP address has been established, the actual voice call can travel in myriad ways. Consider the VoIP network shown in figure 11.1. Alice and Bob are both currently connected via the ISP C using router R1 and ISP D using router R2, respectively. Alice, however, uses VoIP Provider 1, a customer of ISP A, while Bob gets his service from VoIP Provider 2, a customer of ISP B. Both Alice and Bob are traveling and thus are in varying locations; they connect via different ISPs without changing their VoIP providers.[31]

When Alice calls Bob, her VoIP phone sends a message across the Internet to her VoIP provider, which contacts Bob's VoIP provider, and Bob's VoIP provider in turn notifies Bob. (The flow of the call setup messages is shown via dashed lines.) The actual data flow of the phone conversation can be completely different however; there is *no* requirement that the call

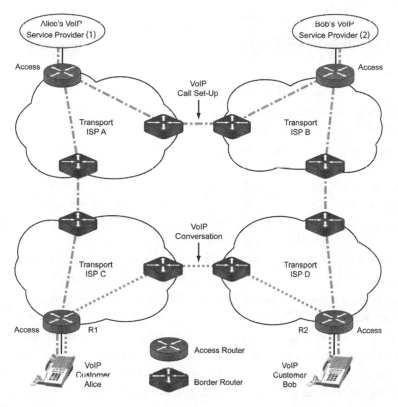

Figure 11.1
Alice and Bob talking via VoIP with multiple providers.
(Illustration by Steve Bellovin and John Treichler.)

go through Alice or Bob's VoIP providers (the call is shown by the two dotted lines).

Suppose we are trying to wiretap Alice's calls to Bob. The obvious points from which to do the tapping are access routers R1 and R2 (these are the Internet analogy—to the extent there is one—of the local telephone switches). However, neither router knows who Alice or Bob are or, for that matter, any information analogous to a telephone number that has a long term connection with Alice or Bob. This is something in the domain of the two VoIP providers. For the tap to succeed, R1 or R2 would have to receive a "start recording" instruction from one of these VoIP providers. But here is where the situation becomes complicated. The VoIP providers can be located at arbitrary places on the Internet, and they need have no business or technical relationship to any ISP other than their own. In fact, they could easily be located in and owned by foreign (and even hostile) countries. How can Alice's ISP trust such a wiretap request?[32]

If Alice's VoIP Provider is owned by her ISP (that is, ISPs A and C are one and the same), the issue is simpler, and for sure, many broadband ISPs do own VoIP operations. This, however, is not required nor even expected to be the norm. Skype, for example, is a non-US company, and is not associated with any ISP. The disassociation of the VoIP provider from the ISP combined with the mobility of the VoIP user makes CALEA applied to VoIP exceedingly complex. As things stand, investigations targeted against people who are constantly on the move are likely either to fail or to violate the privacy of innocent bystanders.

The old models of communications surveillance do not translate easily to the dynamic, packet-routed, signaling-and-content-combined network architecture that forms the Internet. The Internet's flexibility enables a wide variety of VoIP implementations; some, like those picked off by the current FCC order, simply connect an end user on a telephone to a VoIP network and those can relatively easily accommodate CALEA. Others are genuine peer-to-peer applications. Because of the packet-routing nature of the Internet, the path of a VoIP conversation cannot be predicted—and that seriously complicates the placing of the wiretap. It can't be put on the client device, for danger of being discovered and thwarted. But for the wiretap to be placed anywhere else, there must be some way to determine the route the communication will take, and that is not plausible.

Mobility further exacerbates the wiretapping problem. Not only can't you predict which route an Internet communication will take but because of mobility it is not always simple to tell whose packets are the ones of interest, which presents serious problems for minimization. The ease with which new identities can be created on the Internet, a far simpler process than adding a new phone line, also complicates internet wiretapping.

Finally there is the problem of ensuring safe transport of the wiretap information to the law-enforcement facility, a problem far more complex in the packet-routed world than in its telephone counterpart.

A report examining the FCC order observed:

Building a comprehensive VoIP intercept capability into the Internet appears to require the cooperation of a very large portion of the routing infrastructure, and the fact that packets are carrying voice is largely irrelevant. Indeed, most of the provisions of the wiretap law do not distinguish among different types of electronic communications. Currently the FBI is focused on applying CALEA's design mandates to VoIP, but there is nothing in wiretapping law that would argue against the extension of intercept design mandates to all types of Internet communications. Indeed, the changes necessary to meet CALEA requirements for VoIP would likely have to be implemented in a way that covered all forms of Internet communication. (Bellovin 2006a, p. 13)

Such modifications to Internet protocols present a clear risk of introducing vulnerabilities into Internet communications. After all, wiretapping is a legally authorized security breach, and introducing a security breach into a communications network always entails serious risks. In 2000 the Network Working Group of the Internet Engineering Task Force (the IETF designs the protocols for the Internet) studied putting wiretap requirements into Internet protocols and concluded that it could not be done securely (IETF 2000). Their conclusion arises from fundamental engineering principles: complexity is the bane of security. Every function added to a secure program must be evaluated to be sure that it does not contain a vulnerability, when used either alone or in combination with other features (Landau 2006, p. 431).

The industry group examining the issue concluded that, "In order to extend authorized interception much beyond the easy scenario, it is necessary either to eliminate the flexibility that Internet communications allow, or else introduce serious security risks to domestic VoIP implementations. The former would have significant negative effects on US ability to innovate, while the latter is simply dangerous. The current FBI and FCC direction on CALEA applied to VoIP carries great risks." (Bellovin 2006b, p. 2)

Of course, VoIP is not society's only form of mobile telephone communication; cellular phones are also mobile. The mobile telephony of cellphones operates very differently from VoIP and makes accommodating wiretapping easier than for VoIP but not as easy as it is on the wireline system. When a cellphone is operating within its home cell, it behaves much like a wireline phone and wiretapping is simply done at the switch. When the cellphone is roaming outside its normal service area, the situation becomes more complex. At the time the roaming phone is initially turned on, and maybe every fifteen minutes after that, it sends a signaling message back to its home switch.[33] At this point only signaling information has been transferred to the home network. If the roaming cellphone is called, the incoming call passes through its home system during call setup and wiretapping will be initiated at this point. The situation is entirely different, however, when the roaming cellphone makes the call. Once the phone is *registered* with the local switch, if a call is made locally by the cell phone, there is *no* immediate notification about that call to the home switch (or the billing system) and the call is *not* routed through the home switch unless that happens to be the call's destination. This means that no wiretapping can be done on the incoming calls of roaming cellphones by their home switches. It would be possible to route calls artificially back through the target's home system and back again to facilitate wiretapping, but such routing might well be detected by the target due to changes in timing, voice quality, or billing.

There is also the issue of *roving wiretaps*: wiretaps in which, because law enforcement has reason to believe the suspect is using a variety of phones—either because the suspect is trying to avoid a wiretap or because her business is such that she naturally moves around—the telephone number to be tapped is left unspecified in the court order (Solove, pp. 325–326). Law enforcement can, with a single warrant, tap the suspect at Joe's Pizza at 10 and at the bank of phones by Gate 9 East in Penn Station at noon. Roving wiretaps thus appear to mimic VoIP taps. However, minimization, which requires that it is the suspect's call—*and only the suspect's call*—is wiretapped, plays a significant role in distinguishing the two situations. Even if the suspect frequents Joe's Pizza and the bank of telephones at Penn Station, law enforcement cannot have active wiretaps on all the phones at Joe's Pizza or all the telephones by Gate 9 East; rather the wiretap can only be activated when law enforcement knows the suspect is on a particular line. Thus physical surveillance plays a role in roving wiretaps for which there is no analogy in VoIP wiretapping.

The Appeals Court was asked to reconsider its decision in an *en banc* review,[34] but declined. Meanwhile, in a little-noticed policy statement issued the same day as the FCC statement on CALEA applicability, the FCC also announced:

> The Federal Communications Commission today adopted a policy statement that outlines four principles to encourage broadband deployment and pre-

serve and promote the open and interconnected nature of the public Internet: (1) consumers are entitled to access the lawful Internet content of their choice; (2) consumers are entitled to run applications and services of their choice, subject to the needs of law enforcement; (3) consumers are entitled to connect their choice of legal devices that do not harm the network; and (4) consumers are entitled to competition among network providers, application and service providers, and content providers. Although the Commission did not adopt rules in this regard, it will incorporate these principles into its ongoing policymaking activities. All of these principles are subject to reasonable network management. (FCC 2005b)

The meaning of this statement is not clear; indeed, it seems contradictory. "Encouraging broadband deployment" and "promoting an open and interconnected Internet," seem at odds with being able to "run applications and services of their choice," only "subject to the needs of law enforcement." These needs have not been defined, let alone legislated and the statement was, to many, a threatening stake in the ground. At the time of this writing, a year later, there has been no clarification of the policy statement.

The National Security Agency

Current NSA programs

The basic architecture of US surveillance law from the end of the 1970s on had been simple. Interception of communications outside the United States was governed by policies set by the executive branch, largely by the National Security Agency. These policies were subject to high-level review by the intelligence oversight committees of Congress and rarely came to the attention of the courts. Communications interception inside the United States was governed by either the Title III provisions for wiretaps in ordinary criminal cases or by the Foreign Intelligence Surveillance Act. In either case, warrants, specifying individual targets, were required either before or shortly after the start of interception.

After 9/11, NSA, presumably on the urging or outright orders of the White House, began to relax (or bend, depending on your viewpoint) its rules about communications that were in whole or in part US communications. The issues in question concern both communications that originate or terminate (or both) inside the US and communications to which some of the parties are *US persons* wherever they may be located.

It can be argued that the most dangerous of our terrorist enemies are not those abroad but those who have already entered the United States. These people, however, are part of foreign organizations and can thus be expected to "call home" from time to time. On this logic, a communication with one terminal in the US and the other at a known terrorist location outside would seem to be fair game.

The issues are: Who is the target (outside the US)? Where is the target? How is the traffic being collected? Where is it being collected? The Bush administration believed that FISA did not cover all the legitimate cases, but was unwilling to be public about the issue and thus did not seek administrative relief. Instead the administration chose a secretive and unsettling solution: authorizing warrantless wiretaps between targets abroad and those with whom theycommunicate inside the United States. Secrecy might have prevented the program from ever coming under judicial scrutiny but much to the administration's distress, the program was leaked to the *New York Times* (Risen and Lichtblau 2005), which published it in the fall of 2005.[35] Suit was brought against the NSA and its director by the ACLU and numerous co-plaintiffs. In August of 2006, the plaintiffs won their first battle. Judge Anna Diggs Taylor of the Eastern District of Michigan, Southern Division, ruled that the government's program was both a violation of the Foreign Intelligence Surveillance Act (which limits as well as enabling wiretapping) and of the constitution, which prohibits unreasonable searches and seizures (*American Civil Liberties Union et al. v. National Security Agency et al.* (United States District Court Eastern District of Michigan Southern Division, Case No. 06-CV-10204, Hon. Anna Diggs Taylor, *Memorandum Opinion*.)).

One category of communications seems to deserve special consideration. *Transit traffic* is traffic that enters the US only to go on through and come out the other side. Because the communication system of the US is so well developed and because the US is responsible for such a large fraction of the world's commercial activity, it is not surprising that a significant fraction of the communications from Europe to the Far East go via the United States.

How should we treat such communications? Our signals intelligence service is pleased to build antennas on US soil whenever this will give them satisfactory reception, without worrying that once the signal hits the antenna it is in the US and subject to the protection of US law. Are

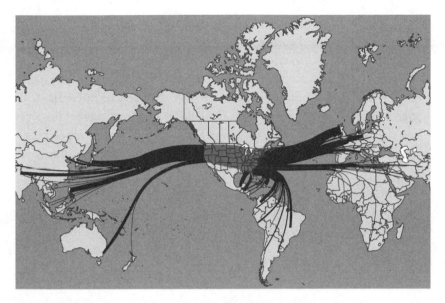

Figure 11.2
Worldwide telecommunications flows.
(Illustration by Nancy Snyder.)

signals that come in on wires different? The NSA did not think so and at some point began collecting the transit traffic.

Another area of contention is information *about* communications, what is called *call detail recording*. The US telecommunications companies have a long history of handing over communications and associated records to NSA and its ancestors. In 2006, a new manifestation of this practice surfaced; most of the country's major carriers were giving NSA the call detail recording (essentially billing) information on their customers, even for calls that originated and terminated entirely within the US. The action did not go unchallenged. The lawyers of the Electronic Frontier Foundation brought suit "on behalf of themselves and all others similarly situated" seeking declaratory and injunctive relief from the federal courts against AT&T (and some 20 "Does") (*Tash Hepting et al. v. AT&T Corporation et al.*, United States District Court for the Northern District of California, Case 3:06-cv-00672-vrw).

Although call detail recording falls far short of content interception for many purposes, for others it is more useful because it provides a window into the past. It parallels the electrical engineering practice of watching a signal with a delay line; triggering observation on the basis of something that happens later; and then looking back at the signal in the delay line to see what happened earlier.

Suppose that on March 1 there is a bombing. On March 2, a well-known radical makes a public statement on the subject, or calls a targeted suspect, or flies out of town. Investigators with access to the radical's old phone records now look to see whom she called in the previous week, two weeks, month, a year. These queries will lead to yet other queries about earlier calls, queries that take into account the location of the bombing, calls to travel agencies or airlines, calls to other people under investigation. It would of course be useful to have all of the communications of the people under investigation but, legal issues aside, it is generally infeasible to record such large volumes of traffic.[36]

Changing US Role in the World

The United States, with only a few percent of the world's population, has long been accustomed to being its dominant economic and military power and one of its major cultural powers. With the rising wealth, as well as population, of China and India, and the increasing unification of the European Union, all this is likely to change.

Despite the ambitious and surprisingly internationalist cryptographic standardization program being undertaken in the US the new generation of American-sponsored cryptographic systems may have less influence today than DES did a generation ago. Europe has undertaken a broad program to develop cryptographic tools to support its information security needs.[37] China has also undertaken the development of new families of cryptographic systems, but if these are public within China, which seems doubtful, they have not made it to the west.

Among the most profound changes that will affect privacy and security policy is the increasing virtualization of computers. At present, the nominal basis of computer usage is the personal desktop or laptop computer computer—a moderately large, moderately expensive and very capable device that can execute billions of instructions a second and store tens of gigabytes of data. The great growth areas in computing, however, are above and below this. On one hand enterprises like Google supply computing services to their customers from computing engines built out of the better part of a million rack-mounted PCs. On the other the tool that people use to access this resource is ever more likely to be a palmtop rather than laptop computer—something small enough to carry everywhere and just powerful enough to serve as your gateway to the Internet.

In a world of virtual computing neither individuals nor organizations will be able to have the sort of security they have been accustomed to in the past. When powerful computations can be done for you by specialized providers more satisfactorily and less expensively than you could do them yourself, use of such services will become a way of life. In order to use them, however, you must accept a breach of security comparable to using an unencrypted radio. If you ask for services, there is no way of concealing what you want done, just as there is no way of getting Google to search on your queries without revealing what your interests are.

The popularity of the MMORPGs points the way to another level of this development. If your major way of interacting with other people is though virtual environments, you will be able to augment yourself with capabilities undreamed of in the past. In a virtual world no one will know whether you are a dog with a prodigious memory or whether you are a dog that is really skillful at typing fast Google queries. In this world a matter that has concerned us since our first chapter comes to the fore. We noted that things that had been facts of life became subject to social policy. In the virtual world, this will be true of almost everything.

Yet What Exactly Is the Terrorist Threat?

Law-enforcement's view of what works in terrorist cases was summed up in 1991 by FBI Director William Sessions: "If a terrorist attack does occur, it is our view that a swift and effective investigation culminated by arrest, conviction and incarceration is a powerful deterrent to future acts of terrorism." (Sessions 1991, p. 72) Thus through the 1990s and into the current decade, the law-enforcement community has pushed for tools that enable investigations that end in "arrest, conviction, and incarceration."

Not everyone in law enforcement agrees with Sessions's approach. During hearings on the USA PATRIOT Act, Associate Deputy Attorney General David Kris said: "And just as a tactical matter, sometimes prosecution is not the right way to go. Other times you just want to monitor these people or do something else. You try to recruit one of them as a double agent. You feed them false information. You disrupt them using some other technique," even while, "In some cases you do want to prosecute." (Kris 2002, p. 28)[38]

The passage of 15 years reveals that Sessions's statement does not encompass the current terrorist threats. Terrorists, even those who do not sacrifice their lives in the act of terror as the 9/11 hijackers did, seem undeterred by the prospect of prosecution. Timothy McVeigh, convicted of the 1995 bombing of the Murrah Federal Building in Oklahoma City, accepted execution with an observation that might have been written by a spokesman for the US military. The body count was in his favor: he had killed 168; they would kill one.

In the disorienting days immediately after 9/11, it was easy to forget that such attacks had been anticipated. In addition to the now-famous memo of August 2001,[39] there were investigators in both the FBI and the CIA who had been tracking bin Laden and other terrorist groups. There were also various studies of the security of the United States in the post Cold War era. Perhaps the best known is the report of the United States Commission on National Security/21st Century, more commonly known as the Hart-Rudman report, after the two former senators who co-chaired the committee.[40] Seven months before the 9/11 attacks, the Hart-Rudman report observed that the US would become "increasingly vulnerable on the America homeland" and that the US military would not be fully able to stop the threats. They went on to say:

> We believe that American strategy must compose a balance between two key aims. The first is to reap the benefits of a more integrated world in order to expand freedom, security, and prosperity for Americans and for others. But second, American strategy must also strive to dampen the forces of global instability so that those benefits can endure and spread.
>
> ... the United States should ... promote pluralism, freedom of thought and speech, and individual liberty. Not only do such aims inhere in American principles, they are practical goals as well. There are no guarantees against violence and evil in the world. We believe, nonetheless, that the expansion of human rights and basic material well-being constitutes a sturdy bulwark against them. On the negative side, these goals require concerted protection against four related dangers: the proliferation of weapons of mass destruction; international terrorism; major interstate aggression; and the collapse of states into internal violence, with the associated regional destabilization that often accompanies it. (Hart et al. 2001, p. 5)

In considering the terrorist threat, the issue at hand is that wiretapping has non-monetary costs. Creation of a surveillance society will quickly be negatively perceived in some immigrant communities both because such societies are what they sought to escape in coming to the US and because they perceive themselves as the explicit targets of surveillance. The effective imposition of surveillance threatens innovation and security by building wiretapping capabilities directly into communications infrastructure. Such costs must be undertaken cautiously and require asking, "What is the value of wiretapping in terrorist investigations?"

The current most serious threat of terrorism, at least in terms of large-scale attacks, comes from violent Islamic fundamentalists, people who look at a war in terms not of years but of centuries (Fallows 2006, p. 60). Investigation of this threat will need the cooperation of domestic immigrant communities (Heymann 1998, pp. 101–102). The threat posed by sleeper cells is particularly serious and the need for community cooperation in uncovering them is acute. As London Police Commissioner Ian Blair put it, "The whole deal here is to engender the trust that one afternoon may allow one of those Islamic leaders to say to the sergeant, 'You know, I'm worried about young so-and-so.'" (Caldwell 2006b, p. 44). The investigation into the 2006 London airline threat began with just such a tip from "several people in Walthamshow," the East London home of some of those accused (Van Natta 2006, p. A8).

Experience with terrorist investigations in Israel and Northern Ireland shows that harsh techniques—massive searches and surveillance, ill-treatment and abuse of prisoners—often backfire (Heymann 1998, pp. 132, 141–142). In Northern Ireland the advantages gained through this type of policing appeared to have been "offset by the effect of stimulating IRA recruitment" (ibid., p. 126). In this light, the US government's early response to 9/11: singling out young male immigrants from Islamic nations and subjecting them to extensive questioning by the Immigration and Naturalization Service when there was no *a priori* suspicion of wrongdoing was counterproductive.[41]

Arab and Muslim immigrant populations in the United States have assimilated—second-generation American Muslims "are culturally and economically Americanized" (Fallows 2006, p. 65)—while European immigrants have not.[42] According to former of the National Security Council member Daniel Benjamin, American Muslims "have been our first line of defense" (ibid., p. 65) since 9/11. They are a resource to be treated with care.

The construction of a surveillance society, and particularly surveillance in Muslim communities, may well be counterproductive. Government policy can go two ways: it can work to create trust[43] or it can build a surveillance society that many in the Muslim community see, with some justification, as targeting them. As Phillip Heymann, former US Deputy Attorney General, has observed, the latter direction has very serious dangers: "In terms of national well-being, the gravest national dangers from a terrorist act (short of an immense escalation of terrorist tactics), are that the interplay of terrorism, public reaction, and governmental response may sharply separate one significant group from the rest of society." (Heymann 1998, p. 2) In such situations, Heymann notes, "the terrorists will find it far easier to secure communication channels, [etc.]" (ibid., p. 13). The lawyer of the LA 8, David Cole, described the FBI pursuit of the seven men and one woman as costing the US government heavily in ways the government could hardly afford to pay: "[T]he L.A. 8 case, seen in Arab-American communities as the prime example of US hostility toward Arab immigrants, has probably done more to undermine that effort than any case in the past 20 years....The vendetta against the L.A. 8 was a critical reason for the Arab community's deep distrust of the government even before 9/11." (Cole 2003)

Laws authorizing wiretapping in law-enforcement investigation were originally passed because of the threat of organized crime. Organized crime works through a small cadre of tightly linked participants —often family members. This makes the organization difficult to penetrate and complicates investigations. Since radical Islamic fundamentalist groups appear to pose similar investigative difficulties, wiretapping would appear to be a particularly tempting tool. Yet there are significant differences between investigating organized crime and violent religious fundamentalists, differences that change the value of wiretapping in investigations.

While the threat of prosecution is a deterrent to members of organized-crime groups, it is much less effective against those espousing violence as a way to achieve a fundamentalist society. Heymann has observed that law enforcement is not a deterrent to terrorists (1998, p. 79). Quite the contrary: violent Islamic fundamentalists often view jails as excellent recruiting grounds—including among the nationals in the country in which the terrorism is to take place.

Let us be clear: we are not arguing that wiretapping and signals intelligence are without value in terrorist investigations. They have proved their value time and again and the very threat of interception has severely impeded al-Qaeda operations. The organization has learned the danger of communicating electronically, and Osama bin Laden is said to rely on handwritten messages delivered by trusted couriers rather than use the telephone. This represents a serious meaconing of al-Queda's communications directly attributable to US intercept capabilities. Many terrorist communications, however, are sufficiently brief that they are difficult to decipher not because of they are electronically encrypted but because the communications are in "code" known to the insiders but not to the eavesdroppers.[44] Our concern is that in as much as wiretaps are not free, either in their impact on immigrant communities or their impact on network architecture, they are also not an investigative panacea.

Traffic flow information is often as valuable as content and it is much harder to conceal. Investigators have been quite successful in tracking terrorists without being able to learn the contents of their messages. In a 2002 case, investigators tracked al-Qaeda members through terrorists' use of prepaid Swisscom phonecards. These had been purchased in bulk —anonymously. But when investigators discovered through a wiretap on an intercepted call that "lasted less than a minute and involved not a single word of conversation" that they were on to an al-Qaeda group, the agents tracked the users of the bulk purchase cards (Van Natta 2004, p. A1). The result was the arrest of a number of operatives and the breakup of al-Qaeda cells.

This example illustrates what the intelligence community has known for years. In the age of electronic communications, analyzing the content of communications is a rich and fruitful investigative tool when you can get it but developments ranging from the low cost of optical fiber to the critical need to secure civilian infrastructure has made the contents of intercepted communications ever more frequently inaccessible.[45] Traffic analysis—who is communicating with whom, for how long, on what kind of channels, and in what volume—is the more fundamental tool. Traffic analysis reveals an organization's structure, its membership, even the roles of its members. Traffic analysis has another, extremely important benefit; it is a tool that aids investigators without requiring security breaches in on the civilian infrastructure.

It is important to apply common sense to the issue of terrorist investigations and to think clearly about which acts can be prevented and which cannot (Heymann 1998, pp. xxi–xxiii). Timothy McVeigh's attack on the federal office building in Oklahoma City was the work of a group of three people. The al-Qaeda attacks of September 11 involved the coordination of a group six times that size but still quite small. Unless the United States moves to a surveillance society on the scale of the former East Germany, the country will never be able to protect itself fully against attacks by "lone warriors" such as McVeigh. We need to factor such common sense into all of our thinking about security.

We also need to have clear understanding of how serious the threat of international terrorism to the domestic United States actually is. Despite government classification of much of the information, there is some data publicly available which can clarify some of these issues. In June 2006 the US Department of Justice released a *Counterterrorism Whitepaper* detailing investigative successes in terrorism cases (ibid. 2006). This report stated that in the period from 9/11 to June 2006, there were 441 defendants charged with terrorism or terrorism-related activity of an international 'nexus' (ibid., p. 13). But this data is not all that it seems. To be sure, the report included a number of serious cases: the indictment (on charges of providing material support to a terrorist organization) of four associates of Sheikh Omar Abdel-Rahman, who had himself been convicted in 1995 for his role in the first attempt to destroy the World Trade Center (ibid., p. 16); the conviction of Zacarias Moussaoui; the indictment and conviction of Richard Reid, the "shoe bomber."[46] But the DoJ report also highlighted a number of lesser cases, including the Florida case of Narseal Batiste and associates, accused of plotting to blow up the Sears Tower in Chicago (ibid., p. 64), but which is now considered a "pipe dream," a "quixotic" effort by the plotters, who were led by an FBI informer (Pincus 2006, p. A1). Other cases detailed in the report include that of a husband in possession of ricin, possibly for use in poisoning his wife because of her extramarital affair (USDoJ 2006, pp. 24–25), a case which began with the premise of foreknowledge of 9/11 by a group of stockbrokers and ended with a simple case of racketeering and securities fraud (ibid., p. 19), and various ones of much lesser import, such as visa violation and marriage fraud.

The DoJ report should be contrasted with the analysis of criminal terrorism enforcement by the Transactional Records Access Clearinghouse (TRAC) in September 2006 (TRAC 2006).[47] The TRAC analysis, based on information from the Department of Justice's Executive Office for United States Attorneys (EOUSA) compiled on a monthly basis, shows quite different results than the official *Counterterrorism Report*. From July 2001 through May 2006, the government prosecuted 335 people as "international terrorists" (ibid.). As we shall see in a moment, few of these cases were actually terrorist cases. In any case, this is a somewhat different number than the DoJ report, which had lumped together international terrorists, such as Abdel-Rahman, Moussaoui, and Reid, with domestic actors (violent anti-abortion activists, members of the Ku Klux Klan, right-wing militias, and the like). While the latter have been and remain a serious concern—until 9/11 the most serious recent[48] act of terrorism within the domestic United States was carried out entirely by domestic terrorists (Timothy McVeigh et al.)—the current area of concern is international terrorism. Because the TRAC numbers are based on EOUSA data that does not include names, it is difficult to directly reconcile the two reports.

The TRAC data is the hard numeric data compiled by US attorneys. In many ways, the difference between the DoJ and TRAC reports parallels the difference between 1994 Freeh testimony to Congress regarding the efficacy of wiretaps and the later analyses of the Wiretap Report produced by the Administrative Office of the US Courts. The former discussed the value of wiretapping in kidnapping cases[49] —a conclusion that a study of the Wiretap Report did not support[50]—and lumped together all electronic surveillance results. Since electronic surveillance includes bugging—and some of the most important convictions, including Gotti's and Stanfa's came about because of electronic bugs and *not* wiretaps— this imprecise information was far less useful—and far more politically motivated—than the results that came from a careful study of the Wiretap Reports.

TRAC's hard data had some very interesting results. There is a surge in federal prosecutions of international terrorism cases in the year immediately after 9/11, but a recent return to pre 9/11 levels. There were 213 convictions in international terrorism cases between July 2001 and May 2006. Of these, 123 individuals received prison sentences, but most convictions were not for serious crimes. Only fourteen individuals received sentences of five years or more, and only six of these received sentences of twenty years or more (TRAC 2006). From the vantage point of 9/11, the bombings in Bali, Madrid, and London, it is clear that there is a serious worldwide terrorism threat from violent Islamic fundamentalists; the data from the TRAC report puts that threat in a clearer perspective. As TRAC asks "Is it possible that the public understanding about the extent of this problem [of international terrorism] is in some ways inaccurate or exaggerated?"

Maybe the focus in communications should be elsewhere? We must be prepared for the possibility that we will not be able to prevent terrorist attacks from occurring in the United States and must, therefore, develop infrastructure to enable recovery—whether it is from natural disasters or man-made ones. In that, the lessons of both 9/11 and Hurricane Katrina demonstrate a clear need for interoperable and robust communications systems. The aim of avowed terrorists to cause the greatest possible disruption of society argues against creating centralized resources whose loss would be crippling. In the physical world this argues for distributed sources of energy, manufacturing, and food. In communications it argues for security in depth. A system in which there is no all pervasive mechanism that can provide or deny security at will, may give an opponent unintended access to communications and computing throughout a national network. These may be the concerns that caused the national-security agencies to support changes in standards and export control regulations to encourage widespread use of strong encryption.

12

Conclusion

Control of society is, in large part, control of communication. From the right to assemble enumerated in the US Constitution to the anti-trust laws prohibiting competitors from agreeing on prices there is a tension between the right to communicate and limitations on communication. As society evolves, particularly as technology evolves, the government's power to control communications changes.

Telecommunication, barely a century and a half old, has so transformed society that, for most people in industrialized countries, it is a necessity, not an option. People move thousands of miles from friends and family, knowing that they can keep in touch by phone and email. People telecommute to work or, having commuted to the office, spend the day doing their work via telephone, email, and Web. People order goods from dealers on the other side of the continent by dialing 800 numbers or opening web pages. For a remarkable range and an increasing number of activities, telecommunication stands on an equal footing with physical communication.

Side by side with the growth of telecommunications there has grown up a major "industry" of spying on telecommunications. Communications interception has played a crucial role in intelligence since World War I, and despite improvements in communication security it continues to grow. The growth of interception is a consequence of the essential fact that the most important effect of the improvements in communications technology on communications intelligence has been to draw more and more valuable traffic into telecommunications channels. As a result, spying on such channels becomes more and more rewarding for governments, businesses, and criminals.

Imagine three versions of an event, one taking place in 1945, one in 1995, and one in 2005. Each involves a major company with physically separated facilities. In 1945 it starts with a brief call, a minute or two. It invites you to an end-of-year project review. You must take a two-day trip to the Pfister Hotel in Milwaukee. It is a nuisance just before Christmas, but there is no alternative. By 1995, the invitation comes not by phone but by email. The project review is to be conducted by conference call and the associated final report will be sent to all the participants by fax or email. In 2005, the invitation again comes by email but now the meeting will take place using web-based collaboration and conferencing tools.

Now consider the significance of the changes from the viewpoint of industrial espionage. A 1945 spy who taps the phone has learned only that interesting information will be available at the Pfister Hotel in Milwaukee a few days hence. The spy knows where to go to get the information, but is still separated from it by substantial cost, work, and risk. On the other hand, the spy of 1995 can expect to have all the information appear on the same phone line the meeting invitation was issued. All that is necessary is to keep listening. The spy of 2005 is in a more complex position. The web conferencing tools operate over the Internet with its combination of high bandwidth and mobility. It is entirely possible for the spy to learn about the meeting—because one participant does email from a cafe with a free and unencrypted wireless connection—but be unable to capture the meeting itself—because the same participant attends from the office. A spy located inside the telecommunication system or, more likely, one who has ways of getting access to intercept facilities built into the telecommunication system, is in a much better position.[1]

The potential impact on privacy is profound. Telecommunications are intrinsically interceptable, and this interceptability has by and large been enhanced by digital technology. Communications designed to be sorted and switched by digital computers can be sorted and recorded by digital computers. Common-channel signaling, broadcast networks, and communication satellites facilitate interception on a grand scale previously unknown. Laws will not change these facts.

Governments have responded to the existence and relative transparency of telecommunications with some willingness to acknowledge rights of communication—in particular rights of private communication —where necessary but have been resistant to developments that could curtail this new ability to watch the citizenry.[2] The result has been an ongoing battle over the legal regulation of communications interception, the inclusion of facilities for interception in communication systems, and the deployment of security measures, particularly by the private sector. The first battleground was cryptography.

When it is not be possible to prevent communications from being intercepted, it may still be possible to protect them. The primary technology for protecting telecommunications is cryptography, which, despite its ancient origins, is largely a product of the twentieth century. For the first 50 years after radio brought cryptography to the fore in World War I, the field was dominated by the military. Then, in the late 1960s and the early 1970s, a combination of the declining cost of digital computation and foreseeable civilian needs brought a surge of academic and commercial interest in the field.

The work of the civilian cryptographers revealed two things. One was that cryptography was not a field that could effectively be kept secret.[3] In the 1930s and the 1950s—both formative periods in American military cryptography—computational capabilities lagged so far behind requirements that building secure cryptosystems took a lot of cleverness and used techniques not applicable elsewhere. By comparison, in a world in which inexpensive digital computing is ubiquitous, cryptography does not usually represent a large fraction of the computing budget.[4]

Today, constructing cryptographic devices and programs is regarded as easy. Developing sophisticated cryptographic hardware is within the abilities of a talented engineer or a small startup company.[5] Developing cryptographic programs is far easier; it is within the means of any competent programmer who possesses a copy of, for example, Bruce Schneier's book *Applied Cryptography*.

In the 1970s, independent cryptographers startled the cryptographic world by demonstrating that privacy can be manufactured "end to end" without the help of any centralized resources. Diffie-Hellman key exchange allows two parties to derive a secret from negotiations in which every bid and every response is public. This changed the basic power relationships in cryptography. Before public-key technology, cryptography always required centralized facilities to manufacture and distribute keys, a feature particularly compatible with the top-down organization of the military. By contrast, public-key cryptography was developed to support the interactions of businesses in a community of equals.

Privacy is the best-known benefit of cryptography; however, it is not the only one, and it may not be the most valuable one. Cryptography also provides authenticity, which enables communicators to be sure of the identities of the people with whom they are communicating.[6] In a business transaction, authentication verifies that the person acting in one instance is the same person who acted in another—that the person who is writing a check, for example, is the same person who opened the account and put the money in it.

The US military responded to the rise of private cryptography by attempting to reestablish control over the technology through Atomic Energy Act-like prior restraint of research and publication.[7] When this effort appeared to have failed (largely as a result of its obvious unconstitutionality), the government attempted to control cryptographic products directly, first through standardization and later through regulation of exports. In 1993, it unveiled the concept of key escrow—cryptography that would provide protection against everyone except the US government. Although the notion was not well received, its proponents (most of them in the government) kept pushing, constantly giving ground to business objections but holding firmly to the view that it is the government's right to take measures to guarantee that citizens cannot encode things so that the government cannot read them.

Despite the government's intransigence, business pressures carried the day. Slightly less than seven years after the announcement of the key-escrow program, the export regulations—the only actual law with much effect on the use of cryptography—were changed.

In relaxing export controls on cryptography in early 2000 and abandoning its attempts to make escrowed encryption the norm, the US government effectively acknowledged defeat in its battle to control cryptography at a direct regulatory level. Cryptography, however, is not a technology that is easy to use on a large scale and those who predicted that ubiquitous cryptography would make wiretapping and signals intelligence things of the past were flatly wrong.[8]

Ever since the explosion in cryptography brought on by the advent of radio in the early twentieth century, the technique has been at its best in protecting the communications of closely knit groups like national military organizations. Cryptography has been less successful when applied to serve the needs of looser groups like coalitions. Before the development of public-key cryptography, the use of cryptography in diverse communities was a non-starter. Cryptography is now a central technique, but many problems of scale are far from solved. It should not be surprising that the decline of regulation was not sufficient to deliver the overnight growth spurt that cryptography required to fulfill its promise.

The government's retreat from the attempt to stifle widespread use of cryptography has not been derailed by anti-terrorist fervor post-9/11; in fact, government promotion of cryptography has grown. The formal adoption of a high-grade cryptographic system as the Advanced Encryption Standard took place on November 26, 2001. A little over a year later, the system was approved for protection of classified information, and in 2005 NSA bestowed that status on a full suite of public cryptographic algorithms. The NSA's actions are seen as serving two ends. The algorithms are expected to lead to widespread commercial incorporation of the approved algorithms and thereby lower government procurement costs. They will also facilitate improved secure communication among the parties to the overnight coalitions that are so active in promoting modern wars.

The deployment of cryptography maintains slow but steady growth and, in the absence of a new regulatory assault, will eventually become ubiquitous. The high profile of the "crypto wars," however, drew attention away from other developments in communications privacy that may prove more important. At present, the battle over communications privacy is moving in new directions, focusing less on the protection of communications and more on their exploitation.

Over roughly a century US law has evolved the concept of wiretapping as a form of search to be controlled by court-issued warrants even more tightly regulated than those required for searches of physical premises.

Although law-enforcement agencies had been intercepting communications since the 1890s, it was not until 1968 that Congress put law-enforcement wiretaps on a solid legal footing. The Omnibus Safe Streets and Crime Control Act, which limited the use of wiretaps to certain crimes and established stringent warrant requirements, was upheld by the courts. As a result, wiretapping has become a generally accepted and ever more widely employed police practice. Law enforcement views the tool as essential, but a closer look at the data shows that things are not so clear cut.[9] Law enforcement spoke freely of its "right to use court-ordered wiretaps" and saw the use of cryptography as a threat to this right.

In discussions of the right to use cryptography, attention focused on the clearly discernible difference between the right to listen and the right to understand what one has heard. The doctrine of wiretapping as a type of search takes for granted the government's ability to practice wiretapping, just as the Fourth Amendment to the Constitution takes for granted the government's ability to break down doors and look under floorboards. It recognizes the power to intercept telecommunication, like the ability to search houses, as having such potential for abuse as to require stringent judicial control. It regulates the right to listen.

Guaranteeing the right to understand is different. To do that, you must regulate the individual to prevent him from taking actions that would otherwise be within his power to protect his communications from being understood. This seems analogous to the ludicrous notion that the government's right to search your house entails a right to find what it is looking for and a power to forbid people to hide things.

There is a important respect in which wiretaps are in conflict with the traditional notion of search in Anglo-American law. Searches have been, by legal intention and usually by physical fact, obvious. It is difficult to search a property and be sure that the search will not be detected. Furthermore, in a tradition dating back to English common law, secret searches were forbidden; where possible, the searchers were expected to knock and to announce their presence.

The no-secret-searches doctrine has been eroded in US law, at first by judicial tolerance and later by congressional action. In the 1970s courts began allowing federal agents to make secret entries into private property in order to plant bugs (Burnham 1996, p. 133). As an outgrowth of the Aldrich Ames case—in which a secret search was conducted, but the legitimacy of the evidence so obtained never enjoyed court scrutiny—the Foreign Intelligence Surveillance Court was given the power under the PATRIOT Act to order secret searches.

Wiretaps, in contrast with searching or bug planting, are inherently difficult to detect. Although it behooves anyone who takes the privacy of communication seriously to assume that every word is being recorded, obtaining confirmation of that fact in any individual instance is usually impossible. Treating wiretaps as searches thus leaves open the possibility that wiretapping may be rampant, may be used as a mechanism of political and social control far beyond the bounds of proper law enforcement, and yet may go unchecked because of public ignorance. Under the "Title III" law of 1968, Congress sought to preclude this possibility by means of stringent reporting requirements. Individuals must be notified that they have been wiretapped, even if they are not prosecuted, and details of all legal wiretapping activity are collected and published in the annual *Wiretap Report*. In 1978, however, Congress created new authority to wiretap, primarily for counterintelligence purposes. Under the Foreign Intelligence Surveillance Act of 1978, only the total numbers of wiretaps are reported. Details need never be made public.[10]

In the shadows of the government's attempts to control the citizens' access to technology for protecting their communications (and thereby guarantee its ability to understand what it intercepts) lurked plans for a dramatic expansion of the basic ability to wiretap. The Communications Assistance for Law Enforcement Act of 1994 (CALEA) requires that telephone companies make their networks "wiretap ready" so that new features in communications do not interfere with government wiretapping.

This expansion of government power to search flies in the face of a gradual acceptance of a basic human right to privacy. Although it was already recognized in ancient times, privacy has come into its own as a legal entity only in recent centuries. In large part this has been a response to the developments of the technological age. Through a series of court decisions (including *NAACP v. Alabama*, *Griswold*, *Katz*, and *Kyllo*), the US Supreme Court expanded the notion of privacy that is implicit, if never called by name, in the Constitution. Though private businesses often intrude upon individual privacy, the consequences of their intrusions pale beside the consequences of government intrusions. Over the past 50 years, government has, on myriad occasions, invaded the privacy of individuals in ways that threaten their fundamental rights. Citizens engaged in peaceful political activity (including the Socialist Workers Party, the civil-rights movement, and Vietnam War protesters in the 1950s and the 1960s, the Committee in Solidarity with the People of El Salvador in the 1980s, and the L.A. 8 in the last two decades), journalists and editors and political leaders (including Supreme Court justices) all have been wiretapped. Members of Congress who disagreed with the president's policies during the Vietnam era were subjects of biweekly FBI reports. Even politically uninvolved citizens who happened to use mail or telegraph to communicate internationally have had their communications intercepted.[11] Information obtained by the government for use in one venue has often been used in another. Census data were used to locate Japanese-Americans so they could be interned during World War II. Some "national-security" wiretaps under various presidents were actually investigations aimed at domestic politics.[12]

The government's record of privacy violations means that any broadening of its snooping powers must be viewed with the gravest concern. CALEA is the basis for a vast expansion of government surveillance powers. Even if the government's record of using its powers were not strewn with tales of abuse, there would be reason to worry.

Intentions can change far more quickly than capabilities. Today the authority of most government officials to use wiretaps is tightly regulated by laws, but laws can change. Were Congress to decide that wiretaps should be usable by any police department without court supervision —much as the police are free to employ stool pigeons without court supervision—the situation would change overnight. The capacity of the telephone system to support wiretaps, by contrast, would not. Although the pre-CALEA phone system was quite capable of supporting the 1500 or so wiretaps that occurred each year, it was not capable of supporting 10 or 100 times as many. Today, more than a decade after the passage of CALEA, this may no longer be the case. The way has been paved for a vast expansion in government surveillance, and only an act of Congress will be required to bring it about.

The push to expand the interception of communications comes at a time when police have experienced an unprecedented expansion of their powers of surveillance in almost every area. Advances in electronics permit subminiature bugs that are hard to detect electronically or physically. Video cameras watch streets, shops, subways, and public buildings. Vast databases keep tabs on the credit, the possessions, and the criminal records of most of the population. Many of these facilities play far greater roles in criminal investigations than wiretaps.

The broadening and deepening penetration of telecommunications into our lives has also shifted the standards of non-governmental surveillance of communications. Although the telephone calls of workers who deal directly with the public are often monitored or recorded for "quality control and training purposes," in other areas of employment some respect for employee privacy seems to prevail. Whether actually required by law, customary, or merely seeming proper to everyone involved, there is still a notion that probable cause is required before an employee's communications can be spied on.

The Internet has changed all of this in two ways: surveillance has become nearly universal, and it is done not by people but by machines. The new instrumentalities of surveillance, moreover, are not passive, like tape recorders; they are active, blocking, censoring, and deleting communications. A number of factors have come together to bring this about.

The most conspicuous are the real dangers of Internet communication to enterprises. Break-ins, denials of service, viruses, and worms are all capable of interfering with enterprise computing, a feature of business that now is just as important as power and light and good employee health. Businesses have responded by installing firewalls that prevent the entry of any malevolent material they can recognize. Not only would it be hard to deny the legitimacy of this action, but in most cases it serves the interests of all parties. Employees are, by and large, grateful when their email is not so cluttered with spam that they miss messages on which doing their jobs depends.

Other measures put employees more at odds with their employers.

Although an employee could tell a company secret to an unauthorized person over the phone, that was not a channel by which an actual copy of a confidential document could be conveyed. Today companies ask "What is to prevent my employees from mailing my most valuable secrets to my competitors?" Responses vary. In extreme cases, like the intelligence agencies, there are separate internal and external networks with only the most tightly controlled connections. More commonly email is recorded so that leaks can be confirmed and analyzed by later investigation if the occasion arises. For companies that do not consider recording sufficient, there are programs that attempt to detect proprietary content in communications crossing the corporate firewall and either alert a security officer or block communications altogether. For a corporate security officer, every day's mail, email, and voice mail brings new pitches from companies claiming to do this more effectively.

Controlling improper use of the Internet by employees also makes up a big piece of the modern information-security pie. From one angle, recreational on the job use of the Internet is a productivity issue. If a job requires Internet use, it is difficulty to tell whether an employee is trying to get the best price for corporate travel or planning an upcoming vacation. In this respect, it is no different for the productivity concern about an employee who spends too much time chatting with friends on the telephone rather than chatting up customers. In another respect, it is far more serious. If an employee is looking at sexy pictures on a workplace display, another employee may be justified in filing a sexual harassment complaint, with devastating consequences for the employer.

Tools used to limit Internet browsing to material employers consider safe combine limitations on the sites that can be visited with scrutiny of the material received. The technology is unsettlingly similar to that employed by parents to control their children's use of the Internet.

Prospects for Intelligence

For thousands of years, a country could strictly limit what other nations could learn about it. Even though it might have difficulty protecting its border, travel was difficult, expensive, and time consuming. Travelers were conspicuous and treated with suspicion. Even when they succeeded in traveling, acquiring information, and returning home with their information, the process might take years. The past century and a half, however, brought the camera, the airplane, and the spy satellite. The interiors of countries are no longer closed to view. They are visible to all the major powers, and with every passing year they are more visible to smaller countries, news media, and commercial interests. In recent years, the development of space technology has served intelligence well by putting cameras and antennas in orbit, where they can collect information about any nation.

If the principal effect of advancing communication technology on communications intelligence is to bring more valuable traffic into telecommunications channels, the secondary effect is increase the complexity of extracting it. Both intentionally protective measures (such as cryptography) and measures that are not primarily protective (such as the use of optical fiber), make the traditional SIGINT practice of starting with an antenna less productive.

In consequence, the character of the COMINT product is changing, improving in some respects and declining in others. Because people are often prone to mourn the loss of something on which they have come to depend and slow to see the possibilities of the unfamiliar, it would not be surprising to find that the change is perceived as decline by many COMINT professionals.

One area in particular in which COMINT has surpassed all other forms of intelligence, with the possible exception of HUMINT, is the discovery of opponents' intentions. Listening to people's communications—particularly when they are speaking or writing candidly out of misplaced faith in their security—can reveal their real objectives and the unspoken desires that underlie their public negotiating positions. This coveted capability is one that COMINT may have to surrender, and a replacement for it seems hard to find.

On the other hand, improvements in communications and increasing human dependence on communications will open new areas of intelligence. Network-penetration techniques will make it possible to capture information that is being stored rather than communicated, and such information is less likely to be encrypted. Even more exciting is the prospect that, in a world with hundreds of countries and thousands of other centers of authority, there will be innumerable agencies responsible for issuing credentials and authorizing acceptance of other agencies' credentials. We will no doubt see numerous cases in which information is leaked to opponents because they are not recognized as opponents. Active network intelligence measures will become the HUMINT of the next century, and it will interact extensively with traditional HUMINT.

In the United States, and perhaps elsewhere, communications intelligence plays less of a role in industrial espionage than in national espionage. Businesses often have a better means of acquiring information: hiring workers away from their competitors. In the world of the Cold War, a world of open hostility between two major coalitions, changing sides was difficult. It did happen, and some people[13] made a big success of it, but it was a risky business and hard to do more than once. In a world of shifting alliances in which international competition is more commercial than military, defection may become as big a feature of national intelligence as of industrial intelligence.

Cryptography is much less successful at concealing patterns of communication than at concealing the contents of messages. In many environments, addresses (and, equally important, procedences) must be left in the clear so that routers will know how packets are to be forwarded. In most environments, the lengths and timings of messages are difficult to conceal. SIGINT organizations are already adept at extracting intelligence from traffic patterns and will adapt to extract more. The resulting intelligence product, abetted by increases in computer power, may not give as detailed a picture in some places but will give a more comprehensive overview.

Some improvements in SIGINT technology cannot easily be categorized as tactical or strategic. They take the form of increased speed and flexibility of the sort that has changed many organizations over the past decades. The current intelligence cycle in SIGINT is a slow one that can be summarized as follows:

- Intelligence consumers formulate requirements.

- The requirements are translated into guidelines about what to intercept.

- Intercepted material is acquired "in the field" and shipped home for analysis and interpretation.[14]

- On the basis of cryptanalysis, interpretation, and political analysis, the information is judged, as are the guidelines under which it is acquired.

- The guidelines are either continued or modified. New intercept facilities may be assigned to a project, new facilities may be built, new instructions on traffic characteristics may be issued, or the project may be dropped.

This process may take weeks, months, or years. Often, significant information will not be acquired simply because it was not being looked for.

Increasing automation and decreasing size and cost of electronic equipment will make for vast improvements in this cycle, resulting in a tighter "target, intercept, analyze" loop. This will be aided by the development of tamper-resistance technology. The secrecy of many SIGINT processes makes intelligence organizations reluctant to use them anywhere but in the most secure areas of their own headquarters. Tamper-resistant chips allow intercept equipment in the field to perform such sensitive operations as cryptanalysis. This permits them to search the contents of ciphertext messages just as they would the contents of plaintext messages.[15]

An example of a SIGINT technology with unfathomed potential is emitter identification. The vanishing cost of signal processors has reduced the cost of this technology and so expanded the range of possible uses.[16] In many cases, emitter identification will counter the concealment of addressing by link encryption.

Not all the growth that can be expected in SIGINT will result from SIGINT technologies. A fast-growing portion of the telecommunications market all over the world is *fixed-position cellular telephony*. The cost of radio technology has dropped to the point that in many rural areas it is cheaper to have a cellular telephone in each house than to run wire. The result is that a whole segment of the telecommunications market that was once effectively out of reach of intelligence organizations is now coming, at least partly, within its grasp.

From a practical viewpoint, it is important to note that nothing will happen overnight. The vast legacy of equipment, services, experience, and investments in communications from the twentieth century will guarantee the future of much of communications intelligence well into the twenty-first.

Prospects for Law Enforcement

The dramatic growth of technology in the twentieth century has given law enforcement a wide variety of technical capabilities, one of which is wiretapping. At present, law-enforcement personnel are worried that advances in communications technology, particularly in cryptography, will lead to a decline in the usefulness of wiretaps. Should this happen, its effect on law enforcement is likely to be modest. Even among tools of electronic surveillance, wiretaps are generally overshadowed by the many kinds of bugging devices used to intercept face-to-face conversations. Electronic surveillance, furthermore, plays a minor role in police investigation by comparison with record keeping, photography, and a broad spectrum of forensic techniques.

Even before CALEA, wiretapping would appear to have gained more than it has lost (and perhaps more than it stands to lose) from modern technology. At one time a wiretap was, literally, a pair of wires attached somewhere between the target's telephone and the telephone office. Its placement and its use entailed a risk of discovery and brought the listeners only disembodied voices. Today, even without the vast wiretapping capacity envisioned by CALEA, wiretaps are "installed" in the software of digital telephone switches. Knowledge about installed wiretaps can be kept to a few telephone-company employees. More important, the taps carry with them extensive call-status information that often makes the identities of the talkers or their locations immediately available.[17]

Law enforcement's gains from advances in technology are not, however, limited to investigation. The police are a mechanism of social control (Manning 1977, p. 23), and their work goes hand in hand with other mechanisms of social control. Improving communication is enhancing "employee supervision" throughout society. In the past, ambassadors and senior military commanders were sent off to the other side of the world with general mission statements and no opportunity to report their successes and failures—let alone ask for advice—for months or years. Today, the president can reach his senior emissaries at a moment's notice anywhere on Earth. At lower levels, employees in many jobs are now monitored by machines. Workers who once had substantial autonomy, such as truck drivers, find that they are subject to the same sort of close monitoring that might have been expected on a factory floor.[18]

Society is also gaining an ability to keep close track of individuals' interests and expertise. Online uses of information resources are intrinsically less private than paper ones. For example, monitoring which documents visitors to libraries consult or what pages they copy would be expensive and, despite the FBI's Library Awareness Program, is probably rare. When people consult sources of information on the Internet, however, monitoring is inexpensive and hard to separate from services the users value. Commercial Web pages record IP addresses and other available information about the "callers" and use it for marketing. Exchange of information among Web sites presents the prospect of a comprehensive profile of each Web user.

The current debate is not, as it was in the 1990s, about the public use of strong cryptography, but rather about communications security and building wiretap capabilities into network infrastructure. At a hearing on the subject of CALEA in 1994, FBI Director Louis Freeh and Senator Larry Pressler had a spirited discussion of the issue (USSH 103 *Digital Telephony*, p. 202).

Asked to state his view of the proper scope of CALEA, Freeh said: "From what I understand ... communications between private computers, PC-PC communications not utilizing a telecommunications common net, would be one vast arena, the Internet system, many of the private communications systems which are evolving. Those we are not going to be on by the design of this legislation." Pressler pressed him: "Are you seeking to be able to access those communications also in some other legislation?" Freeh responded: "No, we are not. We are satisfied with this bill. We think it delimits the most important area and also makes for the consensus, which I think it pretty much has at this point." Pressler then asked: "Yes but in the future, will you be seeking the ability to tap into those other forms of communication?" Freeh gave a prescient response: "It is certainly a possibility. I am sure if, God forbid, somebody blows up the World Trade Center 10 years from now using a PC-PC private communications network, a question would validly be raised in the Congress and by the President as to whether that form of communication now needs to be accessed. But we are not taking that position now. We are not contemplating coming back and asking for additional coverage." Pressler asked for clarification: "So what we are looking for is strictly telephone to telephone, what is said over a telephone?" Director Freeh said: "That is the way I understand it, yes, sir."

In 2001, the World Trade Center was blown up (or at least "knocked down") and, although no one has suggested that the Internet played any significant role, the FBI is indeed seeking to extend CALEA. Is extending built-in wiretapping from the switched telephone network to the Internet a wise precaution or an imprudent risk?[19] In addressing a parallel issue, the National Research Council report on cryptography concluded that "on balance, the advantages of more widespread use of cryptography outweigh the disadvantages" (Dam and Lin 1996, p. 6). Apparently accepting this view, the US government began encouraging the development of strong cryptography in the infrastructure in 2000. We believe the same course would be appropriate here. On balance we are better off with a secure communications infrastructure than with one that builds surveillance into the network fabric. At times this may press law enforcement to exercise more initiative and imagination in its investigations. On the other hand, in a society completely dependent on computer-to-computer communications, the alternative presents a hazard whose dimensions are as yet impossible to comprehend.

Prospects for Security

The world we face now is different from the one many of us envisioned after the demise of the Cold War. Due to various causes—the rising economies of China and India, the rapid rate of globalization—the American hegemony so visible today is likely to have faded by the end of the twenty-first century. The US will undoubtedly remain a major power, but it is unlikely to dominate the world at the end of the twenty-first century as it does at its dawn. Such changes are to be expected and should be part of any national-security planning. What has been less planned for—

or at least less anticipated by the general populace—is the rise of non-state actors and their willingness to perform acts of violence and terror on a grand scale.

Despite poor beginnings, prospects for the security of our information infrastructure are good but only if we accommodate security in our plans from the beginning. Part of the reason for the current poor state of information security is the fear, uncertainty, and confusion created by government opposition to the use of strong cryptography in the 1990s. As the FBI acknowledged, "the use of export controls may well have slowed the speed, proliferation, and volume of encryption products sold in the US" (Dam and Lin, p. 138). Given the enemies we have now, and society's reliance on electronic communications for everything from personal affairs to control of critical infrastructure, it is vital that our computing and communications be properly secured. This means secured against attacks from the outside *and* from the inside.

Time scale is very important. Building interception into our communications system is appealing as a tactical move. The institutions that will have access to this intelligence and law-enforcement resource are institutions that have grown up over the course of the twentieth century and, despite being secretive, are known to the American public. Initially, the new facilities will be far more familiar to those who use them than to those against whom they are used and may be quite efficacious. What will happen to the control of these facilities as the decades pass is hard to assess. Undoubtedly, opponents will become more proficient at employing countermeasures to useful interception. More frightening is the prospect that opponents—particularly opponents within our own society—will learn to turn the new tools to their own advantage.[20] Although a case of this kind has yet to come to light in the United States, there has been one in Greece. For well over a year, interception facilities built into cellular telephone systems were used to tap the phones of over 100 Greek government officials.[21] Who was doing the tapping remains unknown.

In the early 1960s, President John Kennedy promised a new level of control over nuclear weapons. When Don Cotter, director of Sandia National Laboratories, called his senior staff together and told them to start working on this problem, they expressed doubts about what to do when they had only an overall direction and no detailed policies.

"Hardware," Don Cotter told them, "makes policy."[22] In one sense, laws represent a society's highest form of decision making. They are are difficult and expensive to change but not the most difficult or expensive things to change. Long-term investments in infrastructure are even harder to change. Lawrence Lessig put this another way when he titled a book *Code and Other Laws of Cyberspace*. What is committed in design, development, and availability binds everyone, often more firmly than law.

Suppose that the key-escrow program of the 1990s had been successful. Suppose that millions of devices conforming to the Escrowed Encryption Standard had been sold, rather than merely a few thousand. Can there be any doubt that the same junior lawyers in the administration who wrote memos rationalizing the expansion of SIGINT to allow warrantless interception of phone calls between a foreign phone and a domestic one, would argue that the database of escrowed keys should be put at NSA's disposal?

The lesson is simple and unavoidable. By building the machinery for surveillance into the US communication system, we overcome the largest barrier to becoming a surveillance society on a possibly unprecedented scale. By comparison with the years of development and deployment needed to put the system in place, legal decisions to use it in ways that might have been unthinkable when it was approved can be made quickly.

What Kind of Society Do We Want?

In deciding that the Constitution protected Charles Katz against electronic surveillance even though there was no intrusion onto Katz's property, the Supreme Court looked through the propertarian technicality of the Fourth Amendment to its essential objective. As human society changes from one dominated by physical contact to one dominated by digital communication, we will have many opportunities to choose between preserving the older forms of social interaction and asking ourselves what those forms were intended to achieve.

In the societies that have dominated human culture for most of its existence, a general awareness of the pattern of contacts among people was an essential feature of life. In a society dominated by telecommunication, a pattern of contacts is far less visible to the ordinary person and far more susceptible to monitoring by police and intelligence organizations. This produces a fundamental shift of power away from the general population and into the hands of those organizations.

Technology seems to make some losses of privacy inevitable. The capacity to build databases and feed them the details of every credit-card transaction exists, and the result is an excruciatingly detailed portrait of the shopping, traveling, and trysting habits of hundreds of millions of people. Yet, since such databases are an essential component of today's commerce and millions of people work in the industries they support, it seems realistic to accept them. The best we can hope to do is regulate their use in a way that protects individual privacy.

We also seek to preserve both the individual's and the society's security. This is where the government's plans regarding the wiretapping of VoIP and other real-time communications[23] seem remarkably short-sighted. Combining Internet surveillance with inexpensive automated search engines could lead to an unprecedented compromise of American security and privacy (Landau 2006). (The "Titan Rain" exploits described in chapter 4 give a sense of some of the potential problems.) A wiretap is, after all, nothing more than an authorized security breach. This approach is made worse by the direction of the Internet's development. Currently there are millions of devices connected to the Internet, but we are moving to a situation of billions of small devices, such as radio-frequency ID tags and sensors, many of which will communicate via the Internet (ibid.).

Noting the comments of Ayman al-Zahawiri, former leader of the Egyptian Islamic Jihad and second-in-command of al-Qaeda, that "however far our capabilities will reach, they will never be equal to one-thousandth of the capabilities of [our enemies]" (Richardson 2006, p. 232), Louise Richardson observes that we must turn the terrorist threat against itself. We should not take our strengths, which include modern and robust communication systems, and turn them into instruments of surveillance that others can use against us.

Cryptography in Context

The words of the Supreme Court's *Katz* opinion have an importance that transcends the development of American wiretap law. They echo in concrete form Louis Brandeis's view that "time works changes." If there is a right to use cryptography, it must grow from the historical fact of private conversation. Since many conversations today can take place only by telephone, stepping away from other people is no longer a universally applicable security measure. It is not realistic to say to someone "If you don't like the possibility of being tapped, you have the choice of not using the telephone." Stepping away from other people is the expression of a right to keep conversation private in a face-to-face world; use of cryptography is an expression of that right in an electronic world.

In a sense, it is curious that the Constitution regulates the power of the police to search (and, derivatively, their power to conduct electronic surveillance) but leaves activities that are at least as dangerous and disruptive, such as the use of undercover agents and the mounting of sting operations, up to individual detectives or their chiefs.[24]

In light of the curiously small number of prosecutions in which wiretap evidence plays a significant role, it appears that wiretapping is far more valuable as an intelligence tool than as a way of gathering evidence. This utility, however, is not recognized by US law, under which wiretap warrants must name particular suspects and crimes. Police who wish to use wiretaps in the gathering of intelligence are therefore forced into the duplicitous position of representing any wiretap as an attempt to gather evidence. A reform of wiretap law might plausibly recognize the police intelligence applications of wiretapping and give courts the means to supervise it.

Technology might also be applied to streamline the courts' oversight of law-enforcement activities, just as it has made so many improvements in the activities themselves. It seems certain that at some time in the future courts will choose to accept applications and issue warrants electronically, using digitally signed messages. This would reduce law enforcement's logistic overhead and would permit warrants to be more carefully focused. Police might, for example, be more readily granted a warrant limited to communications between two people than a warrant encompassing all the communications of one person. Quick turnaround would permit police to base such warrant requests on the calling patterns of suspects and to get a new warrant promptly when a new link in a conspiracy was identified. Such an arrangement would respond to Brandeis's concern that "whenever a telephone line is tapped, the privacy of the persons at both ends of the line is invaded" (Brandeis 1928, pp. 475–476) by making an effort to target only calls in which both participants were suspects.

Of course, if the utility of wiretaps is no greater than the publicly available evidence suggests,[25] perhaps they should be dropped from police methodology altogether—not because they are an invasion of anyone's privacy, but merely because they are a waste of tax money.

Where Are We Headed?

In the first phase of the communications technology and privacy battle, the central question was very simple: Do people have a right to private communication, and should they be free to express and enforce this right by using cryptography? In the newer phase, the questions are more complex. It is hard to argue that society—including the government and the private sector—has a right to employ communication surveillance to counter imminent threats. On the other hand, there is little question that surveillance has a chilling effect on many activities, from art to politics to personal relations. Can we find a set of rules that give us adequate protection without stripping us of our privacy and autonomy, protected as the protectors see fit, rather than as we see fit.

For many decades a simple dichotomy has served us well in judging the legitimacy of communications interception. Outside the country, national intelligence agencies were allowed to intercept whatever foreign communications they could acquire and considered worth recording. Inside the country, interception followed the reasonable search-and-seizure model of law enforcement. Communications could not be intercepted without probable cause, and that probable cause had to be based on legitimately obtained evidence, typically of some other kind. Although 'inside' and 'outside' are not entirely clear, a workable set of rules has been found. Inside is inside. Communications originating or terminating inside the US, even when they originate with foreign companies or foreign embassies or foreign spies, can be monitored only under the court-regulated model of reasonable searches and seizures. Communications between entities outside the United States can be monitored fairly freely. Moreover, they can be monitored from US territory: the US proper, foreign US military bases, or US embassies.

'Inside' and 'outside' are fairly static notions. A foreign embassy, with its extraterritorial status, may seem to blur the inside-outside distinction, but at least it stays put for years at a time. As travel and communications have become more fluid, determining what is inside and what is outside has become harder. Communications between foreign entities that can be intercepted from antennas at Yakima are unequivocally foreign under the current rules. What about packets entering the United States at the same location and headed for addresses in Europe? Should these be regarded as foreign traffic that happens to have passed close enough to be intercepted—like the radio signals picked up by the antennas—or should they be regarded as domestic traffic because they are traveling over resources provided by US companies and are entitled to protection as invited guests?

Even more provocative are the questions raised by travel. Suppose that intelligence has been monitoring a terrorist traveling abroad and tracks the terrorist onto a flight to New York. Cutting off monitoring at the border would seem particularly foolish if the earlier monitoring showed that the terrorist was on the way to attack a target in the United States. On the other hand, if such emotional cases are allowed to hold sway, we will find ourselves in a world where government can rationalize monitoring anything. Furthermore, if the intelligence model of secrecy about what is being monitored holds sway, as the growing use of FISA rather than Title III wiretaps suggests, the rationale may not have to be explained to very many people.

The task is simple to explain but far harder to achieve. If we do not incorporate adequate security measures in our computer and communications infrastructure, we risk being overwhelmed by external enemies. If we put an externally focused view of security ahead of all other concerns, we risk being overwhelmed by their misuse. We must find a set of rules and a mechanism for overseeing those rules that allows society to defend itself from its genuine enemies while keeping communication surveillance from stifling dissent, enforcing morality, and invading privacy. If we do not, the right to use privacy-enhancing technology that was won in the the 1990s will be lost again.

Notes

Chapter 1: Introduction

1. (p. 4) The Watergate scandal, which forced the resignation of President Richard Nixon in 1974, was initiated by the installation of electronic surveillance in the Watergate offices of the Democratic National Committee. Watergate is only one of a number of cases in which electronic eavesdropping was used by the party in power for *political* purposes and was justified on grounds of "national security."

2. (p. 5) There is one major exception: radio and television are used to deliver a product to consumers.

3. (p. 6) At a 1995 talk in Brisbane, Australia, Ross Anderson of Cambridge University estimated the commercial market at over a billion dollars a year and likely to be augmented by another billion a year from developing industries (Anderson 1995), but the dollar figures do not appear in the printed paper.

4. (p. 7) The exact fraction is difficult to determine, partly because the budget is secret and partly because the most expensive items in the budget are spy satellites (some of which both listen to communications and take photographs).

5. (p. 10) This is one aspect in which telecommunications may forever remain less satisfactory than physical meetings.

Chapter 2: Cryptography

1. (p. 12) A signal sent by satellite can typically be received in an area thousands of miles across. One sent by microwave is hard to pick up more than a few miles from the "line of sight" between the towers. An intriguing approach to secure communication is used in some military satellite systems. Signals are transmitted from one satellite to another at frequencies that are absorbed by oxygen and cannot travel far enough through the Earth's atmosphere to be received on the

ground. This forces an eavesdropper to put up a satellite to spy on other satellites —an expensive proposition.

2. (p. 12) What it means to know someone in this sense is not straightforward. Essentially it means that your expectations about the person you are dealing with are correct. Those expectations may or may not include such identifying information as a name and address. There are some interactions, such as discussing your marital troubles with a stranger in a bar, in which what you are depending on is simply that the other party is, true to appearance, a sympathetic stranger who does not know your spouse and will not carry your tale back home.

3. (p. 13) The principle, alas, is often honored more in the breach than in the observance. For reasons discussed in chapter 3, there is a tradition of secrecy about cryptographic systems, both governmental and commercial. There is, however, a sense in which Kerckhoffs's principle holds, even in regard to secret systems. For example US government cryptographic systems that are used to protect TOP SECRET information are only classified SECRET and the equipment that embodies them is rarely classified higher than CONFIDENTIAL.

4. (p. 14) It is not made any clearer by the practice of using the terms "code" and "coding" for numerous transformations in modern computing and communication that have nothing to do with security.

5. (p. 15) You may also attempt to conceal the fact that what you are conveying is an encrypted message at all. This strategy—called steganography or covert communication—will be touched on from time to time in the text.

6. (p. 16) The police will recognize that they have a plausible candidate for a time-and-place message because the various positions in which only some digits are possible will all assume acceptable values.

7. (p. 17) A mask file is the digital input to a "fab line" that produces a computer chip. It can represent millions of dollars of engineering effort. Furthermore, semiconductors are often fabricated far from where they are designed and so, whether on a disk, by phone, or by Internet, the mask file must be transmitted from one to the other.

8. (p. 17) A one-part code is one in which the plaintext phrases and the code groups are simultaneously in alphabetical (or numerical) order so that the same book can be used for both encoding and decoding.

9. (p. 19) The Venona messages are available from the National Security Agency Web page: http://www.nsa.gov/venona/

10. (p. 21) The alphabets used in the first Vigenère table are in fact related. They were generated algorithmically by starting with a standard alphabet and using the GNU Emacs pseudo-random number generator to pick pairs of letters to be swapped. One thousand swaps were used to produce each alphabet. This process was intended only to produce such examples and is far from secure. The set

of alphabets given can probably be cryptanalyzed to find the parameters of the underlying pseudo-random number source.

11. (p. 21) The classical Caesar Cipher, in which every character is moved forward in the alphabet by three places, is the most famous example of a direct standard alphabet.

12. (p. 22) This form of CAVE is now used only in TDMA (Time Division Multiple Access) a legacy system still supported by Cingular but being phased out in favor of GSM and/or UMTS. (Source: private communication between Diffie and Greg Rose, September 2006).

13. (p. 24) This is usually called a *combiner* if it accepts several bits as inputs and produces one output and an *S-box* if its output is more than one bit.

14. (p. 25) Primary credit for the design of Skipjack appears to go to Paul Timmel, although this has not been officially confirmed. A patent application covering the system remains under secrecy order, presumably because it covers additional ways of using the system that are not needed by the secure email system and remain secret. A study of the published aspects of the escrow protocols makes it clear that the chips needed to encrypt blocks of more than 64 bits in order to escrow the 80-bit key.

15. (p. 26) The vast number of keys was offered as an argument for the unbreakability of ciphers during the Renaissance (Kahn 1967, p. 145) and probably earlier. The more general modern theories, including the theory of *non-deterministic polynomial time* or *NP* computing, (Aho et al. 1974) are far more mathematical but little more satisfactory. More specialized analyses of the vulnerability of particular cryptosystems to particular analytic methods has been more satisfactory.

16. (p. 26) In fact the cryptanalyst must also know enough about the plaintext to be able to recognize it; otherwise a correct solution to the problem cannot be distinguished from an incorrect one.

17. (p. 27) Chosen plaintext can arise in the real world in several ways. Suppose for example that the US Department of State delivers a diplomatique communiqué to a foreign embassy. The embassy must transmit the message home to its own government and is therefore likely to encrypt a message whose contents are known to the US government. Chosen plaintext may also be available to an organization that shares an encrypted channel—for example, a high-bandwidth satellite channel encrypted by the provider—with another organization.

18. (p. 27) This is popularly known as Moore's Law, and has held for the past several decades.

19. (p. 27) This doesn't work the other way around, because a large part of the cost of a computer is in such housekeeping aspects as its case and power supply, whose costs change less quickly.

20. (p. 28) In the US a typical TOP SECRET document is downgraded to UNCLASSIFIED in between 20 and 30 years.

21. (p. 29) Double DES allows a 112-bit key and appears at first to be adequate, but it is subject to an attack called "meet-in-the-middle." First discussed by David Snow at a National Bureau of Standards meeting in September 1976, this attack exploits a matching plaintext-and-ciphertext pair by encrypting the plaintext under all possible keys, decrypting the ciphertext under all possible keys, sorting the results, and looking for a match. Refinements of this technique are applicable in a surprising number of cryptanalytic circumstances.

22. (p. 29) The lifetimes of public-key cryptosystems are harder to quantify because terms such as RSA and Diffie-Hellman denote general techniques; they do not fix key lengths or register sizes.

23. (p. 29) For example, various forms of the German Enigma system were in use from the 1920s until well after World War II and some are probably in use today.

24. (p. 31) Rijndael and AES are not strictly speaking identical. The former includes modes for encrypting larger block sizes that are not part of AES.

25. (p. 31) The name Rijndael, which is pronounced "Rhine Dhal," is a combination of the names of its two designers: Joan Daeman and Vincent Rijmen.

26. (p. 33) As the name suggests, material under two-person control is never handled by one person alone. It would thus require a conspiracy of two people to, for example, make an unauthorized copy. Two-person control is supported by such things as safes with two combinations and areas called *no lone zones* in which at least two people must be present if anyone is present at all. "COMSEC" is short for "communications security."

27. (p. 33) Some keys used in the STU-III secure telephone, for example, have editions. Unless understood in the context of codebooks, the terminology seems peculiar for keys that never exist in any form other than an electronic one.

28. (p. 34) The term usually used for the system about to be described is *key distribution center* (KDC). Because more modern systems do not always distribute keys, the more general term KMF has come into use; we will use it throughout.

29. (p. 34) In fact, sharing a key with the KMF defines membership in the network.

30. (p. 35) The STU-II, an obsolete US military secure phone system was said to suffer from KDC congestion at busy times of the day.

31. (p. 38) Celebrated examples of this are to be found in the cases of Boyce and Lee (Lindsey 1979) and Whitworth (Blum 1987, pp. 280–326). In both cases, keys were saved rather than destroyed after use and were later sold to Soviet intelligence officers.

32. (p. 39) This technique of signing, which might be called a *primitive digital signature* has the disadvantage that in order to maintain the evidence of authorship, the recipient must store the ciphertext and must therefore either store the plaintext as well or decrypt the ciphertext again each time the plaintext is needed. In practice, what is done is to generate a *message digest* using one of several

message digest algorithms (Schneier 1996, pp. 435–455) and compute a primitive digital signature of that.

33. (p. 40) It would be more accurate to call this *key agreement*. Unfortunately, the original name was generally accepted before it was observed that the things exchanged were not keys.

34. (p. 41) In the STU-II telephone system, calls to the KDC were initially too short to trigger billing and the contact had to be lengthened to accommodate telephone company complaints. (source: private conversation between Diffie and Howard Rosenblum circa 1980.)

35. (p. 41) This problem also bedeviled the STU-II. It is the problem solved by Rosenblum's invention of the two-part certificate (Rosenblum 1980).

36. (p. 42) It is possible but difficult to change network protocols. The current version, described here, is IPv4, which has 32-bit (sixteen billion) addresses. In a world with six billion people this no longer seems a generous allotment. A change is underway to IPv6 with 128-bit addresses but has yet to take off.

37. (p. 43) Although the protocol was adopted and published, development continues.

38. (p. 45) This is a codeword of obscure origin, not an acronym.

39. (p. 47) A lawsuit between General Motors and Volkswagen illustrates how valuable information about manufacturing design can be (Meredith 1997).

40. (p. 47) Consider, for example, the 1987 testimony of Cheryl Helsing, chairman of the Data Security Committee for the American Bankers Association, before Congress, "[I]f I were in charge of the Social Security system and concerned about getting those checks out every month, I would be much more concerned about whether those checks were in the correct amounts, made out to the right people, and that they did get out on time, than I would be concerned about an intruder gaining unauthorized access and looking at the files."

41. (p. 47) For example, all transactions over ten thousand dollars must be reported to the Internal Revenue Service. In *United States v. Miller* (425 US 435, 1976, p. 442), the Supreme Court ruled that deposit information does not have an "expectation of privacy" and may be subpoenaed. (A search warrant requires a higher standard of proof.)

42. (p. 47) A Citicorp electronic banking system used directly by corporate customers for funds transfer was penetrated. Posing as a corporate customer, a user from St. Petersburg, Russia, transferred funds to an account in Finland; the money was withdrawn from the bank the next day. The falsified transaction was accomplished with a personal computer and the phone lines. Over the next several months, this scenario was repeated, with the user's locale changing to Rotterdam, San Francisco, St. Petersburg, and Tel Aviv. Citicorp became aware of the thefts after several customers complained of irregularities in their accounts. By August, $12 million had actually been moved, and $400,000 had been stolen.

Relative to the amount that Citicorp transfers daily (about $500 billion) this is relatively little, but of course banks' business is providing security, and in that sense the theft loomed large. Access to the system required a customer's authorization code; it is believed that the perpetrator (allegedly one Vladimir Levin) had an accomplice within the system who supplied these. The bank has now changed its security to using one-time passwords (Carley 1995; Hansell 1995).

43. (p. 48) An Alaskan oil company investigated why it had been losing leasing bids by small amounts to a competitor and discovered that a line between a computer in its Alaska office and one at its home base in Texas were being wiretapped. A competitor was intercepting pricing advice being sent from the Texas office (Parker 1983, p. 322).

44. (p. 48) The most notable of these is the Digital Millennium Copyright Act, Public Law No. 105-304, 112 Stat. 2860.

45. (p. 49) *Annual Report to Congress on Foreign Economic Collection and Industrial Espionage*, National Counterintelligence Center, Washington, D.C., July 1995, as reported in (Dam and Lin 1996, p. 33).

46. (p. 49) Command 9438, para 10, as cited in Fitzgerald and Leopold 1987, p. 148.

47. (p. 50) This directive was issued after a laptop containing information on 26.5 million US veterans and their families was stolen from a VA analyst's home. The data had not been encrypted.

48. (p. 53) One example occurred in the 1970s, when thousands of phone conversations between IBM executives conducted on the company's private microwave network were systematically eavesdropped on by Soviet intelligence agents. NSA informed IBM of the eavesdropping (Landau et al. 1994, p. 1). A similar incident occurred in the 1980s with a different US corporation (Dam and Lin 1996, p. 68).

49. (p. 54) Electronic commerce is still a loosely defined term. At present, examples range from individuals purchasing physical objects over the Internet by sending electronic mail to businesses making automatic purchases from their suppliers using Electronic Data Interchange protocols. The grand conception of large-scale purchase and sale of information over the network using a combination of digital credit cards (which would sign digital sales slips with digital signatures) and electronic cash (which would be anonymous and have many of the properties of physical money) has yet to materialize but electronic commerce using ordinary credit cards and web pages is thriving.

50. (p. 55) Email, which traditionally was done via a user's machine, now can be done through a website; gmail or Yahoo are examples of such a service. This is part of the "Web 2.0" experience.

51. (p. 56) Zfone (Zimmerman 2006) is an add-on security mechanism that can

be used with any Voice over IP system, created by Philip Zimmerman, the author of PGP.

Chapter 3: Cryptography and Public Policy

1. (p. 57) "Probabilistic cryptography," as put forth by Goldwasser and Micali (1984), is a formalization of the long-standing method of using *message indicators* to guarantee that a cryptosystem starts each message in a unique state.

2. (p. 60) For quite some time, communications security was poor in parts of the US government (particularly the Department of State) that lacked access to the services of the military cryptographers.

3. (p. 61) A 2400-bit-per-second mode of operation is one of the *Minimum Essential Requirements* of the third generation secure telephone unit (STU-III), which went into use in 1987.

4. (p. 63) There appear to have been rotor-based voice systems, but these were probably analog scramblers that filtered the signal into several bands employed rotors to shuffle the bands in a constantly changing pattern.

5. (p. 64) A small number of serious books were published in Europe, in particular Eyraud 1959.

6. (p. 64) Papers on points pure of mathematics whose cryptographic inspiration is clearly visible to people familiar with the subject were written by Andrew Gleason, Marshall Hall, W. H. Mills, and presumably others.

7. (p. 64) Kennedy's orders do not mention cryptography, but require that US nuclear weapons be put under *positive control* of the National Command Authority (the President *and* the Secretary of Defense), wherever in the world they may be located. What this came down to was that they could not be armed by anyone unable to send them properly encrypted messages. The key component in this program is the *permissive action link*, which, in effect, issues an encrypted order to a nuclear weapon. Earlier PALs used conventional cryptography; more recent ones use public-key techniques.

8. (p. 65) Source: private conversations between Diffie and Feistel.

9. (p. 65) Eventually, in the late 1960s, the cryptographic system Feistel's group designed was bundled together with the existing modes of operation of the Mark X IFF. The result was called the Mark XII (there never was an XI), and its cryptographic mode was Mode 4. The Mark XII is employed extensively by the military aircraft of the US and its allies.

10. (p. 65) Source: private conversations between Diffie and Carl Engelman of Mitre in the 1970s and between Diffie and Horst Feistel circa 1990.

11. (p. 66) The importance of the Federal Information Processing Standards is illustrated by FIPS 1, the American Standard Code for Information Interchange

or ASCII. The government's adoption of this code, which is ubiquitous today, made it dominant over the rival EBCDIC encoding used by IBM, then the world's largest computer manufacturer.

12. (p. 67) The one in the 2984 is now called the Alternate Encryption Technique. At least two other IBM systems were also called Lucifer. One designed by John Lynn Smith, but never developed into a product, presents the fullest exposition of Feistel's techniques (Smith 1971). Another system called Lucifer was used only as a tutorial device (Feistel 1973).

13. (p. 67) At NSA, Howard Rosenblum, Deputy Director for Communication Security, and Doug Hogan; at NBS, Ruth Davis, Seymour Jeffery, and Dennis Branstad.

14. (p. 67) NOFORN means "No foreign dissemination allowed." This is an odd designation for many NSA algorithms, since several of the most important are NATO standards.

15. (p. 68) In testimony to congress, NSA Director Bobby Ray Inman asserted that public key cryptography had been discovered at NSA 10 years earlier. It appears that Inman preferred to give credit to three Brits with clearances than three Yanks without. The work in question is that of GCHQ employees—James Ellis, Clifford Cocks, and Malcolm Williamson—and was carried out between late 1969 and mid 1976. The precise scope of the British discoveries did not emerge until after James Ellis's death in the fall of 1997, when Ellis's retrospective history of the work and at least some of the original papers were released (CESG). Although GCHQ claimed priority and most of the discoveries it did make (apparently neither digital signatures nor knapsack systems had occurred to them) were earlier than those made in the public world, the two efforts overlap. Ellis's paper in 1969 is several years before any of the outside work but Williamson's secret internal memo on "Diffie-Hellman" comes two months after the idea had been presented at the US National Computer Conference.

16. (p. 69) The term is a misnomer because the items exchanged are not actually keys. In contemporary literature, the more precise terms *key negotiation* or *key agreement* are preferred, but the original terminology persists.

17. (p. 70) In the early 1970s, for example, secrecy orders were placed on some of the inventions of Horst Feistel, nucleus of the cryptographic research group at IBM.

18. (p. 70) Secrecy orders are often helpful to a company because they delay the granting (and thus expiration) of its patents until a time when the invention is more appropriate to the market. In 1939 the famous actress Hedy Lamarr filed for the first patent on frequency hopping radio (Markey 1942). Had this application been kept secret until the 1970s, when spread spectrum technology emerged from military into civilian applications, Hedy Lamarr would have enjoyed a much more comfortable retirement.

19. (p. 72) Uriel Feige, Amos Fiat, and Adi Shamir had discovered a practical im-

plementation of "Zero Knowledge" protocols (Feige et al. 1987). They submitted a US patent application even as Shamir lectured worldwide on the algorithm. The Army requested a secrecy order be placed on the invention. This was classic shut-the-barn-door-after-the-horse-has-fled; for several months the researchers had been giving lectures about the work. Since secrecy orders forbade the discussion of the research with the foreign nationals, and Feige, Fiat, and Shamir were all Israeli citizens, what American law could do in this situation was unclear. Fearing to present the work at an American research conference under the circumstances, Shamir let various colleagues know about the problem. Help came. Shamir's lawyer got an anonymous call from Dr. Richard A. Leibler, retired head of R5, telling him precisely whom to call to get the secrecy order lifted. Shamir publicly thanked "the NSA ... who were extremely helpful behind the scenes ..." (Landau 1988, p. 12)

20. (p. 72) It was rumored that DES was used by the Argentines in the Falklands War and had seriously hampered British SIGINT.

21. (p. 72) CCEP was modeled on the earlier Industrial Tempest Program, begun in the 1970s, which encouraged industry to build electromagnetically shielded versions of their products.

22. (p. 72) The government also appeared to be laying legal framework for broadened availability of cryptographic equipment. For as long as anyone could remember, all cryptographic devices approved for protection of classified traffic had been owned by the government. Now with NSA's COMSEC Instruction 6002 it provided two ways that government contractors could own the equipment and charge the costs back to government contracts in the same way they did with buildings, computers, or safes.

23. (p. 73) Type I equipment is managed through COMSEC accounts and is basically available only to organizations with government contracts. Under the new rules, owners of Type II equipment would not have COMSEC accounts but would need to have the equipment supplied to them by government sponsors. From the point of view of the user, the distinction between having a government sponsor and having a government agency as customer was minor.

24. (p. 74) Development of the STU-III was paid for directly by NSA, beginning by funding five competitors to prepare proposals.

25. (p. 74) Only a year late, and about 50% over the target price.

26. (p. 74) The Type II version, affected by the same fluctuation of availability rules as other Type II equipment, was not a success. In a move incomprehensible to marketing people everywhere, the Type II STU-III, though advertised from the beginning as inferior to the Type I, was always priced higher. This is at least partly because the Type II never achieved the volume of production originally planned and did not benefit from the same economy of scale as the Type I.

27. (p. 74) Perhaps to help NSA avoid the need to pay royalties for public key technology, just as it used secrecy to avoid paying Hebern.

28. (p. 74) "This policy assigns to the heads of Federal Government Departments and Agencies the responsibility to determine what information is sensitive, but unclassified and to provide systems protection of such information which is electronically communicated, transferred, processed, or stored on telecommunications and automated information systems." (Poindexter 1986, p. 542)

29. (p. 76) Ten years later, DES remains a Federal Information Processing Standard.

30. (p. 76) By this time Poindexter was deep in the middle of the Iran-Contra controversy, and the administration was loath to have him appear at any congressional hearing lest the questioning veer to Iran-Contra. Thus Poindexter did not appear when first requested, and the House committee then subpoenaed him (USHH *Hearings on HR 145*, p. 381). A discussion ensued between the White House and the committee, and the committee delayed hearings an additional two weeks, while the White House withdrew Poindexter's directive (Carlucci 1987), hoping to avoid Poindexter's appearance in Congress.

The committee insisted that the former Presidential National Security Advisor appear, which he did, accompanied by counsel. Despite the fact that Representative Jack Brooks, chair of the Committee, promised that questions would be limited to issues related to the NSDD-145 and the Poindexter Directive (USHH *Hearings on HR 145*, p. 399), Poindexter declined to answer any questions and pleaded the Fifth Amendment. The congressmen, having achieved the withdrawal of the Poindexter directive, did not pursue the matter further.

31. (p. 76) "The development of standards requires interaction with many segments of our society, i.e. government agencies, computer and communications industry, international organizations, etc. [NIST] has performed this kind of activity very well over the last 22 years. NSA, on the other hand, is unfamiliar with it." (USHR 100-153 *Computer Security Act*, p. 26)

32. (p. 77) The Committee on Governmental Operations was the subject of a similar attempt by NSA. "In January 1981, the Director of the NSA even went so far as to write this Committee and complain that the Committee had not forwarded to NSA a copy of its investigative report, 'The Government's Classification of Private Ideas,' *prior* to its issuance. As pointed out by Chairman Brooks in reply to NSA, Congress does not submit its reports to Executive Branch agencies for prereview." (USHR 100-153 *Computer Security Act*, pp. 21–22)

33. (p. 78) The Budget Office also noted that the Act would result in savings due to the elimination of fraud and other financial losses (USHR 100-153 *Computer Security Act*, p. 43).

34. (p. 78) Officially known as the Balanced Budget and Emergency Deficit Act of 1985, the act set annual deficit targets for five years, aiming for a balanced budget in 1991. It was never fully implemented.

35. (p. 78) There was an additional $800,000 of 'reimbursable' funds from other agencies; such funds are typically for help in deploying advanced technologies.

36. (p. 79) One example of such deference to NSA was NBS's failure to support its own standard in the International Standards Organization. About 1985 ISO took up consideration of DES as an international standard and approached the American National Standards Institute, which in turn approached its cryptographic committee X3T1, on which NBS sat. NBS cast its vote in X3T1 against recommending DES; ANSI abstained in the international committee, and ultimately ISO did not adopt DES. Another example is Raymond Kammer's decision that NIST would support NSA's decision that NIST abandon RSA as a choice for a public-key signature standard (Source: private conversation between Landau and Kammer, December 19, 1996.)

37. (p. 80) Source: private conversation between Landau and McNulty, December 2, 1996.

38. (p. 81) Source: private conversation between Landau and McNulty, December 2, 1996.

39. (p. 81) Schnorr applied through the EEC for a patent, thus obtaining patents in Germany, United Kingdom, France, the Netherlands, Italy, Spain, Belgium, Switzerland, Sweden, Liechtenstein, and Austria; see Schnorr 1989, Schnorr 1990, and Schnorr 1991.

40. (p. 81) NIST countered this last point by noting that the Kravitz algorithm was roughly 25 times faster than the RSA algorithm in signing (USDoC 1991a).

For most applications speed of verification is more important than speed of signing, since a signature is signed only once and may be verified many times. In some applications such as signing software to protect against the introduction of viruses, the signature may literally be checked billions of times over the lifetime of the product. On the other hand, there is something to be said for making the signing operation more economical, because it is the one that uses the secret information and is best done in an isolated environment like a smart card. At the time, such cards had limited computational power.

41. (p. 82) "The key question during the hearings was: Should a military intelligence agency, NSA, or a civilian agency, [NIST], be in charge of the government's computer standards program?" (USHR 100-153 *Computer Security Act*, p. 19)

42. (p. 82) "Observers—including OTA—consider that [the MOU] appears to cede to NSA much more authority than the act itself had granted or envisioned, especially considering the House report accompanying the legislation." (USC-OTA 1994, pp. 13–14)

The General Accounting Office said: "[T]his Memorandum of Understanding made NSA appear to more influential in NIST's standard-setting procedure relative to cryptographic systems than was intended by the Congress in the Computer Security Act of 1987." (USGAO 1993a, p. 16)

43. (p. 83) Claus Schnorr's patent was licensed from him by RSA Data Security, which provides DSS code in its cryptographic toolkits. On the other hand, many

people have studied Schnorr's patent and maintain that it does not cover DSS. To date the issue of the patent's validity has yet to be litigated.

44. (p. 83) Source: private conversation between Landau and Brooks, January, 17, 1997.

45. (p. 83) Source: private conversation between Landau and Kammer, December 19, 1996.

46. (p. 84) Source: private conversation between Landau and Kallstrom, January 16, 1997.

47. (p. 84) New York City's District Attorney of the fifties, Frank Hogan, was a strong proponent of wiretapping, and in 1955 he testified to Congress: "In these and in many other important prosecutions, the investigative technique of wiretapping was invaluable. In a substantial number I may say, gentlemen, it is indispensable." (USHH 84 *Wiretapping*, p. 322)

48. (p. 84) There were 419 electronic surveillances conducted in New York in 1994 (AO 1995, pp. A26–A33, and p. A90) (AO 1996, pp. 54–60 and pp. 126–140).

49. (p. 84) There were 53 court-ordered surveillances in California during 1994 (AO 1995, pp. A2–A7 and p. A8), (AO 1996, pp. 34–38 and p. 94).

50. (p. 84) Source: private conversation between Landau and Kallstrom, January 16, 1997.

51. (p. 85) Source: private conversation between Landau and Brooks, January 17, 1997.

Chapter 4: National Security

1. (p. 88) On October 24, 1969, President Nixon announced a decision to make narcotics a matter of foreign policy. The CIA was asked to "contribute to the maximum extent possible in the collection of foreign intelligence related to traffic in opium and heroin" (USDoJ 1976, pp. 46–47). President Ford later called the smuggling of opium to the United States a "national-security" issue (ibid., p. 59).

2. (p. 88) In a speech in the Spring of 1997, President Bill Clinton invoked the name of national security in support of education.

3. (p. 88) Cryptography, once so central to information security as to be almost indistinguishable from it is now reduced to the status of one important part.

4. (p. 89) The CIA Foreign Broadcast Information Service publishes transcripts of numerous foreign radio shows.

5. (p. 89) A similar inference might simply have been drawn from looking at the number of cars in the parking lot or the number of lighted windows. The significance of the pizza story lies in showing how difficult operations intelligence

is to counter. Security officers may well have thought to hold meetings in inner offices, or to take measures to avoid having the parking lot look full. Any large scale operation, however, leaves many telltale traces and it is hard to anticipate and cover up all of them. Keeping the kitchens open all night might stem the flow of pizza orders only to create other signs of activity: the amount of raw food being ordered, the quantity of garbage put out for collection, or the hours of the kitchen staff. In many organizations where all professional staff members are required to have security clearances, employees such as cooks, whose activities can be confined to daytime hours and a small part of the building, are not.

6. (p. 89) An amusing example of the CIA's aggressive interest in using travelers as spies is given by the Hemingway scholar Michael Reynolds. In 1975, Reynolds made persistent attempts to get a visa to Cuba in order to study Hemingway's library. In the process, the CIA contacted him on the chance that if he got to Cuba, he might get to meet Fidel Castro—who was known for surprise visits with tourists. Reynolds also gives plausible evidence that the CIA's enthusiasm for his trip to Cuba went so far that they had his phone bugged (Source: Michael S. Reynolds, *Hemingway's Reading*, 1910–1940, Princeton University Press, 1981.)

7. (p. 90) Another new, and very controversial, form of intelligence is: RUMINT, intelligence gathered from unreliable sources, including rumors. Many in the intelligence community discount the value of such intelligence, and, indeed, blame RUMINT for the faulty intelligence the United States used prior to the 2003 war in Iraq (Kristof).

8. (p. 91) These, however, have created embarrassments of their own. Consider the shooting down of an American U2 spy plane over the Soviet Union in 1960 or the capture of the American spy ship *Pueblo* by North Korea in early 1968.

9. (p. 92) This appears to have been a longer-range, and higher-altitude version of the mechanism that seagulls use to detect an impending storm and fly inland. It is an eerie commentary on the success of secrecy in the intelligence community that this project, which is supposed to have been abandoned, was kept secret for nearly 50 years. In 1995 it was offered as the explanation for the Roswell Incident of 1947, which is, despite official denials, believed by many people to have been the crash of an alien spaceship (Thomas 1995).

10. (p. 92) Distinguishing nuclear explosions from other events, such as large lightning bolts or explosions of meteors in the atmosphere, is not easy. The satellite reacts less to the total energy of the blast than to the form of the flash. Nuclear explosions have a characteristic two-humped flash caused by gamma ray induced formation of nitrogen pentoxide (N_2O_5). The time between the humps is called the *bhang metre* and is characteristic of the type of weapon.

11. (p. 93) For example, the Krasnoyarsk radar, which was alleged by the US to violate the ABM treaty, was photographed from space only after its location had been reported by a human source (Richelson 1987, p. 79).

12. (p. 93) The study of radar signals sometimes involves elaborate provocations

designed to create the impression that an attack is in progress and thereby to fool an enemy into using radars that are not meant to "come to life" except under battle conditions. A persistent theory of the strange movements of Korean Airlines 007, which led to its being shot down, is that its purpose was to provoke the radars of the Kamchatka Peninsula air defense system into action so that other aircraft—RC-135's designated Cobra Ball—could observe them (Johnson 1978).

13. (p. 94) It has long been known that, unless a radio is specifically designed to conceal the information, one can discover what station is being received by measuring the frequency of the local oscillator. This, however, depends on knowing the intermediate frequency, and that information may not be available about a radio of unknown type. Around 1960, it was discovered that many radios had a frequency shift in the local oscillator, resulting from an effect of the automatic gain control on the high-voltage supply, that was proportional to the frequency of the received signal. That discovery portended vastly expanded exploitation of unintentional signals emitted by receivers (Wright 1987, p. 93).

14. (p. 94) One function of the US Argos satellites was to monitor telemetry from Soviet missile tests. This fact is believed to have come to Soviet attention as a result of the activities of Christopher Boyce, who was subsequently convicted of spying on CIA projects at a contractor, TRW, in Southern California (Lindsey 1979).

15. (p. 94) In order to prevent tampering with the satellite, control link transmissions are often encrypted. The device used for this purpose by the military, the KI-23, is the main product of Mykotronx, later known as the maker of Clipper and Capstone chips.

16. (p. 95) In the 1950s, British counterintelligence employed a corps of "watchers" to follow hostile diplomats. The watchers communicated with MI5 by radio, and in an attempt to conceal the their activities, they tried communicating in code. Peter Wright recalls in his autobiography *Spycatcher* that this was of little use; the mere occurrence of the traffic was sufficient to reveal to the Russians where the watchers were operating (Wright 1987, pp. 52–53). In a much more recent example of the same phenomenon, Tsutomu Shimomura (Shimomura 1996, Shimomura 1997, p. 76) reports that in tracking Kevin Mitnick, who had broken into Shimomura's computer, Mitnick's use of cryptography "didn't slow them down at all." Quite a different example came to light in conjunction with the Yom Kippur war of 1973: it was said that the Israelis should have been alerted that something was up by the improved communication security on the part of the Egyptians.

17. (p. 95) A remarkable example of this occurred shortly before the Normandy invasion in World War II. The Japanese military attache in Germany demanded a tour of German defenses in Normandy and reported what he had seen to Tokyo. That transmission, presumably encrypted in Purple, was read by the Allies and

supplied them with an expert assessment of German channel defenses (Kahn 1967, p. 508; Boyd 1993). Another example is provided by the Gamma Guppy intercepts of the early 1960s, in which the US embassy in Moscow monitored mobile phone traffic from the limousines of Soviet officials (Bamford 1982, p. 283).

18. (p. 95) Physical taps on lines do have a role, however, and not just in counterintelligence work. In the 1980s a former NSA employee named Ronald Pelton was recruited by the Soviets. One of the things he allegedly told them was that the United States had placed a tap on a cable running under the Sea of Okhatsk. According to a Soviet publication, the tap, which weighed 12 tons, was powered by plutonium and serviced by US submarines.

19. (p. 96) The largest Soviet intercept station outside the USSR was at Lourdes, Cuba. It could pick up satellite transmissions intended for receivers in Washington, New York, and other eastern cities.

20. (p. 96) In the 1970s and the 1980s, there was a war of words between US and Soviet diplomats over Soviet microwave interception activities from a residence the Soviets maintained at Glen Cove, New York (Broad 1982).

21. (p. 98) On the face of it, this incident, in which the Israelis attacked and nearly sank the *Liberty*, is inexplicable. It was claimed that the *Liberty* should actually have been hundreds of miles away, in the waters off Cyprus, but that its orders got delayed. If the *Liberty* was, as publicly claimed, spying on the Arabs, there is no reason for the Israelis to have attacked it. On the other hand, the Israelis' claim that they mistook the *Liberty* for an Egyptian freighter hardly seems credible. Loftus and Aarons (1994) have produced an explanation that, although not supported by overwhelming evidence, is at least sensible. It is their thesis that the *Liberty* was actually listening to traffic from Israeli tanks and manpack radios as part of a secret deal to report weaknesses in the Israeli southern front to Egypt. To do this, it would have to have been quite close.

22. (p. 100) VoIP, with its superb adaptation to mobility, presents related and even more serious difficulties. It is fairly easy to intercept VoIP in a way that gets some part of some of the calls but comprehensive coverage is quite hard.

23. (p. 100) During the shootdown of KAL 007 in September 1983, for example, signals intelligence was severely hampered by the fact that only the transmissions of the interceptor pilots and not those of their ground controllers could be intercepted (Hersh 1986, p. 70).

24. (p. 100) Multiplexing takes three common forms. Two of these are well illustrated by broadcast radio. *Frequency division multiplexing* is the phenomenon by which different stations have different frequencies. To select a station you tune to its frequency. *Time division multiplexing* is the phenomenon that distinguishes programs. Within the frequency of a given station, you listen at a particular time to find the right program. The third form is called *code division multiplexing.* Code division multiplexing is one of the benefits of *spread spectrum communi-*

cation, in which a transmitter uses a wide range of frequencies, often by hopping rapidly from one to another. Using code division multiplexing, multiple transmitters can avoid interference with little prior coordination. The more advanced cordless telephones are perhaps the most common items that use code division multiplexing.

25. (p. 102) The cryptanalysis of World War II systems is discussed in detail in Deavours and Kruh 1985 and in Welchman 1982. Cryptanalysis of classical systems makes up much of the content of *Cryptologia*, the journal of cryptographic history. Cryptanalysis of contemporary cryptosystems can be found in the *Journal of Cryptology* and in the proceedings of numerous annual and biannual conferences, such as Crypto (held in late August in Santa Barbara, California), Eurocrypt (held at a different location in Europe each spring), and Asiacrypt (held each year in the Asia Pacific region). A particularly noteworthy book on the subject is Biham and Shamir 1993. Particularly noteworthy books include the ones by Biham and Shamir (Biham and Shamir 1993) and the encyclopedic reference work by Menezes, van Oorschot, and Vanstone (Menezes) which includes some cryptanalytic material. Sample chapters of the latter are available at: http://www.cacr.math.uwaterloo.ca/hac/ (last viewed 29 August 2006).

26. (p. 102) Most books on the breaking of the German Enigma cryptosystem during World War II focus on the researchers, especially Alan Turing, at Bletchley Park. The actual reading of most of the traffic, however, was done with several hundred special purpose computing machines, called *bombes*, which were operated 24 hours a day by women from the Women's Reserve Navy Service (Welchman 1982, pp. 138–148).

27. (p. 102) Many systems in use today still have either 40-bit keys (which can be searched easily) or 56-bit keys (which can be searched with some difficulty). *Dragging key* (looking through all possible keys) thus has a role to play in contemporary cryptanalysis. A far more subtle, but also universal, cryptanalytic method is the Berlekamp-Massey algorithm (Berlekamp 1968; Massey 1969). It is a fact that any sequence of bits (keystream) whatsoever can be generated by a linear shift register of sufficient length. The Berlekamp-Massey algorithm automatically produces the right register. A major design criterion in modern cryptography is that the "right register" be too long for this approach to be practical.

28. (p. 102) Traffic analysis is fundamentally a matter of discovering the relationships among a number of "address spaces," some observable and others inferred. The call signs, like phone numbers, are the name space of the communications network. Direction finding, emitter identification, and collateral intelligence allow these to be correlated with physical positions, individual pieces of equipment, or command functions.

29. (p. 103) Sanitization goes hand in hand with the desire of intelligence officers to keep raw intelligence out of the hands of their customers. The British learned

this lesson in a particularly blunt fashion at the Battle of Jutland in World War I. Before the battle, a British officer of the line walked into the intelligence center and asked the location of the radio callsign of the admiral commanding the German fleet. He was told, correctly, that it was located in the Jade River. What the officer actually wanted to know was the location of the admiral, who had switched call signs when the fleet had set sail precisely in order to fool the British about his location. British intelligence was not fooled; it knew the German admiral's new location and new callsign. Nonetheless, as a result of the intelligence center's releasing raw intelligence on call-signs, rather than finished intelligence on the locations of forces, the Germans achieved their purpose. The British fleet delayed sailing and the battle, which might have been a major British victory, was indecisive (Beesley 1977).

30. (p. 104) In 1961, William Martin and Bernan Mitchell, two NSA cryptanalysts, defected to Moscow and gave a press conference in which they revealed interception by the US of its allies' communications. According to David Kahn, the loss of intelligence was felt immediately (Kahn 1967, p. 694).

31. (p. 104) In the mid 1970s a panel headed by Nelson Rockefeller concluded that the Soviets were intercepting conversations on microwave telephone channels from Capitol Hill. Even though congressmen are not supposed to discuss classified information over unsecured telephones, the information intercepted from such high-level people, particularly when taken in aggregate, has tremendous intelligence potential. It has been speculated that the Soviet activity was detected because the volume of traffic intercepted was sufficient to permit correlations between fluctuation in the Capitol Hill traffic and communications from the Soviet Embassy to Moscow to be observed.

32. (p. 104) Bobby Inman remarked in an informal discussion after his talk at AFCEA West in Anaheim, California on January 8, 1981 that NSA's product had never been better.

33. (p. 105) Speaking in 1980 at the IEEE Wescon conference in San Francisco, Robert Morris (then at Bell Labs and later Chief Scientist of the National Computer Security Center) said: "We are just leaving a period of relative sanity in cryptography that began shortly after the First World War. During that time people spoke of cryptosystems that were secure for hours, days, weeks, months, and sometimes, years. Before it and after it, they spoke of cryptosystems that were unbreakable."

34. (p. 105) In the 1980s, for example, NSA built two new operations buildings, a new research and engineering building, a chip fabrication facility, and two advanced laboratories away from Fort Meade to be operated by a contractor. Major construction at Fort Meade has subsided since that period but GCHQ, its British cognate, has built a giant round building (called "the doughnut") in Cheltenham.

35. (p. 105) Kim Philby is believed to have had access to information on the

Venona program; the Soviets would thus have learned about it soon after it began.

36. (p. 106) This laboratory is the subject of Aleksandr Solzhenitsyn's novel *The First Circle* (1968) and of a later memoir by Lev Kopelev (who was Rubin in the novel). It is Kopelev (1983, pp. 52–55) who discusses the remarkable technique of assessing the security of *mosaic* or *two-dimensional* (time and frequency) voice scramblers they were developing by printing out a sonogram (a plot of energy and frequency over time) and measuring the time it took to solve the sonogram as though it were a jigsaw puzzle and reassemble it into one representing human voice. In *The First Circle*, which takes place around Christmas 1948, Solzhenitsyn and his fellow workers are under the gun from Stalin to deliver "secret telephony" within about six months. The year I read it was 1974. That year, digitized speech (pre-requisite to high-quality secret telephony or as we call it "secure voice") was the main topic at the ARPA (Advanced Research Projects Agency) Principal Investigators' Conference.—WD

37. (p. 106) After the end of the Cold War Soviet crypto machines began to appear in the collector's market. One of these is a 10-rotor machine called Fialka. Since 'fialka' is a Russian word (meaning violet) and 'Albatross' is a western codeword, the names are of no help in establishing a relationship. Fialka, however, had a number of models spanning the appropriate period. It is interesting to note that although Fialka has the same number of rotors as Sigaba, its rotors are all in one row, compared with Sigaba's two.

38. (p. 108) In the 1980s, US companies were not permitted to export optical-fiber communications systems to the USSR, presumably on the ground that communications carried by fiber would replace radio communications and could not be intercepted.

39. (p. 108) The raw data rate of the V.fast standard is 28 kilobits-per-second, but it incorporates real-time data compression and can often achieve effective throughput of 200 kbps—far more that is available on many current leased line networks.

40. (p. 109) The difficulty of separating the two signals in the communication of autocancelling modems is a function of the size of the *constellation*, the number of combinations of amplitude and phase used in communication. V26ter uses four points, V32bis uses 32 and the more recent V.fast uses 64.

41. (p. 109) Much of dynamic routing technology was developed for another purpose: it increases the survivability of networks against direct attack, a phenomenon that occurs primarily, though not entirely, during open hostilities.

42. (p. 110) AT&T developed a specialized cryptographic device for protecting signaling channels (Myers 1979; Brickell and Simmons 1983, pp. 4–5).

43. (p. 110) The US government's successor to the STU-III, the Secure Terminal Equipment (STE), is primarily an ISDN phone, but is compatible with STU-III. The STE is being manufactured by Lockheed Martin and systematically being

used to replace the aging STU-IIIs. Likewise, the British Brent telephone is an ISDN instrument.

44. (p. 111) Skype can operate between Internet-connected devices or between such devices and more conventional phones. In the latter case, the conventional telephony portion will not be covered by Skype encryption.

45. (p. 111) A precise figure is made difficult to obtain by the problem of deciding what counts as encrypted. At one time, most of the world's encrypted traffic consisted of scrambled pay-tv broadcasts, a good example of the sort of encrypted traffic that either does not interest intelligence agencies or can be accessed without resorting to cryptanalysis.

46. (p. 112) The bombing of communication facilities in France forced the Germans to use radio for their communications with Berlin. The traffic that thereby became available for interception was encrypted with the Siemens and Halske T52 cipher machine. This was especially fortunate because the principles of operation of the T52 are similar to those of the Lorenz SZ40 (an online cipher machine that had earlier been used with radios), and cryptanalytic methods developed to attack the SZ40 proved applicable to the T52. It was to attack these machines, not the Enigma, that the Colossus—arguably the first computer—was built.

47. (p. 112) Photographs of the destruction of a bridge in Baghdad were repeatedly shown during the early days of the attack. The bridge was destroyed, not for its capacity to carry cars and trucks, but to destroy the optical fiber that ran underneath.

48. (p. 113) One development has been the *HARM* or *High-Speed Antiradiation Missile* which is launched from aircraft to home in on the fire-control radars of anti-aircraft weapons and destroy them.

49. (p. 113) The destructive effects of the *Electromagnetic Pulse* or *EMP* was first observed by the United States in a high altitude nuclear test above Johnson Island in the South Pacific. The test damaged electronic equipment as far away as Hawaii. The technique, which has since been refined and can be produced by non-nuclear means, goes under the name *High Energy Microwave* (Van Keuren 1991; Schwartau 1994; AWST 1997a).

50. (p. 113) Jamming describes transmissions intended to interfere with an opponent's communications or other signals such as radar. This is not always a wartime phenomenon. In the mid-eighties HBO was briefly pushed off the air by a more powerful beam carrying a message critical of HBO activities.

51. (p. 113) Communications deceptions are classified as *imitative* if they mimic the communications of an opponent. More subtle communications deceptions are *manipulative*: they do not misrepresent the allegiance of the sender, but convey a false impression of its activities. In the months leading up to the invasion of Normandy in 1944, General George Patton commanded a division, stationed

in southern England, that was pretending—by its communications and other activities—to be an entire army.

52. (p. 114) The distinction between viruses and worms (which might better have been called bacteria) is biologically based. Biological viruses are combinations of genetic material with protective protein coats. They function by invading the genetic material of cells and instructing the cell to produce more viruses. In a similar way, computer viruses incorporate themselves into computer programs. When the program is executed, the virus is executed and exploits the occasion to copy itself into other available programs. A worm, by comparison, is a "free-living" program that invades a computer or a network and tricks its host into running it as a separate process.

53. (p. 114) Viruses first became visible in the 1980s. Their origin is unclear. (I recall discussing the notion of viruses—though not what term was used—with my colleague Jack Holloway in 1970. When I mentioned this to Oliver Selfridge, member of the Baker committee and a longtime advisor to NSA, he told me that the notion had been about in the late 1950s.—WD)

54. (p. 114) Although the claim that viruses were employed against the Iraqis in the first Gulf War appears to be groundless, there are repeated discussions of their development for military applications (AWST 1993; Richardson 1991; Robinson 1993b).

55. (p. 115) A cut out of this sort that prevents the tracing of phone calls is called a *cheese box*.

56. (p. 115) This was at the Air Force IT Conference in Montgomery, Alabama (Onley).

57. (p. 117) Even introducing a small number of errors makes the analysis of data far more difficult, and an error rate of just over 11% reduces the information content of a channel by half. In the mid-1980s, the notion of having the DoD give out false information about weapons developments was publicly mooted (North 1986).

58. (p. 118) Motorola manufactured a device called Ladner to encrypt analog telephone lines. Linkabit, California Microwave, Racal Datacom, and Cylink made high-speed DES-based encryptors to protect the digital ones.

59. (p. 118) AT&T developed DES and public-key-based encryption devices that were subsequently applied to securing common channel interoffice signaling (Myers 1979).

60. (p. 121) As we will see later, this is no longer entirely true.

61. (p. 122) People often refer to high grade cryptographic systems as being "unexportable." In fact, much of the best US cryptographic equipment—for example, the KG-84, general purpose data encryptor—is sold to the governments of NATO countries and other American allies and in some cases even "co-manufactured" in foreign countries. Exports of equipment of this sort are gov-

erned by individually approved export licenses and usually take place under the *Foreign Military Sales Program.*

62. (p. 122) Precisely what the capabilities of intercept equipment are is hard to tell. Under a deal between NSA and the Software Publisher's Association, some cryptographic systems with 40-bit keys could be rather freely exported by the early 1990's, when embodied in "mass market software." Since computers could already execute 2^{40} instructions in an hour at that time, 40-bit keys did not represent very much security from a commercial viewpoint. On the other hand, it is unlikely that intercept devices, which are comparable in price to high-end workstations, could do any better. Since decisions about intercept must be made not in hours, but in fractions of a second, it is prudent to presume that NSA knew how to break the ciphers in question with a workfactor substantially less than 2^{40}.

63. (p. 123) That the true mission of NSA's export-control office is intelligence and not administration is revealed by its organizational designation: G033 (later changed to Z033) rather than Q or D—arguably a failure of operational security.

64. (p. 123) Aside from electronic funds transfers between banks, businesses use telecommunications for a variety of other high value communications. Oil companies routinely prospect at locations scattered around the world. Their analyses of core samples and other data form the basis for bids on drilling rights. Bids by multinational corporations on contracts distant from their headquarters require communication of information that is sometimes valuable enough to affect the company's survival. Internal transfers of equipment and supplies, can rival actual funds transfers in value.

Chapter 5: Law Enforcement

1. (p. 126) Fingerprints serve two related but distinct functions in police work: identifying available people uniquely and identifying unavailable people via *latent* fingerprints on objects at crime scenes. The former function was not new—fingerprints impressed in clay had been used by the Babylonians for identification of written tablets—but before fingerprinting Europeans used the Bertillion system of body measurements. Fingerprints were an improvement both in being more precise and in having a forensic as well as identificational function (Kelling 1991, p. 960).

2. (p. 128) Earlier *stipendiary police*, like bounty hunters and some sheriffs, were paid at least in part through a share of collected fines—a mechanism whose corrupting potential is obvious. In some measure this system has been reintroduced via forfeiture laws that reward police departments, thought not their members directly, with a share of the proceeds derived from selling property confiscated from criminals.

3. (p. 128) The British scholar Sydney Fowler Wright (1929), commented that so

great was the influence of the police over the magistrates' courts that they had come popularly to be called "police courts."

4. (p. 128) The police commonly express the sentiment "We don't make the laws, we merely enforce them." Although it is technically true that laws are made by legislatures, the law-enforcement community exercises substantial influence over the process. Not only do senior law-enforcement officials ranging from the assistant directors of the FBI to the attorney general frequently testify before Congress on pending bills; many bills are first seen by Congress and its staff in the form of drafts prepared by law-enforcement agencies.

5. (p. 130) Even circumspect statements on a wiretapped phone can be quite useful. Fat Ange Ruggerio of the Gambino crime family was not aware his phone was being wiretapped when he told a colleague, "[I'm handling some] H." The FBI was listening, and agents photographed Ruggerio as he made deliveries to three different drug traffickers (Blum 1993, p. 83).

6. (p. 130) Gravano read the government transcripts. He saw the strength of the Federal case and learned that Gotti was angry with him for being too greedy (Blum 1993, pp. 255–257 and pp. 317–318). Fearing that Gotti was developing a strategy to blame him for various crimes, the underboss turned the tables, and testified against Gotti (Blum 1993, pp. 319–326).

7. (p. 130) Although US agents learned of meetings, they never succeeded in tracking Ames to one (Weiner et al. 1995, pp. 229–230, pp. 245–246).

8. (p. 131) A tap of this kind is often called a bug and not clearly distinguished from a microphone listening to the room. Such devices are inexpensive and easy to install. A radio bug built into an RJ11 "octopus plug" has been advertised in *Popular Electronics* by a company called Seymore-Radix. Its price is about $30.

9. (p. 131) For a more detailed exploration of the ways a line can be tapped see Dash et al. 1959 and Fitzgerald and Leopold 1987.

10. (p. 131) On his first visit to Democratic National Committee Headquarters in the Watergate Building, James McCord succeeded in placing a bug in the phone of the chair's secretary. But this elicited very little useful information, so McCord returned a few weeks later for a second—and fateful—try.

11. (p. 133) Apparently because the results were written down with a pen.

12. (p. 133) In Europe this has not been the case. Long-distance bills were instead compiled by means of a tone-based message-unit system that did not reveal the called number.

13. (p. 133) Signaling System 7 (SS7), introduced to support ISDN in the 1980s, passes the identity of the called phone from switch to switch throughout the whole length of the call.

14. (p. 133) Clifford Stoll (1989, p. 68) gives a dramatic account of such an exercise that took place as late as the mid 1980s.

15. (p. 133) Privacy blocking will prevent the ID information from being given to

the receiving telephone but will not conceal it from either a telephone company switch or private branch exchange attached to the network by a DS1 connection.

16. (p. 133) Analysis of billing information during their investigation of the 1993 bombing of the World Trade Center led the FBI from the initial suspect to his co-conspirators (Mashberg 1993; Bernstein 1994). More recently, it has come to light that after the 9/11 attacks the National Security Agency began receiving billing information in vast quantities for similar purposes.

17. (p. 135) In the United States and Canada, 911 is the phone number for emergency services: police, fire, and ambulance.

18. (p. 135) Another conspirator in the 1993 World Trade Center bombing, Eyad Ismoil, was picked up through a matching of telephone records with airline manifests; he was later convicted (McKinley 1995a).

19. (p. 135) Investigators also used photos from several days before the explosion to prove that Timothy McVeigh was the "Robert D. Kling" who, on the afternoon of April 17, 1995, in Junction City, Kansas, rented the Ryder truck used in the bombing. Days and weeks after the bombing investigators meticulously reconstructed McVeigh's movements on April 17. Surveillance photos taken at a McDonald's about a mile from the Ryder agency showed McVeigh at the restaurant at 3:49 and 3:57 PM on that day. Shortly afterward, "Kling" rented the truck. When prosecutors claimed that the McDonald's photo was of McVeigh, his lawyer did not dispute the point. The photo was taken several days before there was any hint it would be useful in a criminal case—and *then the evidence was available when needed* (Brooke 1997a).

20. (p. 136) For decades, state-issued drivers' licenses have been *de facto* identity cards in the US. Congress has until recently rejected the introduction of national identity cards. Now it has changed its mind in a remarkably oblique manner. As the *New York Times* put it, "What Congress [did] instead is to ram through a bill that turns state-issued driver's licenses into a kind of phony national identity card through the mislabeled 'Real ID' provision. And in order to make absolutely sure there's no genuine debate, the sponsors have tied it to a crucial bill providing funds for American troops in Iraq and Afghanistan" (New York Times 2005). (The Real ID Act was introduced as HR 418, but was eventually attached to the emergency Supplemental Appropriations Act for Defense, the Global War on Terror, and Tsunami Relief 2005 (HR 1268).) The Real ID Act required that beginning in 2008, state drivers licenses were to adhere to common machine-readability standards determined by DHS. The licenses were to include name, birth date, sex, ID number, a digital photograph, address—and the data had to be verified with the federal government and other states before a driver's license could be issued. No longer would drivers be allowed to have more than one license, which had been a common practice, for "snowbirds" who spent their winters in Florida and their summers in northern climes, and only citizens and legal residents would be permitted to have such licenses.

21. (p. 137) The provisions were later extended to the other armed forces.

22. (p. 138) Other investigators have reached different conclusions (Burnham 1996, p. 218).

Chapter 6: Privacy: Protections and Threats

1. (p. 142) Article 17 1. No one shall be subjected to arbitrary or unlawful interference with his privacy, family, home or correspondence, nor to unlawful attacks on his honour and reputation. 2. Everyone has the right to the protection of the law against such interference or attacks (United Nations 1985, p. 149).

2. (p. 143) In East Germany clergymen and other religious workers were informants; siblings informed on one another, and there were even husbands who informed on their wives (Kinzer 1992).

3. (p. 143) The main purpose of the Privacy Act of 1974 (Public Law 93-579) was to ensure that federal records on individuals were accurate, timely, complete, and relevant (US-PPSC 1977, p. 17).

4. (p. 145) The term "Secretary of State" must have designated an office more like that of the secretary of state of California (whose duties include certifying election returns) than like that of the US secretary of state, whose position in Britain is called "foreign secretary."

5. (p. 145) "You will know from whom this comes without a signature: the omission of which as rendered almost by the curiosity of the post office. Indeed a period is now approaching during which I shall discontinue writing letters as much as possible, knowing every snare will be used to get hold of what may be perverted in the eyes of the public." (Thomas Jefferson, in a letter to James Thomas Callender, October 6, 1799; see Jefferson, *Works*, Federal Edition, Vol. 9, p. 488).

6. (p. 145) From one post office per forty-three thousand inhabitants in 1790, the US postal system had grown to one post office per slightly over one thousand in 1840 (Ellis, p. 51). "There is an astonishing circulation of letters and newspapers among these savage woods," wrote Alexis de Tocqueville in 1831 (deTocqueville, p. 283).

7. (p. 145) The complaints centered on theft rather than lack of confidentiality.

8. (p. 146) Mail from prisoners of war, and between the Union and the Confederacy, was a different matter; it was routinely opened and censored (Scheele 1970, p. 88).

9. (p. 146) Wiretapping appears to have been rare.

10. (p. 147) See, e.g., *State v. Litchfield* 58 Me. 267 (1870), *National Bank v. National Bank* 7 W. Va. 544 (1874), *United States v. Babcock* 3 Dill 567 (1880), *United States v. Hunter* 15 Fed. 712 (1882), *Ex Parte Jaynes* 70 Cal. 638 (1886),

Re Storrer 63 Fed. 564 (1894), *Western Union Telegraph Co. v. Bierhaus* 8 Ind. App. 563 (1894), as reported in (Seipp 1977, p. 59).

11. (p. 148) In fact, people felt more secure than was justified. Confidence in the sanctity of first class mail was so great that most people were unaware that there were legal circumstances under which it could be opened.

12. (p. 150) Under the Fourteenth Amendment the citizens are protected from intrusions by the states.

13. (p. 152) Specifically, the court held that, "Where, as here, the Government uses a device that is not in general public use, to explore details of a private home that would previously have been unknowable without physical intrusion, the surveillance is a Fourth Amendment 'search,' and is presumptively unreasonable without a warrant." (Kyllo, p. 38).

14. (p. 153) Bork was a candidate for the Supreme Court. During his confirmation hearings, the press reported his video-rental habits, which tended to run to Hitchcock and Cary Grant.

15. (p. 154) For example, the Social Security Administration matches its Supplemental Security Income Benefit with the Internal Revenue Service's tax data so as to avoid paying duplicate benefits (USGAO 1990, p. 24)

16. (p. 154) PL 93-579.

17. (p. 154) Under the Privacy Act, there continue to be notices in the Federal Register about federal systems of records, so it is theoretically possible to gather the aggregate information. In practice, such counts are likely to be inaccurate.

18. (p. 155) In 1900 there were about two-billion people and no database of two-billion items. Today many people could store the names of the world's six-billion people on the multi-hundred gigabyte disks of their laptops and laptops will surely have the capacity to store full dossiers on everyone long before a database of such dossiers is collected.

19. (p. 156) Thus, for example, the doings of the Mississippi Sovereignty Commission, in which, the state, from 1956 to 1977, authorized spying, harassment and intimidation of civil-rights workers in order to delay or halt desegregation, only became public in 1998 (Kettle 1998).

20. (p. 156) Act of June 18, 1929, ch. 28, sec., 11, 46 Stat. 25.

21. (p. 157) Thomas Clark, who later became US Attorney General, was assigned to the Western Command. He recalled that a Census Bureau member had shown him files detailing exactly where Japanese-Americans lived (reported in (Okamura 1981, pp. 112–113)).

22. (p. 157) Postal workers are permitted to open first-class mail, but only with the explicit permission of the addressee or if the employee is trying to determine an address to which to send the mail (39 U.S.C. 3623(d)). Otherwise, a warrant is needed; that has been US law since at least 1878 (*Ex Parte Jackson*, 96 US 727, 1878, p. 733).

23. (p. 158) See Chapter 2 of Charns 1992, for a fuller discussion of this incident.

24. (p. 159) "Evidence indicates that the FBI did not believe that the Communist Party [constituted] as serious a threat as it had in the 1940s" (USSR 94 *Intelligence Activities: Rights of Americans*, p. 66).

25. (p. 159) These are from the following FBI memos: Memo from FBI Headquarters to New York Field Office, July 11, 1960; Memo from FBI Headquarters to New York Field Office, December 16, 1960; Memo from FBI headquarters to New York Field Office, November 3, 1961, as reported in (USSR 94 *Intelligence Activities: Staff Reports*, pp. 363–364).

26. (p. 159) $42,500 for disruption activities by the FBI, $96,500 for surreptitious entries by the FBI, and $125,000 for the FBI's use of informants (*Socialist Workers Party v. Attorney General of the United States*, 73 Civ. 3160, 1986).

27. (p. 161) The justification was "Martin Luther King, Jr., head of the Southern Christian Leadership Conference (SCLC), an organization set up to promote integration which we are investigating to determine the extent of Communist Party (CP) influence on King and the SCLC, plans to attend and possibly may indulge in a hunger fast as a means of protest." (Sullivan 1964)

28. (p. 161) The memo is reproduced on p. 713 of USSH 94 *Intelligence Activities: Huston Plan*.

29. (p. 161) In a letter from the FBI to Vice President Agnew, Ralph Abernathy, President of the SCLC, is characterized as a man "who, although he advocates nonviolence, has invited violence by some of his statements." (USSH 94 *Intelligence Activities: FBI*, p. 494, Exhibit 38-3.) In 1970 the FBI forwarded information on Abernathy's private life to Vice President Agnew. The Church Committee hearing exhibits include a letter the FBI [signature blanked out] to the Vice President, "... In response to your request, there is attached information regarding ... Ralph David Abernathy ... The material also includes information about [his private life] (sic) ..." Exhibit 38-3, in (USSH 94 *Intelligence Activities: FBI*, p. 494).

30. (p. 162) Bond, a Georgia state legislator, and Jackson, executive director of SCLC Operation Breadbasket, were active in the civil rights movement. Baez and Guthrie (son of the legendary Woody Guthrie) were folk singers, Coffin, chaplain at Yale, Spock, a physician and the author of the well-known *Baby and Child Care*, that had been the bible of American parents in the post-war years, were all active in the anti-war movement. Stevenson made it into the files because of his association with Jackson (O'Brien 1971, p. 127). Mikva, a member of the House active in the anti-war movement, said that he learned from Senator John Tunney "how I became eligible for the files. Jesse Jackson is a constituent of mine; Adlai Stevenson is a friend of mine; and my wife used to work for the American Civil Liberties Union." (Mikva 1971, p. 130). Ralph Stein, formerly with US Army, Counterintelligence, in (Stein 1971, p. 266) told of the surveillance of Baez, Bond, Coffin, Guthrie, Jackson, King, and Spock. Stein did not mention

Mikva or Stevenson, but Mikva testified to the existence of Army surveillance files on both, as did O'Brien. (Mikva 1971, p. 136; O'Brien 1971, p. 120 and p. 127)

31. (p. 163) This included Lloyd Norman, a *Newsweek* reporter writing on US military plans in Germany, and Hanson Baldwin, a *New York Times* reporter and military historian who had written on Soviet missile sites (USSR 94 *Intelligence Activities: Rights of Americans*, p. 63).

32. (p. 163) During Johnson's administration Attorney General Nicholas deB. Katzenbach had wrested control of electronic surveillance back from the FBI and imposed certain limitations on its use (USSR 94 *Intelligence Activities: Rights of Americans*, p. 105).

33. (p. 163) As he signed the bill, Johnson said: "Title III of this legislation deals with wiretapping and eavesdropping.

My views on this subject are clear. In a special message to Congress in 1967 and again this year, I called—in the Right to Privacy Act—for an end to the bugging and snooping that invade the privacy of citizens.

I urged that the Congress outlaw 'all wiretapping and electronic eavesdropping, public and private, wherever and whenever it occurs.' The only exceptions would be those instances where 'the security of the Nation itself was at stake—and then only under the strictest safeguards.'

In the bill I sign today, Congress has moved part of the way by

— banning all wiretapping and eavesdropping by private parties;

— prohibiting the sale and distribution of 'listening-in' devices in interstate commerce.

But the Congress, in my judgement, has taken an unwise and potentially dangerous step by sanctioning eavesdropping and wiretapping by Federal, State, and local law officials in an almost unlimited variety of situations.

If we are very careful and cautious in our planning, these legislative provisions could result in producing a nation of snoopers bending through the keyholes of the homes and offices of America, spying on our neighbors. No conversation in the sanctity of the bedroom or relayed over a copper telephone wire would be free of eavesdropping by those who say they want to ferret out crime." [Johnson 1968]

34. (p. 164) Attorney General Edward Levi later wrote Kraft that the FBI's file "did not indicate that [Kraft's] activities posed any risk to the national interest" (Pincus 1976).

35. (p. 165) "This demonstration could possibly attract the largest number of demonstrators ever to assemble in Washington, D.C. The large number is cause for major concern should violence of any type break out. It is necessary for this Bureau to keep abreast of events as they occur, and we feel in this instance ad-

vance knowledge of plans ... would be most advantageous to our coverage and the safety of individuals and property." (Hoover 1969b)

36. (p. 165) These included Columbia University's Mathematics and Science Library, the New York Public Library, the Lockwood Memorial Library at the State University of New York at Buffalo, the Courant Institute of Mathematical Sciences Library, the University of Maryland at College Park Engineering and Physical Sciences Library, the University of Houston Library, and the Engineering and Mathematical Sciences Library at the University of California at Los Angeles (Foerstal 1991, pp. 54–69).

37. (p. 166) The Foreign Agents Registration Ac (22 U.S.C. 611 et. seq.) was passed in 1938 in response to Nazi propagandists working to influence the US government and the public. The law requires those in pay of a foreign government seeking to sway US public opinion through engaging in political activities, acting in a public relations role, soliciting or distributing items of value for a foreign principal, or representing the foreign principal to a member of the US government to register with the Foreign Agent Registration Unit within the Criminal Division of the US Department of Justice.

38. (p. 168) Two of the eight, Khader Hamide and Michael Shehadeh, were permanent residents and thus were charged with being associated with a group that advocated destruction of property, a deportable offense for non-citizens; the others were charged with "technical" violations of their visas (ibid., p. 35).

39. (p. 168) One member of the case was finally granted his petition for citizenship in 2006 (Caldwell).

40. (p. 168) The Uniting and Strengthening America by Providing Appropriate Tools Required to Intercept and Obstruct Terrorism Act of 2001 (Public Law 107-56); see chapter 11 for a discussion of the Act.

41. (p. 168) See chapter 5 for a discussion of the Real ID Act.

42. (p. 169) Despite repeated warnings from the General Accounting Office, these browsings continued. There were 449 unauthorized file searches in 1994, 774 in 1995, 797 in 1996 (USGAO 1997; Richardson 1997).

43. (p. 169) During the investigation of the "sugar lobby" in 1962, ten phone lines of a Washington law firm were wiretapped. Several advisors to Martin Luther King who were lawyers were wiretapped (USSR 94 *Intelligence Activities: Staff Reports*, p. 340).

Chapter 7: Wiretapping

1. (p. 173) "Eavesdrop" does not, as it might appear, mean to hang from the eaves and listen to what is going on in an adjacent room. The eavesdrop is the area within the eaves of a house, what we would today call, the footprint of the

house, and to eavesdrop is to trespass within the eavesdrop in order to look or listen.

2. (p. 175) In 1998 it was revealed that the Los Angeles Police Department had routinely used information gleaned from ongoing—and legal—wiretaps to open new investigations in which, in direct contravention of the law, suspects were never informed of the role that wiretaps had played in their case, even during trial (Krikorian 1998).

3. (p. 177) "General Stuart was always accompanied by his own telegraph operator, who had no difficulty in connecting his portable instrument at any point of the wires, and could thus read off and reply to the messages *in transitu*. One of these on the occasion in question, was addressed to the Quartermaster-General, who had just sent off to the Federal army a large number of mules, all of which had fallen into the hands of Stuart. Accordingly, the following message was despatched [sic] to this official :—"I am much satisfied with the transport of mules lately sent, which I have taken possession of, and ask you to send me soon a new supply.—J.E.B. Stuart."(von Borcke, p. 168).

4. (p. 182) Hoover went to extraordinary lengths to hide the wiretap logs as well as records that would reveal wiretapping had occurred; the FBI Director even hid the name of the filing system in which wiretap records were stored. After the Coplon case, wiretap information went into the "June" files, June being Hoover's codeword for "Top Secret." See Theoharis and Cox 1988, pp. 256–261 for a discussion of Hoover's methods.

5. (p. 184) For many years stories circulated that before Hoover's annual testimony to Congress the FBI Director had wiretaps removed, and then had the taps reinstated afterwards. This way Hoover could minimize the number of active wiretaps reported to Congress. The Church Committee carefully examined the number of wiretaps for the dates in question and concluded this story was apocryphal (USSR 94 *Intelligence Activities: Staff Reports*, p. 302).

6. (p. 185) He certainly had no objection to doing so when the president made such requests (FBI 1975a; USSR 94 *Intelligence Activities: Staff Reports*, p. 313–314).

7. (p. 185) The wiretapped justices include Hugo Black, Stanley Reed, William O. Douglas, Abe Fortas, and Potter Stewart (Charns 1992, pp. 17, 25, 87); (Hoover 1970).

8. (p. 194) The Judiciary Committee Report on the act said that "each offense was chosen because it was intrinsically serious or because it is characteristic of the operations of organized crime," (USSR 90-1097 *Omnibus Safe Streets and Crime Control*, p. 97) and that "the last provision [interstate transport of stolen goods] is included to make it possible to strike at organized crime fencing" (USSR 90-1097 *Omnibus Safe Streets and Crime Control*, p. 98).

9. (p. 194) In an emergency, a wiretap may be placed without a warrant; however, if a warrant is not obtained within 48 hours, the information produced

—like any electronic communication intercepted in violation of Title III—may not be received in evidence or even divulged (Omnibus Crime Control Act 1968 §515).

10. (p. 194) The fax and computer provisions were added by the Electronic Communications Privacy Act.

11. (p. 194) The stringent requirements for obtaining a wiretap order do not, however, mean that such surveillance may only be done as a "last resort." (*United States v. David Smith.* 893 F. 2nd 1573 (9th cir. 1990))

12. (p. 194) This requirement was codified in a supplementary law enacted in 1970 (Omnibus Crime Control and Safe Streets Act §2518(4)).

13. (p. 197) Aside from Morton Halperin, there were:

- National Security Council members Helmut Sonnenfeldt, Daniel Davidson, Richard Sneider, Winston Lord, and Tony Lake;
- State Department members Richard Pedersen and Richard Moose, Ambassador William Sullivan;
- Department of Defense member Colonel Robert Pursley;
- White House staff John Sears, William Safire, and James McLane, and;
- correspondents Henry Brandon (*London Sunday Times*), Hedrick Smith (*New York Times*), and Marvin Kalb (CBS News).

14. (p. 197) The *Post* began to publish the papers after the *Times* was served with an injunction barring publication.

15. (p. 198) Ellsberg had also been picked up on the Halperin wiretaps; during the 21 months, Ellsberg had been overheard on 15 occasions (Hersh 1983, p. 325). Halperin was circumspect in his conversation, but Ellsberg was not; he talked about taking "trips" and carrying "stuff" to a friend's house—clear allusions to drugs. The wiretap transcripts, including these comments, went to the White House.

When Ellsberg's role in leaking the Pentagon Papers was discovered, Kissinger, who had earlier hired Ellsberg as a consultant to the National Security Council, tried to distance himself from the leaker. He disparaged Ellsberg to the president by calling him a drug abuser. When Nixon queried Kissinger about this, the National Security Advisor replied "There is no doubt about it." (Hersh 1983, p. 384) Thus we see the insidiousness of wiretaps; the Halperin-Ellsberg wiretapped conversations, *which had never showed any evidence of national-security leaks* (ibid., p. 397), were forwarded to the White House, where the private discussions between two colleagues became ammunition for character assassination and worse.

16. (p. 202) The Church Committee observed that certain types of surveillance carried out by the intelligence agencies had been illegal at the time (USSR 94 *Intelligence Activities: Rights of Americans*, pp. 12–13). The members recommended

legislation to regulate "domestic security activities of the Federal Government" (ibid., p. 295).

17. (p. 202) *Recommendation 6.*—The CIA should not conduct electronic surveillance, unauthorized entry, or mail opening within the United States for any purpose (USSR 94 *Intelligence Activities: Rights of Americans*, p. 302).

18. (p. 202) *Recommendation 15.*—NSA should take all practicable measures consistent with its foreign intelligence mission to eliminate or minimize the interception, selection, and monitoring of communications of Americans from the foreign communications.

Recommendation 16.—NSA should not be permitted to select for monitoring any communication to, from, or about an American without his consent, except for the purpose of obtaining information about hostile foreign intelligence or terrorist activities, and then only if a warrant approving such monitoring is obtained in accordance with procedures similar to those contained in Title III of the Omnibus Crime Control and Safe Streets Act of 1968 (USSR 94 *Intelligence Activities: Rights of Americans*, p. 309).

19. (p. 202) *Recommendation 52.*—All non-consensual electronic surveillance should be conducted to judicial warrants issued under authority of Title III of the Omnibus Crime Control and Safe Streets Act of 1968.

The Act should be amended to provide, with respect to electronic surveillance of foreigners in the United States, that a warrant may issue if:

(a) There is probable cause that the target is an officer, employee, or conscious agent of a foreign power.

(b) The Attorney General has certified that the surveillance is likely to reveal information necessary to the protection of the nation against actual or potential attack or other hostile acts of force of a foreign power; to obtain foreign intelligence deemed essential to the security of the United States; or to protect national security information against hostile foreign intelligence activity.

(c) With respect to any such electronic surveillance, the judge should adopt procedures to minimize the acquisition and retention of non-foreign intelligence information about Americans.

(d) Such electronic surveillance should be exempt from the disclosure requirements of Title III of the 1968 Act as to foreigners generally and as to Americans if they are involved in hostile foreign intelligence activity (except where disclosure is called for in connection with the defense in the case of criminal prosecution) (USSR 94 *Intelligence Activities: Rights of Americans*, pp. 327–328).

20. (p. 202) For the purposes of FISA, a "United States person" is a citizen, a permanent resident alien, a group of such people, or a US corporation.

21. (p. 202) There are two exceptions to this rule. After a declaration of war, the president, through the attorney general, can authorize a wiretap for foreign intelligence purposes for up to 15 days without a court order. A court order is

also unnecessary if communications are exclusively between foreign powers or involve intelligence other than spoken communications from a location under the exclusive control of a foreign power.

22. (p. 202) Originally there were seven judges on the FISA Court, but the USA PATRIOT Act increased the number to eleven.

23. (p. 203) An approved application may result in several wiretap orders, since the target may be using several communication devices.

Chapter 8: Communications in the 1990s

1. (p. 205) This was the best-case scenario; the worst-case showed no access to intercepted communications by 1995 (Advanced Telephony Unit 1992).

2. (p. 206) In 1994 the US Telephone Association estimated that the costs for call forwarding information alone would come to as much as $1.8 billion (Neel 1994b, p. 101). New equipment and software for wiretapping were estimated to be another $450 million (Neel 1994a, p. 60).

3. (p. 206) In particular, by 1991 virtually all mid-size and large companies were using PBXs, with more than 25 million lines (NTIA 1992).

4. (p. 207) "Everything considered, would you say that you approve or disapprove of wiretapping?" In 1994 76% of Americans said they disapprove (USDoJ 1994b, p. 173).

5. (p. 207) All the missing reports occurred in state electronic surveillance cases.

6. (p. 207) In the case of multiple crimes, the *Wiretap Report* lists only the most serious crime as the reason for the order.

7. (p. 207) Since a case takes several years to wend its way through the courts, convictions are usually reported several years later.

8. (p. 208) 355 F. Supp. 523, 542 (S.D. Calif. 1971), cited in Schwartz 1974, p. 194.

9. (p. 208) The case involved bid-rigging in the window-replacement industry. Lawyer Benjamin Brafman argued that his client had been entrapped, "Others on the tapes would refer to my client as 'the kid.' " The defendant was acquitted (Marks 1995).

10. (p. 209) "In attacking Mr. Fortier today, the defense played recordings of a series of his telephone conversations that were wiretapped by Federal agents in the weeks immediately after the bombing, when Mr. Fortier himself was considered a possible suspect. In those recordings, turned over to the defense as part of a pretrial process, he boasted that he could mislead Federal agents and make a million dollars through book rights from his connection to Mr. McVeigh." (Brooke

1997b) Fortier was the lead witness against McVeigh, who was charged with bombing the federal building in Oklahoma City.

11. (p. 210) Elimination of supplies from one area leads to increased cultivation elsewhere (Reuter 1985, pp. 90–93). Increased law enforcement in northern California led to marijuana growing shifting to Kentucky (Kleiman 1993, p. 284). The success of the "French Connection" case caused a significant reduction in the flow of heroin into the United States from Turkey but within three years this was replaced by heroin from Mexico and Southeast Asia (Moore 1990, p. 136). When US efforts eliminated Mexican marijuana, the drug was replaced almost instantly by hemp from Colombia that turned out to be significantly more potent than the Central American variety (Reuter 1985, pp. 91–92).

Simple arithmetic makes clear an additional reason for the failure of interdiction: drugs that are valued at $2000 per pound where they are grown end up costing well over $100,000 per pound on American streets (Rydell 1994, p. 11). The profit margin is sufficiently high and the demand by addicts sufficiently inelastic that seizures have little effect on the commerce in drugs.

12. (p. 211) This is to achieve a 1% reduction in current annual consumption (Rydell, p. xiii). Treating all heavy users once each year would reduce US consumption of cocaine by half in fifteen years and by less than half in earlier years (Rydell, p. xix).

13. (p. 211) Over the years, FBI Directors and Attorneys General have been eloquent in their appeals for electronic surveillance.

"I dare say that the most violent critic of the FBI would urge the use of wire tapping techniques if his child were kidnapped, and held in custody." [Hoover 1950, p. 230]

"[E]very Attorney General over the last twenty-two years has favored and authorized wire tapping by Federal officials in security cases and other heinous crimes such as kidnapping. . . ." Attorney General Herbert Brownell, in [Brownell 1954a, p. 201]

"By way of background, telecommunications systems and networks are often used to further organized crime, racketeering, extortion, kidnapping . . ." Assistant Attorney General Lee Rawls to Speaker of the House Thomas Foley in 1992 [Rawls 1992]

"Wiretapping is used in the most important life and death cases—terrorism, espionage, drug trafficking, organized crime, kidnapping, and a variety of other crimes" FBI Director Louis Freeh, testifying to Congress in hearings on the digital telephony bill [USS 103b, p. 6]

Freeh was speaking of digital telephony and wiretapping when he told the committee, "I sat last week with Polly Klaas' father [Polly Klaas was the victim of a kidnapper], who came from California to talk to me, and he said to me, 'Mr. Freeh, the FBI did everything in that case to find my little girl.' I do not want to be in a position where I am going to tell some father I could not do everything I would normally do because I could not get the access that I have today." [USS 103b, p. 13]

In fact, as the Director well knew, wiretapping would not have prevented Klaas's murder.

14. (p. 211) Approximately at the time that the FBI began its hard push on CALEA, the use of electronic surveillance in kidnapping cases saw a sharp increase. During the period 1968–1993, there were a total of 69 court orders for electronic surveillance. In 1994 the number jumped to 11 kidnapping cases using electronic surveillance and in 1995 to 25. Much is heard about how carefully the courts review electronic surveillance applications but it is striking to note that of the 11 court orders in kidnapping cases, there were 2 surveillances that were never installed and two that were installed but yielded no incriminating intercepts. Furthermore, 2 of the kidnapping cases (one of which did not have an intercept installed) were related to other cases investigated through wiretaps.

The 1995 kidnapping cases show the same pattern: 1 intercept was not installed, 10 had no incriminating intercepts and 2 of the cases (1 installed, 1 not) were related to gambling cases already being investigated through wiretaps. Thus of the 25 so-called kidnapping electronic surveillances, at most 13 yielded any information in a kidnapping case. Installed surveillances that yield no incriminating intercepts are rare, and the kidnapping cases before 1994 do not show this pattern.

15. (p. 211) According to the FBI, the precise numbers are 1990:624; 1991:481; 1992:495; 1993:401; 1994:418 (Source: Michael Kortan, Unit Chief, FBI National Press Office, private communication to Susan Landau, August 7, 1995).

16. (p. 211) This is an area in which improvements in telecommunications have made investigation easier and kidnappers' lives riskier. The familiar process of trying to keep the caller talking long enough that the line can be traced is often made unnecessary by caller-ID mechanisms that reveal the calling number immediately. When this is coupled with *911 databases* that give police information about the locations of phone numbers, it means that even a kidnapper who calls from a phone booth must talk fast and leave quickly.

17. (p. 211) Administrative Office of the United States Courts, *Wiretap Report*, Government Printing Office, Washington, DC, for the years 1988–2005.

18. (p. 212) Source: private conversation between Landau and Iglehart, January 16, 1997.

19. (p. 212) The Los Angeles police had been engaging in 'hand-offs,' in which the first set of investigators, when they discover illegal activity from a wiretap, pass the information on to a new set of investigators without revealing the source. The second set of officers then establish probable cause in order to obtain a new wiretap warrant. In this procedure, the accused would *not* be informed that their case originally developed from wiretapped information (NYT 1998a). When this

practice was uncovered in 1998, the Los Angeles Police Department (LAPD) said it would adopt an interim policy of notifying defendants if their cases involved wiretaps (NYT 1998a).

20. (p. 212) The Administrative Office of the US Courts releases annual reports on wiretaps. In late spring of each year, data become available on the electronic surveillances of the previous year (except for those still in use) as well as new information on previous surveillances. It typically takes about four years for cases involving wiretaps to wend their way through the court system; thus we have picked 1988 as a year to study, since that leaves a sufficiently long window.

21. (p. 213) The Wiretap Reports show arrests through 1996 in cases involving wiretapping investigations in 1988. After 1996, there were no additional arrests or new court cases.

22. (p. 213) These statistics are based on the raw data provided in the *1994 Wiretap Report*. Although the "Reports by Judges on Court Authorized Intercepts" are supposed to be exact data, some of the reports appear instead to be estimates (presumably supplied by the prosecutors). For example, on pages A36–A37, cases AO 471*, 472*, 473*, and 474*, list 2000, 1500, 300, and 1000 intercepts and 100, 300, 200, and 200 incriminating intercepts respectively. This seems unlikely. On pages A38–A39, AO 475* lists 2000 intercepts, of which 500 are recorded as being incriminating. Similarly, on pages A90–A91, cases AO 13, 14 and 15 list 6200, 1200, and 200 intercepts, and 180, 80, and 60 incriminating intercepts respectively. There are a number of other such anomalous figures in the *1994 Wiretap Report*.

23. (p. 213) For this statistic, we are including court authorizations that are solely for wiretaps and not for combination wiretap and electronic bug surveillance.

24. (p. 215) (Public Law 106-197), Continued Reporting of Intercepted Wire, Oral, and Electronic Communications Act.

25. (p. 215) One possibility is that encoding mechanisms that would not be though of as cryptography by security professionals have been reported as such. This would seem more plausible, however, if law enforcement were reporting that it had had difficulty with encryption rather than that it had not. If a data-compression encoding, for example, had been mistaken for encryption but subsequently unscrambled, the initial misunderstanding should not have found its way into the subsequent report. Perhaps, the encryption in question has been done by hand rather than by machine—emails or phone calls containing code words for activities. Another possibility is that the commercially available encryption tools are poorly implemented and permit the plaintext to be recovered without confronting the encryption directly. When encryption is used to protect files on disks, great care is needed to avoid leaving accidental plaintext copies unexpunged. Using encrypted email, it is possible to encrypt the message to some addressees and fail to encrypt those to others.

The most interesting possibility is that there is an unadvertised law-enforcement program for dealing with encrypted communications. This might take several forms. Secure telephony can always be bypassed by installing bugs in or near the telephone or on the line near the telephone. Both original research and scrutiny of government programs (Kuhn) have shown that interception of compromising emanations, particularly by active techniques, is a richer field than is generally imagined. Similarly, the use of encrypted email can frequently be bypassed by installation of spyware, keyboard loggers, and local packet sniffers.

Carrying this speculation a step further, it is possible that the available tools have been compromised either in individual instances or *en masse*. Even where security products are open-source, adequate security evaluations are difficult to conduct initially and difficult to maintain as the products evolve. Typical users "upgrade" their software when upgrades or packages are offered, without even thinking of the possibility that they may have been targeted for a Trojan horse.

26. (p. 216) Table 6 in the appropriate *Wiretap Report*. In using the summary tables of the *Wiretap Reports* one loses specificity; the numbers cited are the sum of all surveillances (phone, electronic, and combination) that are not purely microphone.

27. (p. 216) Annual Department of Justice letter to the Chair of the House Judiciary Committee, as reported in (Burnham 1996, p. 159).

28. (p. 216) Most calls, regardless of distance, now "arrive" with indication of the calling number. This information is often blocked from going to the subscriber but it is available to the local telephone switch and thus to law enforcement.

29. (p. 216) At the time of Foster's death, Clinton's friends and advisors had been scattered across the continent. Had they instead been down the hallway, these conversations—five-minute discussions—might have disappeared into dust. But with hard records of when phone conversations took place, political Washington drew all sorts of conclusions.

30. (p. 216) An exception occurs if there is an emergency; in that case, a court order authorizing the tap must be approved within 48 hours, or all oral and wire communications intercepted in violation of Title III cannot be used in evidence—or even divulged (Omnibus Crime Control and Safe Streets Act, §2515).

31. (p. 217) Source: private conversation between Diffie and Charney, 1975.

32. (p. 217) One possibility is to relax the minimization requirement but increase the reporting requirement by requiring recording of all conversation on a tapped line and making the entire body of material available to the wiretap victims at the close of the investigation, whether or not that investigation leads to a prosecution.

33. (p. 218) Source: private conversation between Diffie and Charney, 1975.

34. (p. 219) Communications Assistance for Law Enforcement Act, Public Law 103-414.

35. (p. 220) One such example is Freeh's reference to the Tina Isa murder (Bryant 1993), which had been recorded by an FBI bug—not wiretap.

36. (p. 220) "Coincidentally, Director Freeh, with your testimony today the *Philadelphia Inquirer* has a major story on 'FBI Nets Stanfa in Mob Sweep,' and the subheadline is 'FBI's Rich Harvest is a Tale of the Tape,' which could not come at a more opportune time to underscore the kind of need of which you are testifying," Specter said (USSH 103 *Digital Telephony*, p. 46).

But the Stanfa case is described in detail in several newspaper articles and the surveillance used is microphone bugs, including one planted in Stanfa's lawyer's office in Camden, New Jersey (Anastasia 1994; Hinds 1994). The case corresponds to AO number 230 in the *1993 Wiretap Report*; the surveillance is explicitly listed as a microphone bug.

37. (p. 220) CALEA applied only to telecommunications carriers and did not affect companies supplying information services, including electronic mail, and Internet services.

38. (p. 220) This includes not only wiretaps but also dialing and signaling information, including "redirection numbers" (call forwarding, call transfers) and call attempts (including unanswered calls).

39. (p. 221) The FBI did not release information indicating which geographic areas corresponded to which categories.

40. (p. 222) Of course, in 1968, you could not typically trace a call in less than several minutes. Furthermore, tracing a call only tells you the calling phone number. The location of the phone is now available to law enforcement from the databases constructed to support the 911 service but these did not exist in 1968.

41. (p. 222) "The proposed legislation does not seek to expand the current laws." (Freeh 1994b, p. 29)

42. (p. 222) The growth factor varies depending on whether the number is "actual" or "maximal" and whether the interception is for wired or wireless communications. For "actual" wired communications the growth factor is 1.259, for "actual" wireless is 1.707, for "maximal" wired communications is 1.303, and for "maximal" wired communications is 1.621 (FBI 1997b).

43. (p. 222) "Actual" means the number of simultaneous communications intercepts, pen registers, and trap-and-trace devices, that the Attorney General anticipates will be simultaneously conducted in 1998, "maximal" means the maximum number (FBI 1997b).

44. (p. 223) The funding was approved in the "Omnibus Consolidated Appropriations Act," and it provided for funding through a combination of money supplied by various intelligence agencies, as well as $60 million in direct funding. An additional $12 million was provided through unspent Department of Justice funds.

45. (p. 223) *Antiterrorism and Effective Death Penalty Act*, Public Law 104-132. This added subsection (f) to Title 18, §2703.

46. (p. 223) The 1996 Antiterrorist and Effective Death Penalty Act (PL 104-32) empowered the Attorney General to determine whether a group constituted a foreign terrorist group and made this designation immune to subsequent judicial review.

47. (p. 224) Some of the evidence was merely circumstantial, as when demonstrators, who used only a telephone to communicate the particulars, appeared at a march or rally and discovered a police presence, or when members of the government knew about tactics that union officials had decided on a short time earlier (Fitzgerald and Leopold 1987, pp. 27–28). But the ubiquity of such wiretapping was confirmed by the General Treasurer of the Post Office Engineering Union—until 1980 the British Post Office ran the telephone system—who in 1980 said there was much evidence to confirm that the Security Services monitored the calls of union officials during work actions (ibid., p. 29). The activities that have come to light occurred before 1985, when Britain codified the procedure for obtaining a wiretap. However, the British green movement, whose most disruptive tactics consist of blocking road-building projects, has been investigated by the Anti-Terrorist Squad. Like labor activists before them, environmental protesters have found police waiting for them at demonstrations whose venues had been relayed only by telephone (Monbiot 1996).

48. (p. 224) In preparing for the discrimination case, the Chief Constable had wiretaps put on the Assistant Chief Constable's private and work phone lines (the private line being at work but a private line). The case went to the European Court of Human Rights, which ruled, "The Court, bearing in mind that the interception of calls made by Ms. Halford on her office telephones at Merseyside police headquarters, not subject to any regulation by domestic law, appears to have been carried out by the police with the primary purpose of gathering material to be used against her in sex-discrimination proceedings" *Halford v. The United Kingdom*—20605/92 [1997] ECHR 32 (25 June 1997) and awarded the ten thousand pounds in damages plus twenty-five thousand pounds in costs. The more important aspect of this decision was that it brought attention to the lack of codes of practice for police wiretapping (Donohue, p. 1167). The result, however, was most disturbing. Instead of establishing safeguards as required by the European Convention on Human Rights, the government used the opportunity to expand police wiretapping powers (Donohue, p. 1167–1168).

49. (p. 224) It was only in 1985, in response to a European Court ruling that objected to the lack of a clear warrant procedure for wiretaps, that Britain adopted wiretap legislation. Before that wiretaps proceeded through a combination of warrants, executive orders, and even informal requests.

Malone challenged the legality of the wiretap, arguing that (i) telephone users had privacy rights, (ii) the wiretapping violated the European Convention on

Human Rights, and (iii) in the absence of a specific wiretap law, the interception was illegal (Fitzgerald and Leopold 1987, pp. 134–135). The British High Court rejected Malone's arguments but the European Convention on Human Rights, after determining the case was admissible, referred it to the European Court on Human Rights, whose rulings can require governments to correct deficiencies in the law. The European Court ruled that under British wiretap law "it cannot be said with any reasonable certainty what elements of the powers to intercept are incorporated in legal rules and what elements remain within the discretion of the executive ... the minimum degree of legal protection to which citizens are entitled under the rule of law in a democratic society is lacking...." (Bailey et al. 1991, p. 803)

Current British wiretap law is significantly less specific than US law governing wiretaps and permits the interception of lines not specified in the warrant if it is believed these are *likely* to make contact with lines that are specified in a warrant (Fitzgerald and Leopold 1987, p. 146). In a surprisingly broad view of circumstances that can justify wiretap surveillance, the 1985 British law allows interception in cases that the Foreign Secretary deems to be necessary "to safeguard the economic wellbeing of the country." (Command 9438, paragraph 10, as cited in (ibid., p. 148))

An Independent Commissioner provides an annual report on wiretapping activity to Parliament, but this report is relatively superficial, not even providing the number of intelligence wiretaps (Donohue, p. 1159).

50. (p. 224) See "Interception of Communications," Report to COREPER, ENFOCO 40, 10090/93, Confidential, Brussels, 16.11.93, as reported in "European Union and FBI Launch Global Surveillance System," (Statewatch, London).

51. (p. 225) "Memorandum of Understanding concerning the lawful interception of telecommunications," ENFOPOL 112, 10037/95, Limite, Brussels, in "EU and FBI" (see preceding note).

52. (p. 225) Correspondence with Ministers, 9th Session 1995–1996, HL 74, pp. 26–29, in "EU and FBI."

53. (p. 225) Draft letter to non-EU participants in the informal international Law Enforcement Telecommunications Seminar regarding the Council Resolution, ENFOPOL 180, 11282/96, Limite 6.11.96, in "EU and FBI."

54. (p. 225) "Legally permitted surveillance of telecommunications systems provided from a point outside the national territory," Report from the UK delegation to the Working Group on Police Cooperation, ENFOPOL 1, 4118/95, Restricted, 9.1.95, Report from the Presidency to the Working Group on Police Cooperation, ENFOPOL 1, 4118/2/95 REV 2, Limite, 2.6.95, in "EU and FBI."

55. (p. 227) For example, Louis Freeh testified to Congress on May 19, 1994 that "The proposed [Digital Telephony] legislation relates solely to advanced technology, not legal authority or privacy. It has nothing to do with the separate, but important, 'Clipper Chip' technology." (Freeh 1994c)

56. (p. 227) This took place at a conference on Global Cryptography in Washington, D.C.

57. (p. 227) The memo is dated January 17, 1991 but from context it is clear that the correct date is January 17, 1992.

Chapter 9: Cryptography in the 1990s

1. (p. 231) This committee has since been reconstituted as the Information System Security and Privacy Advisory Board with expanded scope.

2. (p. 231) Raymond Kammer, acting director of the National Institute of Standards and Technology, and Clinton Brooks, assistant to the director of the National Security Agency, briefed the FBI on the dangers encryption posed to wiretapping technology; see Chapter 8.

3. (p. 231) For example, the NSA Director wrote to Dr. Willis Ware, chair of NIST's Computer System Security and Privacy Advisory Board that, "The National Security Agency has serious reservations about a public debate on cryptography." (McConnell 1992)

4. (p. 232) According to one of the NSA "flag badges," NSA's Deputy Director for Operations went looking for the Deputy Director of Information Security with a TSD (telephone security device) in hand. At an encounter in the hall, he rammed the TSD firmly into his opposite number's stomach as though he were passing a football and said: "What are you trying to do to me?" (Subsequent to first publication, I have been told on equally good authority that this could not have happened because the DDO just wasn't the sort of person to do such a thing.—WD)

5. (p. 232) The original producer was RCA's COMSEC division in Camden, New Jersey, which was bought by GTE and later absorbed into Lockheed Martin.

6. (p. 232) Use of the STU-III in secure mode is controlled by an "ignition key," a 64-kilobit storage device packaged in the form of a small plastic key. One key may authorize its holder to use as many as 8 different phones, and as many as 32 distinct keys may be used in any one phone.

7. (p. 233) In addition to new developments, some STU-IIs remain in use, and Clipper phones have official status if not much market share. STU-IIs are also used in parts of NATO and some of our allies have secure-phone systems of their own: Brent in the UK, and Speakeasy in Australia, for example.

8. (p. 233) In fact it had a second signal processor dedicated as a modem, but this was a big improvement on earlier secure phones, some of which had seven.

9. (p. 235) Originally, this was more candidly entitled the Law Enforcement Exploitation Field (LEEF), a phrase consistent with standard SIGINT terminology. The less accurate term "access" was adopted for marketing reasons.

10. (p. 235) The natural question arises: Why not escrow Type I keys? Such a proposal is in line with the standard command and control objective, so carefully sought in the nuclear field, of denying the use of captured weapons to an opponent and may ultimately be undertaken. At present, however, there are hundreds of thousands of Type I devices in the field and any prompt conversion is out of the question. All known forms of key-escrow, moreover, harbor potential vulnerabilities. Introducing key escrow technology first in Type II equipment provides a less sensitive environment in which to refine the techniques.

11. (p. 236) Although the standard was announced on April 16, 1993, it was first published in the *Federal Register* on July 30. The public comment period ran through September 28 (USDoC 1993).

12. (p. 236) The Department of Energy, the US Agency for International Development, and the Nuclear Regulatory Commission all submitted letters opposing the adoption of the Clipper standard.

13. (p. 237) "Authorized implementations may be procured by authorized organizations for integration into security equipment." (USDoC 1994b, p. 6004)

14. (p. 237) There is also be a vulnerability associated with each individual chip. No tamper-resistant technology seems likely to be immortal. At some point, recovery of the device unique key from an individual chip may become economical, rendering each device a threat to all the past traffic it was used to transmit.

15. (p. 237) NSA, however, doesn't seem worried. In early 1996, Fortezza cards were authorized for SECRET traffic and NSA officials used the TSD 3600 to stay in touch with their offices while traveling. (I have subsequently been told that authorization was only for compartmentation in an already-adequately-protected system, so perhaps NSA's faith was all that great.—WD)

16. (p. 238) Federal procurement practices generally combine a standard with a process for approving exceptional requirements. The object is to lower costs through volume purchases resulting from conformance to the standards. Getting approval for exceptions can therefore be very tedious.

17. (p. 239) Source: private conversation between Diffie and AT&T personnel.

18. (p. 239) This lack of confidentiality led to the embarrassing problem Speaker of the House of Representatives Newton Gingrich faced in January 1997 (Lacey 1997).

19. (p. 239) By the early fall of 1992, industry groups working on secure computing had been promised a "Type IIE" cryptosystem—a system with an 80-bit key that would be certified for protecting sensitive government information, but would also be exportable—and had been told the names Skipjack and Capstone. This appears to have been the main program and was probably planned to handle voice traffic among other things. The more limited Clipper program seems to have been pushed forward to accommodate the needs of AT&T's new secure telephone.

20. (p. 239) The name Tessera was taken from a form of "ID" used by the Roman empire to identify subject peoples. Many people considered this a fitting name, but it seems to have been dropped due to an unforeseen trademark infringement.

21. (p. 240) These two principles, taken together, were widely regarded as a show stopper, because a receiving email agent does not transmit and thus has no way of making up for the sender's failure to include an escrow field.

22. (p. 240) The Department of Justice representative even said they would probably have to have SECRET facility clearances.

23. (p. 241) This idea was first put forth by Silvio Micali at Eurocrypt '94 in Italy. His point was that if you want to get something from the user (the escrowing of his key) you have to demand it at a point where the user is getting something he cannot do for himself. Since privacy can be manufactured on an end-to-end basis by a pair of users and authenticity cannot, the service that provides users with letters of introduction, the key management infrastructure, is an appropriate place to attach the string.

24. (p. 242) There was an announcement in early 1997, however, that the earlier form of key escrow was being removed and replaced with the commercially oriented "key recovery" techniques (O'Hara 1997).

25. (p. 242) These were Top Secret Special Intelligence or TS/SI clearances. The three who chose not to go through the process were Colin Crook, Leslie Gelb, and Raymond Ozzie.

26. (p. 243) The other panelists were Lee Bollinger, Colin Crook, Samuel Fuller, Leslie Gelb, Ronald Graham, Martin Hellman, Julius Katz, Peter Neumann, Raymond Ozzie, Edward Schmults, Elliot Stone, and Willis Ware.

27. (p. 243) Even opponents of publicly available strong unescrowed encryption agreed that this was the case. The FBI testified to the NRC panel "the use of export controls may well have slowed the speed, proliferation, and volume of encryption products sold in the US" (Dam and Lin 1996, p. 138).

28. (p. 244) The degree of success that the US lobbying has achieved in Britain should not be surprising. Cryptologic cooperation between the two countries which began during World War II and was later codified into the UK-USA Treaty (Richelson 1985).

29. (p. 245) The hostility to privacy in British law has spread beyond cryptography. A recent law vastly expands police powers of search and virtually removes judicial oversight.

30. (p. 245) Members of the OECD are: Austria, Australia, Belgium, Canada, the Czech Republic, Denmark, Finland, France, Germany, Greece, Hungary, Iceland, Ireland, Italy, Japan, South Korea, Luxembourg, Mexico, New Zealand, the Netherlands, Norway, Poland, Portugal, Spain, Sweden, Switzerland, Turkey, the United Kingdom, and the United States.

31. (p. 245) OECD recommendations form the basis for privacy laws in more than a dozen European and Pacific Rim nations (Rotenberg 1996, p. 5).

32. (p. 245) Source: private conversation between Landau and Deborah Hurley, April 3, 1997.

33. (p. 246) In an ironic twist, in early 1997 the German company Brokat Informationssysteme proposed that it ship its strong cryptography systems to the United States, where they would be embedded in products to be exported. Arguing that the shipment from the United States would simply constitute returning the strong encryption to its original markets, Brokat sought to circumvent the restrictive export controls imposed by the US government (Andrews 1997).

Chapter 10: And Then It All Changed

1. (p. 249) The overtly foreign submissions were: LOKI97 from Australia, Rijndael from Belgium, CAST-256 and DEAL from Canada, FROG from Costa Rica, DFC from France, Magenta from Germany, E2 from Japan, CRYPTON from Korea and Serpent from the U.K., Israel, and Norway. The domestic submissions were: MARS, RC6, SAFER+, Twofish, and the Hasty Pudding Cipher, the only purely US entry.

2. (p. 250) IBM, which designed the Data Encryption Standard, built the algorithm to be secure against differential cryptanalysis as later described by Don Coppersmith (Coppersmith). DES is not optimal against linear cryptanalysis, developed by Mitsuru Matsui (Matsui) in 1994, which appears not to have been anticipated by IBM. The earlier history of linear cryptanalysis is not clear. It is essentially the technique the British used to attack the high-grade (above the level of Enigma) German systems during World War II and is implicit in NSA work in the 1960s (Rothaus) but the NSA evaluators do not seem to have imposed it on the design of the DES S-boxes.

3. (p. 252) Neal Koblitz has been a radical since college and was one of the protesters who sat in at the Communications Research Division (now called the Center for Communications Research) of the Institute for Defense Analyses, a research organization that works entirely for NSA, when he was a graduate student in mathematics at Princeton. A few years after his cryptographic discovery, Victor Miller left IBM to take a position at this same laboratory.

4. (p. 252) Elliptic-curve cryptography does not provide a direct replacement for the RSA cryptosystem; instead its key management and signature functions are performed by an Elgamal-type signature and elliptic-curve Diffie-Hellman.

5. (p. 253) In practice, what is required is that it be impossible to alter a message so that its message digest remains unchanged—the *second preimage* problem. For safety, message digests are only considered secure if there is no known way of finding any two messages with the same digest.

6. (p. 253) A feat repeated by the public community shortly thereafter (Chabaud 1998).

7. (p. 253) The work required to find two messages that hash to the same digest can never be greater than that of the workfactor of a cryptosystem whose key is half the size of the hash algorithm's output. SHA-1 with its 160-bit output was designed to have a workfactor of 2^{80}. The later algorithms were designated SHA-256, SHA-384, and SHA-512 and were designed to have workfactors of 2^{128}, 2^{192}, and 2^{256} respectively.

8. (p. 254) Although less than one might have hoped for or expected in half a century.

9. (p. 255) *Daniel Bernstein v US Department of State*, 922 F. Supp. 1426, 1428–30 (N.D. Cal. 1996)

10. (p. 255) *Bernstein v US Department of State* 176 F. 3d 1132, 1141, rehearing en banc granted, opinion withdrawn, 192 F. 3d 1308 (9th Cir. 1999)

11. (p. 255) Since commercial communications play a large and growing role in government communications (both military and nonmilitary), they are a legitimate target of traditional national intelligence collection. The US-government position is that it does not provide covert intelligence information to US companies, but will make use of such information in helping them to counter what it considers foreign corrupt practices.

12. (p. 256) The Administration's anti-cryptography policy was inimical to Silicon Valley, whose support was seen as crucial for the Vice President's bid for President.

13. (p. 257) Information in transit is not considered valuable; if it gets lost, you resend it. If the recipient can't decrypt it, you may have to renegotiate keys and resend. Communications keys are destroyed on a schedule that takes account of what might have been encrypted in them. By contrast, if you depend on cryptography to protect stored data, the keys are just as valuable as the data they were used to encrypt and are valuable for as long as you want the data. Robust availability of the key is therefore the single most important point in management of storage keys.

14. (p. 258) The *Electronic Key Management System* or *EKMS* is a key-management system built by NSA to serve the needs of national security communications. Keys are manufactured at the *central facility* in Finksburg Maryland. When cryptographic equipment is first installed, keys are distributed to it in physical form. After that devices are generally keyed electronically, a process called OTAR or Over the Air Rekeying.

The keying process for the STU-III telephone (currently being phased out in favor of an ISDN phone called the STE) is typical. Shortly after a user receives a new phone, the user receives a small memory device containing a *seed key* usable only for communicating with the central facility. The user makes a secure call to a special number and the seed key is replaced by operating key, which can be used to call other STU-IIIs. Subsequent changes of key are done in the same way.

The EKMS makes extensive use of public-key technology. In the terms of the commercial world, it plays the role of a certifying authority. Because the military world is smaller and more centralized than the commercial, the EKMS operates as a single level keying hierarchy in which the central facility is the only certifying authority.

In testimony before the Senate in May of 1994 (McConnell 1994) John Michael McConnell, Director of NSA, stated that the Key Management Facility had cost fourteen million dollars to build and would cost sixteen million dollars a year to run.

15. (p.258) The fact that the memorandum from the Office of Management and Budget was a "recommendation" rather than a "requirement" might look like a weak action but, in fact, given that noncompliance would result in action from an agency's Inspector General, the force of the OMB recommendation is actually as strong as a requirement.

16. (p.259) How widespread a cryptosystem is depends some on how you count and four obvious measures come to mind. The first is the number of devices. If this is the measure, either SSL in browsers or cryptography in smart cards seem likely to win. Another is total investment, perhaps a smaller number of expensive devices (high-speed trunk-line encryptors, for example) cost more than many cheap ones. A third possibility is the number of bits encrypted. Once again it is possible that a smaller number of highspeed devices will exceed the total traffic volume of a larger number of slow ones. Finally, one might ask if the devices are really used. SSL is in every browser but comparatively few servers operate securely, so the use of SSL is not as great as the browser base suggests.

17. (p.259) Drugs are an example of a product with a high cost of development and a low marginal cost of production but a high marginal cost of reproduction.

18. (p.260) http://www.dvdcca.org

19. (p.262) Ferguson's fears may have been reasonably founded, considering what happened to Dmitry Sklyarov, who was arrested while visiting the US to talk about flaws in the security of e-books at Defcon (McCullagh 2001a).

20. (p.262) Under California law, because the research was done by state employees in the normal conduct of their duties, the legal team would defend the researchers in any civil action.

21. (p.262) DMCA §1201 (g) (2) (C).

22. (p.262) Private conversation between Landau and David Wagner, August 7, 2006.

23. (p.263) Attestation appears to be due to John Manferdelli of Microsoft but does not seem to have been published outside of the TCG documentation. (Source: conversation between Diffie and Manferdelli, November 9, 2004)

24. (p. 264) Tighter control of enterprise networks will have the socially significant effect of reducing the power of the employees who use them. In the era of timesharing, all control was central. PCs empowered users of all kinds with the ability to configure their machines as they saw fit and to run what programs they wished. Gradually, the PCs owned by corporations have been brought more tightly under the control of corporate IT departments or replaced by centralized servers providing "second generation timesharing."

25. (p. 264) The authors are grateful to Scott Rotondo of Sun Microsystems for pointing out this particular example.

26. (p. 266) English, at 312 million people, is the native language of 30% of Internet users, while Chinese, at 132 million, is the native language of just under 13% (http://internetworldstats.com/stats7.htm, last viewed 18 July 2006). The percentages will undoubtedly shift in the direction of Chinese as more members of that populous nation go online.

27. (p. 266) XML makes up for one of HTML's most glaring defects: HTML has no definitions; a sequence repeated over and over in an HTML document must be repeated over and over; it cannot, as in most computer languages, be abbreviated into a macro or routine.

28. (p. 267) In particular, see (Saltzer), "The function in question can completely and correctly be implemented only with the knowledge and help of the application standing at the endpoints of the communications system. Therefore, providing that questioned function as a feature of the communication system itself is not possible."

29. (p. 268) When I arrived at Sun in 1991, I was asked to choose a name for my workstation, a task I took unreasonably seriously. I departed the day after I was hired for the Crypto conference in Santa Barbara and during the trip stayed with friends in Los Angeles. My hosts obligingly connected to Sun and printed out a list of all the computers on its network, so that I would know what names had been taken. This would not be so easily done today.—WD

30. (p. 269) Information about Carnivore became public as a result of an FBI effort to install Carnivore at the ISP Earthlink. Originally Earthlink was served with an order for a pen register to be installed. Despite the ISP's efforts to carry out the order—Earthlink provided the FBI with headers of incoming mail and some headers of outgoing mail—the FBI was not satisfied and sought to install Carnivore at Earthlink. Earthlink opposed this and went to court, but lost (*In the Matter of an Application of the United States of America for an Order Authorizing the Installation of a Pen Register and Trap and Trace Device*, United States District Court, Central District, Western Division - California, Criminal No. 99-2713M). Part of the difficulty was that the installation of Carnivore crashed Earthlink's remote servers (Wingfield). Earthlink's lawyer, Robert Corn-Revere, testified to the House Judiciary Committee on the issue (Corn-Revere). As a result of his testimony, Carnivore became public.

31. (p. 269) Carnivore was part of the "Dragon Ware" suite of FBI computer programs for Internet surveillance, which included "Packeteer" and "CoolMiner." Packeteer took the data from the raw packets and reconstructed the original format of the communications, while CoolMiner organized the information in a user-friendly manner, so that an investigator could see a target's steps browsing the web, sending email, chatting via ICQ, Yahoo Messenger, AIM, IRC, etc.

32. (p. 270) Bellovin et al. noted that in pen register mode, Carnivore captured lengths of various communications. This allows a certain type of traffic analysis, namely, "[I]n the case of a user visiting a web site, knowing the length of the objects returned can often be used to identify which web page he was visiting (at least for static HTML content), and this is clearly not authorized in pen mode" (Bellovin 2000, itemized comment on 4.2.8).

33. (p. 270) Earthlink objected; see note 30 above.

34. (p. 271) The setting should have been entered in the Carnivore filter set by FBI agents who were detailed to work on Carnivore and who were the only FBI personnel given logical access to the Carnivore appliance. The agents in charge of an investigation should not have had access to the appliance itself but should have been required to make written requests for such changes to the "Carnivore agents."

35. (p. 271) A group of computer-security researchers reviewed the IIT review, and found it sorely lacking (Bellovin 2000). They objected to the lack of systematic review of system issues (flaws that arise when two complex systems interact), lack of systematic search for bugs (especially for string buffer overflows, a well-known problem), and an "inadequate discussion of audit and logging."

36. (p. 271) The Electronic Privacy Information Center discovered in 2002 that this design flaw caused serious consequences. Apparently while the FBI's UBL unit —UBL is the US government's abbreviation for Usama bin Laden—was conducting FISA surveillance, "The software was turned on and did not work properly. The FBI software not only picked up the E-mails under the electronic surveillance of the FBI's target [redacted] but also picked up E-mails on non-covered targets. The FBI technical person was apparently so upset that he *destroyed* all the E-mail take, including the take on [redacted] under the impression that no one from the FBI [redacted] was present to supervise the FBI technical person at the time" (FBI 2000).

37. (p. 273) In principle anything that runs over TCP, or Transmission Control Protocol, which is the Internet's transport-layer reliable transport protocol, can run on Tor.

38. (p. 273) Source: private conversation between Landau and Roger Dingledine, August 11, 2006.

39. (p. 273) Beginning in 2004, work on Tor has also been supported by other government agencies and nonprofits including the Electronic Frontier Foundation.

40. (p. 273) See the discussion regarding data retention in chapter 11.

41. (p. 273) Source: private conversation between Landau and Roger Dingledine, August 11, 2006.

42. (p. 274) http://www.projectliberty.org

43. (p. 275) Disclosure: both authors have worked on the Liberty protocols.

Chapter 11: Après le Déluge

1. (p. 279) Available at http://www.opsi.gov.uk/acts/acts2000/20000023.htm
   (last viewed October 14, 2006).

2. (p. 280) 50 USC 1801 et seq.

3. (p. 280) Because the CIA had determined that Khaled al-Mihdhar and Nawaf al-Hazmi participated in a meeting in Malaysia with planners of the USS Cole plot, the agency linked the two men to al-Qaeda in 2000 but did not inform the FBI of this. The men were not placed on a "watch list" and were allowed to enter the United States, which is where they were during the summer of 2001. Both men were among the hijackers of American Airlines flight 77 on September 11.

4. (p. 280) USA PATRIOT Act, Pub. L. 107-56. This act is often referred to as the Patriot Act.

5. (p. 280) §218 of PATRIOT Act modified 1804(a)(7)(B).

6. (p. 281) Zacarias Moussaoui was a suspicious French national of Moroccan descent who took flying lessons in Minnesota in 2001. After Moussaoui behaved oddly at flight school, his instructor called the FBI, who arrested Moussaoui on a visa violation. FBI Special Agent Colleen Rowley sought a FISA warrant to search Moussaoui's personal belongings (including his computer), but the FBI was unwilling to apply for a FISA warrant. On September 11, a criminal warrant was issued and Moussaoui was discovered to have connections with al-Qaeda. Moussaoui was later convicted in federal court of conspiring to kill Americans and sentenced to life in prison.

7. (p. 281) As discussed in chapter 8, the Administrative Office of the US Courts publishes an annual *Wiretap Report* that details all electronic surveillance orders of the previous year, including the date surveillance was authorized, how long the surveillance was, who the prosecutor was, who the presiding judge was, how many conversations were surveilled, how many incriminating conversations were surveilled, and whether there were arrests or convictions.

8. (p. 283) In this context "law enforcement officials" refers both to FBI agents and criminal prosecutors (USFISC 2002c, p. 2).

9. (p. 285) Letters containing deadly anthrax spores were sent to news agencies and two US Senators, Tom Daschle and Patrick Leahy (both Democrats, from

South Dakota and Vermont, respectively). Twenty-two people, including postal workers, developed anthrax and five died. For a number of weeks Congressional office buildings were closed while being tested and cleaned.

10. (p. 286) Because information about FISA taps is not public, it is impossible to determine how often a FISA wiretap has been used in what turned into a criminal investigation. We do know that the Justice Department uses the PATRIOT Act tools for investigations of non-terrorists, including: "suspected drug traffickers, white-collar criminals, child pornographers, money launderers, spies, and corrupt foreign leaders and to pursue a broad law-enforcement agenda" for the department has said so (DoJ 2006, p. 56).

11. (p. 286) Offenses under Section 1030 include: intentionally access[ing] a computer without authorization or exceeds authorized access, and thereby obtains[ing] information contained in a financial record of a financial institution, or of a card issuer [or] information from any department or agency of the United States or information from any protected computer if the conduct involved an interstate or foreign communication; intentionally, without authorization access[ing] any nonpublic computer of a department or agency of the United States, access[ing] such a computer of that department or agency that is exclusively for the use of the Government of the United States; knowingly and with intent to defraud, access[ing] a protected computer without authorization, or exceeds authorized access, and by means of such conduct furthers the intended fraud and obtains anything of value, unless the object of the fraud and the thing obtained consists only of the use of the computer and the value of such use is not more than $5,000 in any 1-year period.

12. (p. 288) P. L. 109-160, which extended the PATRIOT Act provisions until February 3, 2006, and P. L. 109-170, which extended the provisions until March 10, 2006.

13. (p. 288) §108(a) of P.L. 109-177 amending 50 U.S.C. 1804(a)(3) and 50 U.S.C. 1805(c)(1)(A).

14. (p. 288) This may be extended to sixty days under appropriate circumstances.

15. (p. 288) §108(b)(4) P.L. 109-177, adding 50 U.S.C. 1805(c)(3).

16. (p. 289) The Y2K problem was one of errors caused by computer programs that used 2-digit dates and could not distinguish the year 1900 from the year 2000. Massive checking of legacy programs seems to have found the errors because there were few problems. The disaster many people expected on January 1, 2000 never materialized.

17. (p. 289) Title X, Subtitle G, of P.L. 106-398.

18. (p. 290) Public Law 107-347.

19. (p. 290) In a rather odd situation, there were actually three FISMAs: the Treasury and General Government Appropriations Act for Fiscal Year 2003 FISMA amendment, "FISMA A," and "FISMA B."

The Appropriations Act FISMA had an amendment to extend GISRA past its one-year limit, but by the time the appropriations bill passed, the two other FISMAs had already become law. There was no need for the amendment, which was not folded into the Consolidated Appropriations bill.

FISMA A, Title X of the Homeland Security Act of 2002 (P.L. 107-296), extended GISRA, but put defense agencies in charge of its implementation.

FISMA B, Title III of the E-Government Act was quite similar to FISMA A, but included a provision (§11331 (d)) that placed the Office of Management and Budget, the federal agency responsible for coordinating executive branch management procedures, in charge of implementation.

In signing the E-Government Act, the president designated which bill was operative through a "signing statement" clarifying the implementation of the two laws: "Title III of this Act is the Federal Information Security Management Act of 2002. It is very similar to Title X of the Homeland Security Act of 2002, which also bears the name Federal Information Security Management Act of 2002 and which I signed into law on November 25, 2002. I am signing into law the E-Government Act after the enactment of the Homeland Security Act, and there is no indication that the Congress intended the E-Government Act to provide interim provisions that would apply only until the Homeland Security Act took effect. Thus, notwithstanding the delayed effective dates applicable to the Homeland Security Act, the executive branch will construe the E-Government Act as permanently superseding the Homeland Security Act in those instances where both Acts prescribe different amendments to the same provisions of the United States Code." (Bush 2006)

20. (p. 290) Though cryptography standards are the activity that draw the most public attention, the Computer Security Division's responsibilities are much broader that that. Under FISMA, CSD's role includes developing guidelines for secure system implementation, security research on emerging technologies, and evaluation of security testing labs.

21. (p. 291) For example, the CSD provided cryptographic standards to the banking industry (ISPAB, p. 6) and how NIST provided core technologies that enabled "the [pharmaceutical] industry to create new value" (ISPAB, p. 7).

22. (p. 291) Source: private conversation between Landau and William Barker, September 18, 2006.

23. (p. 292) EZPass is a system in which a user prepays an account using a credit card, personal check, or cash, and receives a small electronic tag for their vehicle, which the driver places on the windshield. This enables the car to drive through specially-equipped toll lanes without stopping. The toll is automatically deducted from the owner's account. EZPass is a system in use throughout the northeastern United States. There are similar systems in other parts of the country and the world.

24. (p. 294) During 1994 hearings, FBI Director Freeh made clear his under-

standing that the bill was limited to telephony systems. In particular, he said, "From what I understand ... communications between private computers, PC-PC communications not utilizing a telecommunications common net, would be one vast arena, the Internet system, many of the private communications systems which are evolving. Those we are not going to be on by the design of this legislation." Senator Larry Pressler asked, "Are you seeking to be able to access those communications also in some other legislation?" Freeh responded, "No, we are not. We are satisfied with this bill. We think it delimits the most important area and also makes for the consensus, which I think it pretty much has at this point." (USSH 103 *Digital Telephony*, p. 202).

25. (p. 294) Communications Assistance for Law Enforcement Act, §102 (8)(C).

26. (p. 294) Communications Assistance for Law Enforcement Act, §103 (b)(2).

27. (p. 295) "CALEA applies to facilities-based Internet access providers and interconnected VoIP service providers" (FCC 2005a, p. 24).

28. (p. 295) It was not only the petitioners who viewed it this way. In his dissenting opinion, Senior Circuit Judge Harry T. Edwards wrote:

"In determining that broadband Internet providers are subject to CALEA as 'telecommunications providers,' and not excluded pursuant to the 'information services' exemption, the FCC apparently forgot to read the words of the statute." *American Council on Education, Petitioner v. Federal Communications Commission and United States of America*, No. 05-1404 et al. (D.C. App. June 9, 2006, Edwards, dissenting).

29. (p. 295) The petitioners did not completely lose the case; the Appeals Court ruled, for example, that a CALEA exemption remained for private networks, such as those maintained by educational institutions.

30. (p. 295) The authors were part of an industry study (Bellovin 2006a) of the impact of applying CALEA to VoIP and what follows draws heavily on that report.

31. (p. 296) This diagram and discussion first appeared in (Bellovin 2006a).

32. (p. 298) We are glossing over the difficulty of Alice's VoIP Provider even knowing who Alice's ISP is, let alone the location or identity of R1. These are non-trivial issues in and of themselves.

33. (p. 300) In fact the signaling message is sent to the home location register, a database containing the identity and service profile of the subscriber.

34. (p. 300) In spirit an *en banc* review is a review by the full court but in practice, it is review by a panel of judges significantly larger than the original panel of three but smaller than the whole court.

35. (p. 302) The *New York Times* had uncovered the story in 2004, but the White House requested that the newspaper not publish on the subject, "arguing that it could jeopardize continuing investigations and alert would-be terrorists that

they might be under scrutiny." (Risen) The *Times* reporters investigated the issue for another year and then published. The Bush administration began investigations to determine who had leaked the story to the press. Meanwhile the two reporters who broke the story, James Risen and Eric Lichtblau, won a Pulitzer Prize for "carefully sourced stories on secret domestic eavesdropping that stirred a national debate on the boundary line between fighting terrorism and protecting civil liberty." (Pulitzer)

36. (p. 304) The falling cost of storage holds promise of changing this.

37. (p. 304) NessiE material can be found at www.cryptonessie.org.

38. (p. 306) Senator Charles Grassley added, "I worry down the road that ... some prosecutors who do not have experience dealing with terrorists and spies may be tempted to order an arrest for a reason other than national security. That prosecutor may, for instance, want a convicted terrorist on his record, even though it is smarter to watch the suspect and learn about his plans and and conspirators. The intelligence agencies on the case may still be looking for other terrorists in the cells, but they get overruled by the prosecutor.... I am worried that prosecution is not always the best decision in terms of national security." (Grassley, p. 28)

39. (p. 306) On August 6, 2001, the President's Daily Brief contained a two-page memo titled "Bin Laden Determined to Strike in US" that described "patterns of suspicious activity in this country consistent with hijackings" (White House 2001, p. 2)

40. (p. 306) Gary Hart was a Democrat from Colorado, Warren Rudman a Republican from New Hampshire.

41. (p. 307) The policy was dropped because of the lack of measurable success in fighting terrorism (Bernstein 2004).

42. (p. 307) South Asian immigrants in Britain are three times as likely to be unemployed as white Britons and indeed, 40 percent of Pakistani women in Britain are unemployed, as are 28 percent of Pakistani men (Bernstein 2006). The situation in the United States is markedly different: incomes of people of Pakistani origin are close to the median in New York and slightly exceed the median in New Jersey (there is, however, a large underclass of South Asians) (ibid., p. A1). The pattern of Arab Americans follows the pattern of most immigrant groups in the US that have been here for a few generations: Arab-Americans have a higher rate of college and post-college education. The median income of Arab Americans is higher than the US median (Fallows, p. 65).

43. (p. 308) Queensborough Public Library in New York City, which serves an immigrant population, took steps to preserve privacy of patrons, including delinking of electronic book/patron info when book is returned, daily destruction of Internet usage sign-up sheets, etc. (NRC 2006). The message from the Queensborough library is clear: you are part of our society, deserving of our protections.

44. (p. 309) A case in point is the 9/11 hijackers. Mohamed Atta described a nuclear facility as "electrical engineering" to his fellow pilots (National Commission, p. 245). Khalid Sheikh Mohammed used the code of send "the skirts" to "Sally" to instruct another al-Qaeda member to send funds to Zacarias Moussaoui (National Commission op. cit. at 246). The targets were discussed as if the participants were students at a university: the Pentagon was "arts," the World Trade Center, "architecture," the Capitol, "law," and the White House, "politics" (National Commission, p. 248).

45. (p. 309) This realization undoubtedly contributed to NSA acquiescence to the change in cryptographic export-control regulations in 2000.

46. (p. 310) Reid attempted to blow up an American Airlines flight from Paris to Miami by lighting his shoe tongue. Reid's shoes contained an explosive. The attempt was thwarted by a flight attendant. The plane was diverted and Reid was arrested upon arrival in Boston (USDoJ 2006, p. 26).

47. (p. 311) TRAC is a research center at Syracuse University devoted to data collection, analysis and distribution, about federal government staffing, spending, and enforcement activities.

48. (p. 311) On July 30, 1916, in the midst of World War I—but before US participation in the war—a munitions storage depot in New York Harbor was destroyed by saboteurs, destroying over two million pounds of explosives. The blast burst windows in Jersey City, Manhattan, and Brooklyn, and was heard over 100 miles away in Philadelphia. (Landau 1937, pp. 77–91). This was the largest terrorism act during this period, but there were numerous other explosions at industrial plants, all laid at German saboteurs. They were estimated to have caused over $150 million in damage to essential war goods (Sayers, p. 11). There was even an attack at the US Capitol switchboard (Landau 1937, pp. 305–307).

49. (p. 311) In 1994, when FBI Director Louis Freeh testified before Congress in support of the "Digital Telephony" bill (later passed as CALEA), he emphasized the importance of wiretaps in solving kidnappings and in preventing terrorist actions (Freeh 1994b).

50. (p. 311) At the time of Freeh's testimony (1994b), Title III was just turning 25 years old. If wiretaps are an important tool of law enforcement, there should be enough clear-cut cases in 25 years to allow a persuasive case to be made. Freeh's account of the value of wiretapping is remarkably vague. He refers to numerous convictions, without mentioning the name of a single defendant, court, presiding judge, case name, or docket number. This makes the information difficult to verify or explore. It is one thing to say that you can't give the details of ongoing investigations or that cases ended in plea bargains or that crimes were prevented without any trials resulting; it is another to fail to identify cases that must be the results of public trials.

Credence is further strained by the inclusion of at least one identifiable case

—the Tina Isa case—in which the surveillance was a microphone planted in the living room in which a teenage girl was murdered (Bryant 1993).

Chapter 12: Conclusion

1. (p. 314) It has long been an asset of private detectives, often retired police, to have friends in the department who will give them non-public information—looking up addresses from license plate numbers, for example. In the twenty-first century, well-connected detectives may be among the beneficiaries of CALEA and its descendents.

2. (p. 315) Although widespread wiretapping is an abomination, government surveillance is not in all respects undesirable: the government's ability to serve its citizenry is, after all, dependent on sufficient understanding of the population's activities to know the population's needs (Bogard).

3. (p. 315) At the first of the American public cryptographic conferences, Crypto '81, which was held at the University of California at Santa Barbara, one of the NSA people said to me: "It's not that we haven't seen this territory before, but you are covering it very quickly."—WD

4. (p. 315) In the AT&T TSD3600, for example, encryption represents approximately 1–2% of the computation. It is 3% of the cycles of the main processor, but this is assisted by a dedicated modem chip. In short, the TSD spends almost all of its effort either preparing the speech to be encrypted or preparing the cipher text to be sent over the phone line. The rest goes to encrypting it.

5. (p. 315) For example, the GSP8191 secure telephone, was designed by one person, Eric Blossom, in a little more than 2 years. Skype, the secure VoIP system was done by a dozen in a similar length of time.

6. (p. 316) Establishing that the person you are communicating with now is the same person you were communicating with at some previous time is socially fundamental; it is the way acquaintances develop. The ability to assume a persona and to sign email provides a mechanism by which people can meet on the Internet and have some confidence that they are communicating with the same person each time, without exchanging any absolute information about their identities.

7. (p. 316) The Atomic Energy Act of 1954 created the notion that ideas in atomic energy were "born secret" and were to remain secret unless the government said they could be disclosed.

8. (p. 317) In 1992, the FBI's Advanced Telephony Unit warned that by 1995 no more than 40% of Title III wiretaps would be intelligible and that in worst case all might be rendered useless (Advanced Telephony Unit 1992). In 1994 Assistant Attorney General Jo Ann Harris admitted that, a year after the introduction of the Clipper proposal, the FBI had yet to encounter a single instance of encrypted voice communications (Harris 1994). Further data in the *Wiretap Report* for the

years 2000–2005 bears out that encrypted voice communications are simply not a problem.

9. (p. 318) FBI directors have always emphasized the use of wiretaps in kidnapping investigations, and Louis Freeh was no exception. In fact wiretaps were used on average in only two to three kidnapping cases a year in the period 1968–1993. Terrorist actions were likewise cited as an important reason for wiretaps,despite the fact that there were no Title III wiretaps in terrorist cases in the period 1988–1994.

In pressing for various wiretapping capabilities, FBI Assistant Director James Kallstrom argued: "... just for the FBI alone, we have used court-authorized electronic surveillance to capture terrorists intent on blowing up buildings and tunnels in New York, to detect and capture pedophiles who intended to brutally murder their intended victim, to arrest and convict various organized crime leaders like John Gotti, and to successfully investigate a spy whose espionage cost many their lives" (Kallstrom 1997). However, the Rahman case ("terrorists intent on blowing up buildings and tunnels in New York") turned not on wiretaps, but on other forms of electronic surveillance, including a body wire (which does not require a warrant); the valuable evidence in the Gotti case came from an electronic bug (Less than eight months after Kallstrom's remarks, FBI Director Louis Freeh testified to a Senate Judiciary committee hearing: "John Gotti never implicated himself on a telephone conversation with one of his confederates." (USS 105d).); and the wiretap in the Ames case ("a spy whose espionage cost many their lives") served in a tangential fashion, enabling the government to pressure Ames to reveal information in order that his wife—whose knowledge of his spying activities was revealed on the wiretaps—receive a reduced sentence. (The value of the wiretap in the Ames investigation should be placed in context: a recent Department of Justice (USDoJ 1997) report severely castigated the FBI for inadequately investigating the FBI and CIA spy losses years earlier and thus allowing Ames to inflict further damage on US intelligence.)

10. (p. 319) This is a point that Congress might do well to consider amending.

11. (p. 320) Mail covers operated from the 1940s til the early 1970s; copies of telegrams were also sent to NSA during that period. See chapter 6.

12. (p. 320) See chapter 6, and discussions on Kennedy, Johnson, and Nixon.

13. (p. 324) A good example is Anatoli Golitsin, a Soviet defector who initiated a decade long search for "moles" in the CIA.

14. (p. 325) A common division of responsibility in this respect has been that field stations do signal processing on received material, but leave all cryptanalytic operations to be done by headquarters.

15. (p. 325) One fascinating possibility is that the cryptanalysis of some popular cryptographic algorithm such as DES or 40-bit RC4 might be achieved and embedded in a tamper-resistant chip. Intercept equipment with explicit "counter-DES capability" or "counter-RC4 capability" might thereby become available.

16. (p. 325) Some of the uses are commercial. Emitter identification is being used to detect cloned cellular phones (AWST 1997b).

17. (p. 326) The technology that makes this available is the same as that of caller ID.

18. (p. 327) Systems such as *Teletrack* keep track of the locations of fleet vehicles and report this information automatically to a dispatcher. They may even have profiling capabilities that allow them to warn the dispatcher when a vehicle is out of its expected area, behind schedule, etc.

19. (p. 328) By wiretap law—Title III, FISA, and subsequent amendments—communications on these networks are subject to wiretap. The issue is making the network architecture subject to CALEA.

20. (p. 329) Effects of this kind are seen in microcosm in the case of Robert Hanssen, a spy within the FBI, who was able to tap into counterintelligence databases in order to detect whether he was being investigated (United States of America v. Robert Philip Hanssen, Affidavit in Support of Criminal Complaint, Arrest Warrant and Search Warrants.) (United States District Court for the Eastern District of Virginia, Alexandria Division).

21. (p. 329) Source: Greek government press briefing, February 2, 2006. English translation provided by George Danezis http://homes.esat.kuleuven.be/~gdanezis/intercept.html (last viewed October 16, 2006).

22. (p. 330) Personal reminiscence of Gustavus J. Simmons, retired Senior Research Fellow at Sandia National Labs, told to Diffie in 1991.

23. (p. 331) The current government efforts are focused on VoIP. However, there is a draft Department of Justice bill that would apply the CALEA requirements to any real-time communications. This would include Instant Messaging, MMORPGs, etc.

24. (p. 332) French law requires the police to consult the courts when initiating any investigation of a citizen (Kelling 1991, p. 965).

25. (p. 333) Textbooks on criminal investigation devote approximately 1% of their pages to the subject.

Glossary

Advanced Encryption Standard (AES)—a symmetric-key encryption algorithm working on 128-bit block size data with 128-, 192-, or 256-bit keys that is a Federal Information Processing Standard replacing DES, the Data Encryption Standard.

AFCRC—Air Force Cambridge Research Center.

ANSI—American National Standards Institute.

ASCII—American Standard Code for Information Exchange—The government's adoption of this code, which is ubiquitous today, made it dominant over the rival EBCDIC encoding used by IBM.

analog scrambler—cryptographic device operating on continuous (analog) signals, rather than on discrete elements like bits or letters.

asymmetric cryptography—See: public-key cryptography.

authentication—the process of determining whether an individual is whom they claim to be. Weak authentication methods might be a user name and password; strong authentication methods typically include several factors, e.g., something you are (a biometric), something you have (a physical token), and something you know (a password).

bit—short for "binary digit," the smallest unit of data stored in a computer. Bits have a single binary value, either a "0" or a "1."

block cipher—a cryptosystem on a block of symbols that sequentially repeats an internal function, called a *round*.

bps—bits per second.

CCEP—Commercial COMSEC Endorsement Program—A program set up

in the mid 1980s by NSA in an attempt to get industry to invest in building a new generation of communication security equipment.

cipher text—Unintelligible text or signal produced by a cryptographic system. (adj.: ciphertext)

Clipper—name of a chip implementing the Skipjack encryption algorithm with key recovery.

CALEA—Communications Assistance For Law Enforcement Act—a 1994 law requiring that digitally switched telephone networks deployed after 1998 be built wiretap enabled according to standards defined by the US government.

COMINT—Communications intelligence—the extraction of information for opponents communications.

COMSEC—Communications security—protection of communications against communications intelligence.

crypto wars—political battle in the US in the 1990s over the freedom to use cryptography for personal and commercial applications.

Diffie-Hellman key exchange—a process using exponentiation in modular arithmetic for negotiating a shared secret key between two parties without concealing any of the messages from opponents.

DS1—Telephone communication standard for transmitting 24 voice channels, together with signaling information, simultaneously at a rate of 1.544 megabits per second.

EBCDIC—Embedded Binary Coded Decimal Interchange Code—IBM encoding used for representing characters.

electronic surveillance—the use of electronic equipment to surveil private conversations.

EES—Escrowed Encryption Standard—an originally classified algorithm (Skipjack) that was to be implemented on tamper-resistant chips (Clipper) with escrowed keys.

Enigma—a three-rotor encryption machine developed for commercial use in the 1920s and widely used by the Nazis during World War II.

FBI—Federal Bureau of Investigation—the main criminal investigatory agency of the US government, the FBI is part of the US Department of Justice. The FBI is responsible for investigating all federal crimes except those specifically assigned to other agencies (e.g., the Secret Service is responsible for investigating cases of counterfeiting).

FCC—Federal Communications Commission—the federal agency responsible for regulating interstate and international communication, whether by radio, television, wire, satellite, or cable.

Federal Information Processing Standard (FIPS)—an information processing guideline set for federal government departments and agencies by the National Institute of Standards and Technology, but often having wider applicability.

FISA—Foreign Intelligence Surveillance Act—a 1978 law providing for interception of communications within the United States for intelligence purposes.

Freedom of Information Act (FOIA)—A 1966 federal law establishing the public's right to obtain information from federal government agencies. The law applies to Executive Branch departments, agencies, and offices.

IP—Internet Protocol—the Internet Protocol is the method by which data travels from one computer to another over the Internet.

IP address—the IP, or Internet Protocol, address, is a unique number that devices use to communicate across a computer network.

ISDN—Integrated Services Digital Network—ongoing replacement of existing 'analog' telephone service by digital service.

KMF—Key Management Facility—a network resource that assists users in acquiring keys needed to establish secure communications.

LEAF—Law Enforcement Access Field; the law-enforcement access to the keys of the Clipper system.

mail cover—the process of recording information on the outside cover of mail as well as contents of second-, third-, and fourth-class mail, and international parcel post mail without the consent of the recipient.

message digest—a cryptographic function that takes input data (often a entire message) and outputs a short, fixed length result.

MI5—Military Intelligence 5, more accurately known as the Security Service—the British counterintelligence organization. The closest US cognate is the counterintelligence function of the FBI.

MI6—Military Intelligence 6, more accurately known as the Secret Intelligence Service—the British foreign intelligence service. The US cognate is the CIA.

minimization—In law enforcement, the practice of limiting interception to those portions of communications that are or may be of legitimate investigative interest. In intelligence, the more limited practice of limiting interception to exclude forbidden material such as the communications of the citizens of the host country. In general, minimization may be by channel, person, time, or subject matter.

MOU—Memorandum of Understanding—a legal contract determining the obligations of two governmental entities (or between the government and a contractor) regarding joint work.

NAACP—National Association for the Advancement of Colored People—the largest civil-rights organization in the United States, founded in 1909 with the mission of "ensur[ing] the political, educational, social and economic equality of rights of all persons." The NAACP played a central role in the civil-rights movement of the 1950s and 1960s.

NBS—National Bureau of Standards—US government bureau whose responsibilities included the development of computer security standards for civilian federal agencies; renamed National Institute of Standards and Technology in 1988.

NIST—National Institute of Standards and Technology—US government bureau whose responsibilities include the development of computer security standards for civilian federal agencies.

NRC—National Research Council—research arm of the National Academy of Sciences and the National Academy of Engineering.

NSA—National Security Agency—the US government agency responsible for spying on foreign communications and for protecting military, diplomatic, and intelligence communications of the US government and its contractors.

NSDD—National Security Decision Directive—a directive issued by the president (usually classified) on military, intelligence, and security matters. Such directives, unlike Executive Orders, often change name from one administration to the next and in recent decades have also been called "national security directives," "presidential directives," and "national security action memoranda."

OCR font—Optical Character Recognition font—fonts designed to be read easily by machines.

PBX—Private Branch Exchange—a telephone switch belonging to a business or other organization rather than the phone company.

PCMCIA—Personal Computer Memory Card International Association.

PGP—Pretty Good Privacy—A program originally written by Philip Zimmerman for encrypting computer files and email. The name of the company formed to develop and market the program.

plain text—Intelligible text or signals—text or signals that have not been encrypted. (adj.: plaintext)

private key—In public-key cryptography, the key that is known only to the recipient and is used for decryption and signing.

public key—In public-key cryptography, the key that is widely available and is used by the sender to encrypt and by the receiver to verify signatures.

public-key cryptography—cryptography in which communications are controlled by two keys, one of which can be made public without revealing the other. Public key cryptography makes it possible to separate the capabilities for encrypting and decrypting.

RC2—a block cipher designed by Ron Rivest of MIT and marketed by RSA Data Security as an exportable replacement for DES.

RC4—a fast stream cipher designed by Ron Rivest of MIT and marketed by RSA Data Security.

Real ID Act—a law requiring the issuing of driver's licenses by US states to conform to federal standards that effectively create a national ID card.

realtime—something operating in "real time," i.e., without the opportunity to calculate for as long as needed.

rotor machine—a cryptographic device consisting of a machine with several rotors, a disk that implements a cipher alphabet. Each disk face has a number of electrical contacts corresponding to the letters of the alphabet, and each contact on the front face is wired to exactly one contact on the rear face. As an electrical signal passes through the rotor, the signal is carried to a new alphabetic position, just as a letter looked up in a cipher alphabet changes to another letter. Rotor machines typically have at least three rotors.

RSA—the Rivest-Shamir-Adelman public-key cryptosystem. The security of the RSA system is based on the difficulty of factoring large numbers.

shift register—an electronic device made up of a number of *cells* or *stages*, each of which holds a single 0 or 1 of information. As the shift register operates, the data shift one or more places along the register at each tick of the clock. In addition to moving left or right, some of the bits are modified by being combined with other bits.

smart card—a plastic card, typically the size of a credit card, with an embedded microchip.

SSL—Secure Socket Layer—the transport-lay security mechanism used in Web browsing to support the secure form of the Hypertext Transfer Protocol, HTTPS.

STU-III—third-generation secure telephone unit—A US government secure telephone system constructed during the 1980s and using public-key cryptography.

symmetric cryptography—cryptography in which the capability to encrypt and the capability to decrypt are inseparable, in contrast to asymmetric cryptography.

Title III—Title III of the Omnibus Crime Control and Safe Streets Act of 1968 (18 USC §2510–2521) established the basic law for interceptions performed in criminal investigations.

traffic analysis—the study of the patterns of communication. An opponent can learn a great deal about the activities of an organization without being able to understand any individual message.

transactional information—information revealed during the conduct of a transaction, e.g., the time you left the (paid) parking lot, or the source and destination of an email.

Triple DES—a block employing DES three times in a row with different keys; surprisingly this has a workfactor of 2^{108} to break (rather than the expected 2^{168}).

TSD—Telephone security device—a device that is not a complete telephone but provides encryption when installed in conjunction with a telephone. Especially, the AT&T TSD-3600.

TWG—Technical Working Group set up between NIST and NSA to implement joint work on cryptography.

Type I—A category of equipment certified only for the protection of "unclassified sensitive information" that was available without the administrative controls that applied to equipment for protection of classified information.

Type II—Equipment certified for protection of sensitive information.

Venona—a long-running project by NSA (along with other organizations such as the British Government Communications Headquarters (GCHQ) and MI5) to exploit a set of Soviet messages sent in the 1940s and the early 1950s. The messages, made vulnerable by the reuse of "one-time" keying material revealed much information about Soviet spies working against Britain and the United States.

VPN—Virtual Private Network—a dynamic network constructed from encrypted tunnels through the Internet.

VoIP—Voice over IP—transmission of voice calls over the Internet.

workfactor—the number of operations needed to break a cryptosystem.

911—In the United States and Canada, the phone number for emergency services: police, fire, and ambulance.

9/11—September 11, 2001—the day on which 19 terrorists hijacked four large passenger planes, crashing two into the World Trade Center in New York and one into the Pentagon; the fourth plane crashed in the Pennsylvania countryside.

Bibliography

In this book, congressional documents are often cited as sources of memoranda, of previously published articles, and of other materials. Besides making the bibliographic scheme more efficient, and revealing where materials that otherwise would be nearly unobtainable can more easily be found, this illuminates how the original materials have been used in congressional investigations of privacy-related matters. *Data on the original materials can be found in the cited documents.*

Some sources in the text are cited by abbreviations. The explanation for these abbreviations is:

AO	Administrative Office of US Courts
ACE	American Council of Education
EP	European Parliament
OMB	Office of Management and Budget
USC-OTA	United States Congress. Office of Technology Assessment.
USDoC	United States Department of Commerce
USDoD	United States Department of Defense
USDoJ	United States Department of Justice
USDoW	United States Department of War
USGAO	United States General Accounting Office
USGSA	United States General Services Administration
USH	United States Congress. House.
USS	United States Congress. Senate.

Reports from the House of Representatives are cited in the text as: USHR xxx–yyy. The number xxx refers to the Congressional session; the yyy number is the report number. Thus USHR 102 refers to a House Report from the 102nd session of Congress. Similarly Senate reports are referred to as USSR xxx–yyy with the same meanings for xxx and yyy. Hearings are cited as USHH (House of Representatives) and USSH (Senate) followed by the Congressional session number.

Academy on Human Rights. (1993). *Handbook of Human Rights.*

Administrative Office of the United States Courts. (1972). *1971 Wiretap Report,* Government Printing Office, Washington, D.C., 1972.

Administrative Office of the United States Courts. (1978). *1977 Wiretap Report,* Government Printing Office, Washington, D.C., 1978.

Administrative Office of the United States Courts. (1989). *1988 Wiretap Report,* Government Printing Office, Washington, D.C., 1989.

Administrative Office of the United States Courts. (1990). *1989 Wiretap Report,* Government Printing Office, Washington, D.C., 1990.

Administrative Office of the United States Courts. (1991). *1990 Wiretap Report,* Government Printing Office, Washington, D.C., 1991.

Administrative Office of the United States Courts. (1992). *1991 Wiretap Report,* Government Printing Office, Washington, D.C., 1992.

Administrative Office of the United States Courts. (1993). *1992 Wiretap Report,* Government Printing Office, Washington, D.C., 1993.

Administrative Office of the United States Courts. (1994). *1993 Wiretap Report,* Government Printing Office, Washington, D.C., 1994.

Administrative Office of the United States Courts. (1995). *1994 Wiretap Report,* Government Printing Office, Washington, D.C., 1995.

Administrative Office of the United States Courts. (1996). *1995 Wiretap Report,* Government Printing Office, Washington, D.C., 1996.

Administrative Office of the United States Courts. (1997). *1996 Wiretap Report,* Government Printing Office, Washington, D.C., 1997

Administrative Office of the United States Courts. (1998). *1997 Wiretap Report,* Government Printing Office, Washington, D.C., 1998.

Administrative Office of the United States Courts. (1999). *1998 Wiretap Report,* Government Printing Office, Washington, D.C., 1999.

Administrative Office of the United States Courts. (2000). *1999 Wiretap Report,* Government Printing Office, Washington, D.C., 2000.

Administrative Office of the United States Courts. (2001). *2000 Wiretap Report,* Government Printing Office, Washington, D.C., 2001.

Administrative Office of the United States Courts. (2002). *2001 Wiretap Report,* Government Printing Office, Washington, D.C., 2002.

Administrative Office of the United States Courts. (2003). *2002 Wiretap Report*, Government Printing Office, Washington, D.C., 2003.

Administrative Office of the United States Courts. (2004). *2003 Wiretap Report*, Government Printing Office, Washington, D.C., 2004.

Administrative Office of the United States Courts. (2005). *2004 Wiretap Report*, Government Printing Office, Washington, D.C., 2005.

Administrative Office of the United States Courts. (2006). *2005 Wiretap Report*, Government Printing Office, Washington, D.C., 2006.

Advanced Telephony Unit, Federal Bureau of Investigation. (1992). "Telecommunications Overview" briefing.

Aho, Alfred V., John E. Hopcroft, Jeffrey D. Ullman. (1974). *The Design and Analysis of Computer Algorithms*, Addison-Wesley, Reading, Massachusetts.

American Bar Association. (1988). Special Committee on Criminal Justice in a Free Society of the American Bar Association Criminal Justice Section, Samuel Dash, chair, *Criminal Justice in Crisis: A Report to the American People and the American Bar on Criminal Justice in the United States: Some Myths, Some Realities, and Some Questions for the Future.*

American Council on Education, Public Cryptography Study Group. (1981). *Report of the Public Cryptography Study Group*, February 1981.

Agee, Philip. (1975). *Inside the Company: CIA Diary*, New York: Stonehill.

American National Standards Institute, Committee X9, Financial Services, Working Group X9.E.9. (1984). "Financial Institution Key Management (Wholesale)" American National Standard X9.17.

American National Standards Institute. (1998). ANSI X9.52-1998, Triple Data Encryption Algorithm Modes of Operation.

Anastasia, George. (1994). "FBI Tapes Provide Information Overflow on Mob," *Bergen Record*, July 31, 1994, p. N04.

Anderson, Betsy and Todd Buchholz. (1992). Memo for Jim Jukes, May 22, in EPIC 1994.

Anderson, Christopher. (1996a). *Jack and Jackie, Portrait of an American Marriage.* New York: William Morrow.

Anderson, Ross. (1995). "Crypto in Europe—Markets, Law and Policy" *Cryptography Policy and Algorithms Conference*, July 3–5, 1995, Brisbane, Australia, pp. 41–54.

Anderson, Ross. (1996b). *Information Hiding*, Proceedings of the First International Workshop, Cambridge, U.K., May/June 1996. Springer Lecture Notes in Computer Science No. 1174.

Anderson, Ross. (2001). *Security engineering: A Guide to Building Dependable Distributed Systems*, New York: Wiley 2001.

Andrews, Edmund. (1997). "U.S. Restrictions on Exports Aid German Software Maker," *New York Times*, April 7, 1997, p. D1.

Arbaugh, William, David J. Farber, Jonathan M. Smith. (1997). "A Secure and Reliable Bootstrap Architecture," Proceedings 1997 IEEE Symposium on Security and Privacy, pp. 65–71.

Ashcroft, John. (2003). Testimony in Commerce, Justice, State, and Judiciary Subcommittee, Senate Appropriations Committee, Hearings on April 1, 2003, One Hundred Eighth Congress, First Session.

Associated Press. (2006). "Cell phones won't keep your secrets," http://www.cnn .com/2006/TECH/ptech/08/30/betrayed.byacellphone.ap/index.html CNN.com, August 30, 2006. (Last viewed September 6, 2006).

AWST. (1986). "Contractors Ready Low-Cost, Secure Telephone for 1987 Service Start" *Aviation Week and Space Technology*, January 20, 1986, pp. 114–115.

AWST. (1993). "Infectious Electronics," *Aviation Week and Space Technology*, October 18, 1993, page 29.

AWST. (1997a). "Put Out Their Lights," *Aviation Week and Space Technology*, March 31, 1997, p. 19.

AWST. (1997b). "Not Your Average Calliope," *Aviation Week and Space Technology*, March 31, 1997, p. 19.

Bailey, S. H., D. J. Harris, and B. L. Jones. (1991). *Civil Liberties: Cases and Materials*, Butterworths, London.

Baker, Stewart. (1997). "Summary Report on the OECD Ad Hoc Meeting of Experts in Cryptography," http://www.steptoe.com/276908.htm, March 1997, to appear, *International Lawyer*.

Bamford, James. (1982). *The Puzzle Palace*, Houghton Mifflin.

Banisar, David. (1993). Statistical Analysis of Electronic Surveillance, presentation at the National Institute of Standards and Technology, Computer System Security and Privacy Advisory Board, June 3, 1993.

Banks, William, and Peter Raven-Hansen. (1994). *National Security Law and the Power of the Purse*, Oxford University Press, New York and Oxford.

Barbaro, Michael and Tom Zeller Jr. (2006). "A Face Is Exposed for AOL Searcher No. 4417749," *New York Times*, August 9, 2006, p. A1.

Barrett, P. J. (1994). "Review of the Long Term Cost Effectiveness of Telecommunications Interception," Draft report to the Security Committee of Cabinet (Australia), March 1.

Bates, David. (1907). *Lincoln in the Telegraph Office*, Century Company, New York.

Baumgardner, Frederick. (1964). Memorandum to William Sullivan, King FBI File, January 28, 1964.

Bayh, Birch. (1978). "Unclassified Summary: Involvement of NSA in the Development of the Data Encryption Standard," (United States Senate Select Committee on Intelligence), *IEEE Communication Society Magazine*, Vol. 16, No. 6, November 1978, pp. 53–55.

Beesley, Patrick. (1977). *Very Special Intelligence: The Story of the Admiralty's Operational Intelligence Center*, Hamilton, London.

Bellovin, Steven M., Matt Blaze, David Farber, Peter Neumann, and Eugene Spafford. (2000). "Comments on the Carnivore System Technical Review," December 3, 2000.

Bellovin, Steven, Matt Blaze, Ernest Brickell, Clinton Brooks, Vinton Cerf, Whitfield Diffie, Susan Landau, Jon Peterson, and John Treichler. (2006a). "Security Implications of Applying the Communications Assistance to Law Enforcement Act to Voice over IP," June 2006, http://www.itaa.org/news/docs/CALEAVOIP report.pdf (last viewed August 21, 2006).

Bellovin, Steven, Matt Blaze, Ernest Brickell, Clinton Brooks, Vinton Cerf, Whitfield Diffie, Susan Landau, Jon Peterson, and John Treichler. (2006b). "Security Implications of Applying the Communications Assistance to Law Enforcement Act to Voice over IP: Executive Summary" June 2006, http://www.itaa.org/news/docs/CALEAVOIPreport.pdf (last viewed August 21, 2006).

Berlekamp, Elwyn R. (1968). *Algebraic Coding Theory*, McGraw Hill.

Bernstein, Richard. (1994). "Testimony in Bomb Case Links Loose Ends," in *New York Times*, January 19, 1994, p. B2.

Bernstein, Nina. (2004). "A Longer Wait for Citizenship and the Ballot," *New York Times*, June 11, 2004, p. A24.

Bernstein, Nina. (2006). "In American Mill Towns, No Mirror Image of the Muslims in Leeds," *New York Times*, July 21, 2005, p. A1.

Berson, Tom. (2006). "Skype Security Evaluation," Oct. 18, 2005, http://www .skype.com/security/files/2005-031%20security%20evaluation.pdf (last viewed August 25, 2006).

Best, Jo. (2005). "Europe passes tough new data retention laws," news.com, December 14, 2005, http://news.com.com/Europe+passes+tough+new+data+ retention+laws/2100-7350_3-5995089.html (last viewed October 4, 2006).

Beth, Th., M. Frisch and G. J. Simmons, editors. (1992). *Public Key Cryptography: State of the Art and Future Directions*, Lecture Notes in Computer Science No. 578, Springer-Verlag.

Biddle, Francis. (1941). Confidential Memorandum for Mr. Hoover, October 9, 1941 in USSR 94 *Intelligence Activities: Staff Reports*, p. 281.

Biham, Eli, and Adi Shamir. (1993). *Differential Cryptanalysis of the Data Encryption Standard*, Springer-Verlag.

Blaze, Matt. (1994). "Protocol Failure in the Escrowed Encryption Standard," May 31, 1994.

Blaze, Matt, Whitfield Diffie, Ronald Rivest, Bruce Schneier, Tsutomu Shimomura, Eric Thompson, and Michael Wiener. (1996). "Minimal Key Lengths Symmetric Ciphers to Provide Adequate Commercial Security: A Report by an Ad Hoc Group of Cryptographers and Computer Scientists," Business Software Alliance. See also (http://www.bsa.org/policy/encryption/cryptographers.html).

Blum, Howard. (1987). *I Pledge Allegiance...*, Simon & Schuster, New York.

Blum, Howard. (1993). *Gangland: How the FBI Broke the Mob*, Simon & Schuster, New York.

Bogard, William. (1996). *The simulation of surveillance: hypercontrol in telematic societies*, Cambridge University Press, 1996.

Boyd, Carl. (1993). *Hitler's Japanese Confidant*, University Press of Kansas.

Brandeis, Louis. (1928). Dissenting opinion in (*Olmstead v. United States*, 277 U.S. 438, 1928.)

Brickell, Ernest F., and Gustavus J. Simmons. (1983). "A Status Report on Knapsack Based Public Key Cryptosystems" Congressus Numerantium, Vol. 7, pp. 3–72.

Brickell, Ernest, Dorothy Denning, Stephen Kent, David Maher, and Walter Tuchman. (1993). "SKIPJACK Review: Interim Report, The SKIPJACK Algorithm," July 28, 1993, available electronically from http://www.cpsr.org.

Broad, William. (1982). "Evading the Soviet Ear at Glen Cove," *Science*, Vol. 217 (3), September, pp. 910–911.

Brooke, James. (1997a). "Prosecutors in Bomb Trial Focus on Time Span and Truck Rental," *New York Times*, May 10, 1997, p. A1 and A10.

Brooke, James. (1997b). "McVeigh Lawyers Rest his Defense in Bombing Trial," *New York Times*, May 29, 1997, p. A1.

Brooks, Clinton C. (1992). Memo, April 28, 1992, in EPIC 1996, pp. C8–C13.

Brooks, Jack. (1987). Opening Statement, in USHH *Hearings on HR 145*, pp. 1–2.

Brownell, Herbert. (1954a). "The Public Security and Wire Tapping," in *Cornell Law Quarterly*, pp. 195–212.

Brownell, Herbert. (1954b). Memo to J. Hoover, May 20, 1954, in USS 94e, pp. 296–297.

Bryant, Tim. (1993). "4 Indicted Here as Terrorists; FBI: Tina Isa Killing Part of Conspiracy" *St. Louis Post-Dispatch*, April 2, 1993, p. 1A.

Burnham, David. (1980). *The Rise of the Computer State*, Random House, New York.

Burnham, David. (1996). *Above the Law*, Scribners, New York.

Burrows, William E. (1987). *Deep Black*, Random House.

Busey IV, James B. (1994). Imaging Advances Impact Criminal Detection *Signal* June 1994, p. 13.

Caba, S. (1994). "FBI Nets Stanfa in Mob Sweep," *Philadelphia Inquirer*, March 18, 1994, Sec. A.

Caldwell, Tanya. (2006a). "Palestinian in L.A. Wins Ruling in Fight to Stay in U.S.," *Los Angeles Times*, June 24, 2006.

Caldwell, Christopher. (2006b). "After Londonistan," *New York Times Magazine*, June 25, 2006, pp. 41–75.

Campbell, Duncan. (1999). "Interception 2000: Development of Surveillance Technology and Risk of Abuse of Economic Information," Report to the Director General for Research of the European Parliament, Luxembourg, April 1999.

Campen, Alan D., editor. (1993a). *The First Information War*, AFCEA International Press.

Carley, William, and Timothy O'Brien. (1995). "Cyber Caper: How Citicorp System was Raided and Funds Moved Across the World," *Wall Street Journal*, September 12, 1995, p. 1.

Carlucci, Frank. (1987). Letter to Chairman Jack Brooks, March 12, 1987, in USHH *Hearings on HR 145*, p. 386.

Carney, Dan. (1997). "Promoting Technology in America's Rural Areas," *New York Times*, January 20, 1997, p. D5.

Carroll, Maurice. (1967). "Federal Agencies Open Unified Drive to Control Mafia," *New York Times*, February 22, 1967, p. 1.

Center for Social and Legal Research. (1994). *Privacy and American Business*, Volume 1(3), Hackensack, New Jersey, 1994.

Certicom. http://www.certicom.com/index.php?action=company,press_archive& view=582 (last viewed October 14, 2006).

Chabaud, Florent and Antoine Joux. (1998). "Differential collisions in SHA-0," Crypto, 1998, p. 56.

Charns, Alexander. (1992). *Cloak and Gavel; FBI Wiretaps, Bugs, Informers, and the Supreme Court*, University of Illinois Press, 1992.

Chartrand, Sabra. (1994). "Clinton Gets a Wiretapping Bill That Covers New Technologies," *New York Times*, October 9, 1994, p. A27.

Cinquegrana, Amerigo. (1978). "The Walls (and Wires) Have Ears: The Background and First Ten Years of the Foreign Intelligence Surveillance Act of 1978," 137 *University of Pennsylvania Law Review* 793 (1989), pp. 814–815.

Clark, Tom. (1946). Letter to the President, July 17, 1946, as reported in USSR 94 *Intelligence Activities: Staff Reports*, p. 282.

Clark, Ramsey. (1967). in USHH 90 *Anti-Crime Program*, pp. 285–321.

Clark, Ramsey. (1968). Communications between Clark and J. Edgar Hoover, footnote 372 in USSR 94 *Intelligence Activities: Staff Reports*, p. 349

Clark, Ramsey. (1970). *Crime in America*, Simon and Schuster, 1970.

Clark, Ronald W. (1977). *The Man Who Broke Purple*, Boston: Little Brown and Company, 1977.

Clark, Tom. (1981). Recollections in "Japanese-American Relocation Reviewed," Earl Warren Oral History Project, Bancroft Library, University of California, Berkeley as cited in Okamura, p. 113.

Clark, David Earl. (1986). Testimony in *United States vs. Jerry Alfred Whitworth*, No. CR 85–552 JPV, in the United States District Court, Northern District of California, March 25, 1986. Reporter's transcript of proceedings, Volume 11, page 1345.

Cole, David. (2003). "9/11 and the LA 8," *The Nation*, October 27, 2003.

Communications Electronics Security Group
http://www.cesg.gov.uk/site/publications/media/ellis.pdf
http://www.cesg.gov.uk/site/publications/media/notense.pdf
http://www.cesg.gov.uk/site/publications/media/cheapnse.pdf
http://www.cesg.gov.uk/site/publications/media/secenc.pdf
http://www.cesg.gov.uk/site/publications/media/possnse.pdf
(last viewed October 15, 2006).

Committee on National Security Systems. (2003). CNSS Policy No. 15, Fact Sheet No. 1, National Policy on the Use of the Advanced Encryption Standard (AES) to Protect National Security Systems and National Security Information, June 2003.

Computer Professionals for Social Responsibility, David Banisar and Marc Rotenberg, editors. (1993). *1993 Cryptography and Privacy Sourcebook: Primary Documents on U.S. Encryption Policy, the Clipper Chip, the Digital Telephony Proposal and Export Controls*, 1993.

Congressional Quarterly Weekly. (1967). Vol. 25, Washington, D.C., February 10, 1967.

Congressional Quarterly Weekly. (1968a). Vol. 26, Washington, D.C., June 28.

Congressional Quarterly Weekly. (1968b). Vol. 26, Washington, D.C., July 19.

Congressional Record. (2001). October 25, 2001 (Senate) [Pp. S10990-S11060], from the Congressional Record Online via GPO Access (wais.access.gpo.gov) (last viewed August 27, 2006).

Congressional Research Service, Memorandum. (2006). Subject: Presidential Authority to Conduct Warrantless Electronic Surveillance to Gather Foreign Intelligence Information, January 5, 2006.

Corn-Revere, Robert. (2000). "The Fourth Amendment and the Internet," testimony in USHH 106 *Fourth Amendment and the Internet*, pp. 63–76.

Corson, William R. and Robert T. Crowley. (1989). *The New KGB*, New York: William Morrow and Company, 1985.

Corson, William R., Susan B. Trento, and Joseph J. Trento. (1989). *Widows*, New York: Crown Publishers, 1989.

Council of European Union. (2005). From: Working Party on Cooperation in Criminal Matters, to: Article 36 Committee, "Draft Framework Decision on the retention of data processed and stored in connection with the provision of publicly available electronic communications services or data on public communications networks for the purpose of prevention, investigation, detection and prosecution of crime and criminal offenses including terrorism," 6566/05, Copen 35, Telecom 10, Brussels, February 24, 2005.

Craver, Scott A., Min Wu, Bede Liu, Adam Stubblefield, Ben Swartzlander, Dan

W. Wallach, Drew Dean, and Edward W. Felten. (2001). "Reading Between the Lines: Lessons from the SDMI Challenge," *Proc. of 10th USENIX Security Symposium*, August 2001.

Crosby, Scott, Ian Goldberg, Robert Johnson, Dawn Song, and David Wagner. (2001). "A Cryptanalysis of the High-bandwidth Digital Content Protection System," ACM Workshop on Security and Privacy in Digital Rights Management 2001.

Gersho, Allen, editor. (1982). "Crypto '81," U.C. Santa Barbara Dept. of Elec. and Computer Eng., Tech Report 82-04, 1982.

Crystal, David. (2004). *The Language Revolution*, Polity Press, Cambridge, U.K., 2004.

Dam, Kenneth. (1997). Letter to Orrin Hatch, in (USSH 105 *Encryption, Key Recovery*).

Dam, Kenneth and Herbert Lin, editors. (1996). National Research Council, Commission on Physical Sciences, Mathematics, and Applications, Computer Science and Telecommunications Board, Committee to Study National Cryptography Policy, *Cryptography's Role in Securing the Information Society*, National Academy Press, 1996.

Danby, Herbert. (1933). *The Mishnah*, Oxford University Press, Oxford, 1933.

Darrow, David. (1981). *The FBI and Martin Luther King Jr.*, W. W. Norton, New York, 1981.

Dash, Samuel, Richard Schwartz, and Robert Knowlton. (1959). *The Eavesdroppers*, Rutgers University Press, 1959.

Dash, Samuel. (2004). *The Intruders: Unreasonable Searches and Seizures from King John to Ashcroft*, Rutgers University Press, New Brunswick, New Jersey, 2004.

Davis, Robert. (1973). "Confidentiality and the Census, 1790–1929" in USHEW.

Deavours, Cipher A. and Louis Kruh. (1985). *Machine Cryptography and Modern Cryptanalysis*, Artech House, 1985.

Delaney, Donald, Dorothy Denning, John Kaye, and Alan McDonald. (1993). "Wiretap Laws and Procedures: What Happens When the U.S. Government Taps A Line," September 23, 1993, available electronically from http://www.cpsr.org.

DeLoach, Cartha. (1966a). Memorandum to Clive Tolson, January 10, 1966, as reported in USSR 94 *Intelligence Activities: Staff Reports*, p. 308.

DeLoach, Cartha. (1966b). Memorandum to Clive Tolson, January 21, 1966, as reported in USSR 94 *Intelligence Activities: Staff Reports*, p. 310.

DeLoach, Cartha. (1968a). Memorandum to Clive Tolson, October 30, 1968, as reported in USSR 94 *Intelligence Activities: Staff Reports*, p. 315.

DeLoach, Cartha. (1968b). Memorandum to Clive Tolson, November 4, 1968, as reported in USSR 94 *Intelligence Activities: Staff Reports*, p. 314.

Dempsey, James and David Cole. (1999). First Amendment Foundation, Los Angeles, 1999.

Denning, Dorothy, Mike Godwin, William Bayse, Marc Rotenberg, Lewis Branscomb, Anne Branscomb, Ronald Rivest, Andrew Grosso, and Gary Marx. (1996). "To Tap or Not to Tap," in *Communications of the ACM*, Vol. 36 (3), March 1993, pp. 24–44.

Denning, Dorothy. (1994). "Encryption and Law Enforcement." February 21, 1994. Available electronically from http://www.cpsr.org.

de Tocqueville, Alexis. J.P. Mayer (trans). (1971). *Journey to America*, Doubleday & Co., New York, 1971.

Diffie, Whitfield and Martin E. Hellman. (1976). "New Directions in Cryptography," *IEEE Transactions on Information Theory*, Vol. IT-22, pp. 644–654, Nov. 1976.

Diffie, Whitfield. (1982). "Cryptographic Technology: Fifteen Year Forecast," in Simmons 1982.

Diffie, Whitfield, Leo Strawczynski, Brian O'Higgins, and David Steer. (1987). "An ISDN Secure Telephone Unit" in *Proc. National Communications Forum 1987*, pp. 473–477.

Diffie, Whitfield. (1988). "The First Ten Years of Public Key Cryptography," *Proceedings of the IEEE*, Vol. 76 No. 5, May 1988, pp. 560–577.

Diffie, Whitfield, Paul van Oorschot, and Michael Wiener. (1992). "Authentication and Authenticated Key Exchanges," in *Designs, Codes, and Cryptography*, Volume 2, Number 2, 1992, pp. 107–125.

Dingledine, Roger, Nick Mathewson and Paul Syverson. (2004). "Tor: The Second-Generation Onion Router," *Proceedings of the 13th USENIX Security Symposium*, August 2004.

Dingledine, Roger and Nick Mathewson. (2006). "Anonymity Loves Company: Usability and the Network Effect," in *The Fifth Workshop on the Economics of Information Security (pre-proceedings)*, June 26–28, 2006, pp. 533–544.

Donohue, Laura. (2006). "Anglo-American Privacy and Surveillance," *North-*

western School of Law Journal of Criminal Law and Criminology, Vol. 96, No. 3 (2006), pp. 1059–1208.

Duke, Steven B., and Albert C. Gross. (1993). *America's Longest War: Rethinking Our Tragic Crusade Against Drugs*. Putnam.

Eberle, Hans. (1992). "A High-speed DES Implementation for Network Applications," DEC SRC Research Report 90, September 23, 1992.

Edwards, Hazel. (1994). As reported in USC-OTA 1994, p. 3.

Ehrlichman, John. (1973). Testimony in USSH 93 *Watergate*, p. 2535.

Electronic Frontier Foundation. (1998). *Cracking DES: Secrets of Encryption Research, Wiretap Politics, and Chip Design*, O'Reilly & Associates, 1998.

Electronic Frontier Foundation. (2006). Tor: An anonymous Internet communication system, http://tor.eff.org/index.html.en (last viewed August 9, 2006).

Eggen, Dan and Susan Schmidt. (2002). "Secret Court Rebuffs Ashcroft," *Washington Post*, August 23, 2002, p. A1.

Electronic Industries Association. (1992). "Dual Mode Cellular System" TR45.3 IS-54 Rev. B, Appendix A, "Authentication, Message Encryption, Voice Privacy Mask Generation, Shared Secret Data Generation, A-Key Verification, and Test Data," February 1992.

Electronic Privacy Information Center, David Banisar, editor. (1994). *1994 Cryptography and Privacy Sourcebook: Primary Documents on U.S. Encryption Policy, the Clipper Chip, the Digital Telephony Proposal and Export Controls*, Diane Publishing Co., Upland, Penn., 1994.

Electronic Privacy Information Center. (1995a). *1995 EPIC Cryptography and Privacy Sourcebook: Documents on Wiretapping, Cryptography, the Clipper Chip, Key Escrow and Export Controls*, Diane Publishing Co., Upland, Penn., 1995.

EPIC. (1995b). Letter to Telecommunications Industry Liaison Unit, Federal Bureau of Investigation, November 13, 1995, in EPIC 1996, pp. B14–B20.

Electronic Privacy Information Center. (1996). *1996 EPIC Cryptography and Privacy Sourcebook: Documents on Wiretapping, Cryptography, the Clipper Chip, Key Escrow and Export Controls*, Diane Publishing Co., Upland, Penn., 1996.

ElGamal, Taher. (1985). "A Public-Key Cryptosystem and a Signature Scheme Based on Discrete Logarithms," *IEEE Transactions on Information Theory*, IT-31 (1985), pp. 469–472.

Ellis, Kenneth. (1958). *The Post Office in the Eighteenth Century: A Study in Administrative History*, Oxford University Press, London, 1958.

Elsey, G. (1950). Memorandum to the President, February 2, 1950, as reported in USSR 94 *Intelligence Activities: Staff Reports*, pp. 282–283.

Espiner, Tom. (2005). "Security experts lift lid on Chinese hack attacks," CNET News.com, ZDNet News, November 23, 2005, http://news.zdnet.com/2100-1009_22-5969516.html (last viewed on October 4, 2006).

European Commission. (1997). *Towards a European Framework for Digital Signatures and Encryption*.

European Parliament. (2005). "EP rejects initiative on data retention," September 27, 2005, http://www.europarl.europa.eu/news/expert/infopress_page/019-669-270-9-39-902-20050921IPR00560-27-09-2005-2005–true/default_en.htm (last viewed October 4, 2006).

European Union, European Parliament. (2002). "Directive 2002/58/EC of the European Parliament and of the Council Concerning the Processing of Personal Data and the Protection of Privacy in the Electronic Communications Sector (Directive on Privacy and Electronic Communications," PE-CONS 3636/02, Brussels, July 12, 2002.

Executive Office of the President. (1996). Office of Management and Budget, Interagency Working Group on Cryptography Policy, Bruce W. McConnell and Edward J. Appel, Co-Chairs. *Enabling Privacy, Commerce, Security and Public Safety in the Global Information Infrastructure (draft)*, May 20, 1996.

Executive Office of the President. (2005). Office of Management and Budget, Joshua Bolten, Memo for the Heads of All Departments and Agencies, Implementation of Homeland Security Presidential Directive 12—Policy for a Common Identification Standard for Federal Employees and Contractors, August 5, 2005.

Eyraud, Charles. (1959). *Precis de Cryptographie Moderne*, Editions Raoul Tari, Paris, 1959. This was the second edition; an earlier one appeared in 1953.

Faiola, Anthony. (2005). "Cyber Attacks Target Japan Sites," *Washington Post*, May 22, 2005.

Fallows, James. (2006). "Declaring Victory," *The Atlantic Monthly*, September 2006, pp. 60–73.

Federal Bureau of Investigation. (1949). SAC Letter No. 69, Ser. 1949, June 29, 1949, FBI 66-1372-1, as reported in Theoharis, pp. 460–461.

Federal Bureau of Investigation. (1956). "Racial Tensions and Civil Rights,"

March 1, 1956, statement used by the FBI Director at Cabinet briefing, March 9, 1956, as reported in USSR 94 *Intelligence Activities: Rights of Americans*, p. 50.

Federal Bureau of Investigation. (1960a). FBI Manual Section 87, pp. 5–11, as reported in USSR 94 *Intelligence Activities: Rights of Americans*, p. 48.

Federal Bureau of Investigation. (1960b). Memo from FBI Headquarters to New York Field Office, July 11, 1960, as reported in USSR 94 *Intelligence Activities: Staff Reports*, pp. 363.

Federal Bureau of Investigation. (1960c). Memo from FBI Headquarters to New York Field Office, December 16, 1960, as reported in USSR 94 *Intelligence Activities: Staff Reports*, p. 364.

Federal Bureau of Investigation. (1961a). FBI Summary Memo, February 15, 1961, as reported in USSR 94 *Intelligence Activities: Staff Reports*, p. 329.

Federal Bureau of Investigation. (1961b). Memo from FBI Headquarters to New York Field Office, November 3, 1961, as reported in USSR 94 *Intelligence Activities: Staff Reports*, p. 364.

Federal Bureau of Investigation. (1962). FBI Summary Memo, June 15, 1962, as reported in USSR 94 *Intelligence Activities: Staff Reports*, p. 346.

Federal Bureau of Investigation. (1968). Teletypes from the FBI to the White House Situation Room, October 30, 1968, October 31, 1968, November 1, 1968, November 2, 1968, November 3, 1968, and November 4, 1968 in USSR 94 *Intelligence Activities: Staff Reports*, p. 314.

Federal Bureau of Investigation. (1975a). FBI Summary Memo, February 3, 1975, as reported in USSR 94 *Intelligence Activities: Staff Reports*, pp. 313–314.

Federal Bureau of Investigation. (1975b). Staff summary of FBI Review, August 22, 1975, as reported in USSR 94 *Intelligence Activities: Staff Reports*, p. 344.

Federal Bureau of Investigation. (1987). Memo from FBI New York to FBI Headquarters, September 21, 1987, as reported in Foerstal 1991, p. 55.

Federal Bureau of Investigation. (1989). *FBI Inspection Division Report*, as reported in USSR 101-46 *FBI and CISPES*, pp. 2–6.

Federal Bureau of Investigation. (1992a). "Benefits and Costs of Legislation to Ensure the Government's Continued Capability to Investigate Crime with the Implementation of New Telecommunications Technologies," in EPIC 1994.

Federal Bureau of Investigation. (1992b). "Digital Telephony Proposal," in CPSR 1993.

Federal Bureau of Investigation. (1995). Federal Bureau of Investigation, "Implementation of the Communications Assistance for Law Enforcement Act," in *Federal Register* Vol. 60, No. 199, October 16, 1995, pp. 53643–53646.

Federal Bureau of Investigation. (1997a). FBI, "Electronic Surveillance—Report on the FBI Publication of the Second Notice of Capacity," Press Release, January 14, 1997.

Federal Bureau of Investigation. (1997b). FBI, "Implications of Section 104 of the Communications Act for Law Enforcement," in *Federal Register*, Vol. 62, Number 9, January 14, 1997, pp. 192–1911.

Federal Bureau of Investigation. (2000). Memo from [redacted] to Spike (Marion) Bowman, Subject: [redacted], April 5, 2000, http://www.epic.org/privacy/carnivore/fisa.html (last viewed August 18, 2006).

Federal Bureau of Investigation. (2003a). "Carnivore/DCS-1000 Report to Congress," February 24, 2003, available at http://www.epic.org/privacy/carnivore/foia_documents.html (last viewed August 18, 2006).

Federal Bureau of Investigation. (2003b). "Carnivore/DCS-1000 Report to Congress," December 18, 2003, available at http://www.epic.org/privacy/carnivore/foia_documents.html (last viewed August 18, 2006).

Federal Communications Commission, Industry Analysis Division, Common Carrier Bureau. (1996). "Telephone Lines and Offices," in EPIC 1996, p. B-19.

Federal Communications Commission. (1998a). "Memorandum Opinion and Order," September 10, 1998.

Federal Communications Commission. (1998b). *Petition for the Extension of the Compliance Date under Section 107 of the Communications Assistance for Law Enforcement Act by AT&T Wireless Service Inc, Lucent Technologies Inc. and Ericsson Inc., Memorandum Opinion and Order*, FCC 98-233, September 11, 1998.

Federal Communications Commission. (1998c). "Engineering and Technology Action: FCC Proposes Rules to Meet Technical Requirements of CALEA (CC Docket 97-213)," October 22, 1998.

Federal Communications Commission. (1998d). *In the matter of the Communications Assistance for Law Enforcement Act*, CC Docket No. 97-213, Further Notice of Proposed Rulemaking, FCC 98-292, November 5, 1998.

Federal Communications Commission. (1998). *In the matter of the Communications Assistance for Law Enforcement Act*, CC Docket No. 97-213, Third Report and Order, 14 FCC 16794 (1999).

Federal Communications Commission. (2002). *In the matter of the Communications Assistance for Law Enforcement Act*, CC Docket No. 97-213, Order on Remand, FCC 02-108, April 11, 2002.

Federal Communications Commission. (2005a). *First Report and Order and Further Notice of Proposed Rulemaking*, ET Docket No. 04-295, FCC05-153, September 23, 2005.

Federal Communications Commission. (2005b). *In the Matters of Appropriate Framework for Broadband Access to the Internet over Wireline Facilities; Review of Regulatory Requirements for Incumbent LEC Broadband Telecommunications Services; Computer III Further Remand Proceedings: Bell Operating Company Provision of Enhanced Services; 1998 Biennial Regulatory Review—Review of Computer III and ONA Safeguards and Requirements; Inquiry Concerning High-Speed Access to the Internet over Cable and Other Facilities; Internet Over Cable Declaratory Ruling; Appropriate Regulatory Treatment for Broadband Access to the Internet Over Cable Facilities*, CC Docket No. 02-33, CC Docket No. 01-337, CC Docket Nos. 95-20, 98-10, GN Docket No 00-185, CS Docket No. 02-52, Policy Statement, FCC 05-151, September 23, 2005.

Feige, Uriel, Amos Fiat, and Adi Shamir. (1987). "Zero Knowledge Proofs of Identity," in proceedings *19th ACM Symposium on Theory of Computing*, May 1987, pp. 210–217.

Feistel, Horst. (1973). "Cryptography and Computer Privacy," *Scientific American*, Vol. 228, No. 5, May 1973, pp. 15–23.

Ferguson, Niels. (2001). "Censorship in action: why I don't publish my HDCP results," August 15, 2001, http://www.macfergus.com/niels/dmca/cia.html (last viewed August 3, 2006).

Festa, Paul. (1998). "DOJ Charges Youth in Hack Attacks," News.Com, http://news.com.com/2100-1023-209260.html?legacy-cnet, March 18, 1998.

Fisher, David. (1995). *Hard Evidence*, Simon and Schuster, 1995.

Fitzgerald, Patrick and Mark Leopold. (1987). *Stranger on the Line: the Secret History of Phone Tapping*, Bodley Head, London, 1987.

Flaherty, David. (1989). *Protecting Privacy in Surveillance Societies: the Federal Republic of Germany, Sweden, France, Canada, and the United States*, University of North Carolina Press, 1989.

Flynn, Laurie. (1996). "Company Stops On-Line Access to Key Social Security Numbers," in *New York Times*, June 13, 1996, p. B11.

Foerstal, Herbert. (1991). *Surveillance in the Stacks: the FBI's Library Awareness Program*, Greenwood Press, New York, 1991.

Franklin, Benjamin. (1907). *Writings*, ed. Albert H. Smyth, Volume 5, New York, 1907.

Freeh, Louis. (1994a). Speech to the Executives' Club of Chicago, February 17, 1994.

Freeh, Louis. (1994b). Testimony in USSH 103 *Digital Telephony*, pp. 5–51.

Freeh, Louis. (1994c). Speech to American Law Institute, May 19, 1994, in EPIC 1994, p. 6 of speech.

Freeh, Louis. (1996). Statement to Committee on Commerce, Science, and Transportation, U.S. Senate, July 25, 1996.

Fried, Joseph. (1995). "In Conspiracy Case, Much Evidence that Egyptian Sheik Has a Healthy Ego, at Least," *New York Times*, April 30, 1995, p. A38.

Friedly, Michael and David Gallen. (1993). *Martin Luther King, Jr. The FBI File*, Carroll and Graf Publishers, Inc., New York, 1993.

Friedman, Lawrence. (1973). *A History of American Law*, Simon and Schuster, New York, 1973.

Froomkin, A. Michael. (1995). "The Metaphor Is the Key: Cryptography, the Clipper Chip, and the Constitution," *University of Pennsylvania Law Review*, Vol. 143 (1995) p. 709–897.

Frost, Mike. (1994). *Spyworld: Inside the Canadian and American Intelligence Establishments*, Doubleday, Canada, 1994.

Galbraith, Peter W. (2006). "Mindless in Iraq," *New York Review of Books*, August 10, 2006, pp. 28–31.

Gardner, Martin. (1977). "A New Kind of Cipher That Would Take Millions of Years to Break," *Scientific American*, Vol. 237, No. 2, Aug. 1977, pp. 120–124.

Gennaro, Rosario, Paul Karger, Stephen Matyas, Mohammad Peyravian, Allen Roginsky, David Safford, Michael Willett, and Nev Zunic. (1997). "Secure Key Recovery" IBM TR 29.2273.

Gentry, Curt. (1991). *J. Edgar Hoover. The Man and the Secrets*, Norton, New York, 1991.

Girardin, Ray. (1967). Letter to Herman Schwartz, as reported in USHH 90 *Anti-Crime Program*, p. 1013.

Goldwasser, S. and S. Micali. (1984). "Probabilistic Encryption," *Journal of Computer and System Sciences*, Vol. 28, No. 2, April 1984, pp. 270–299.

Gombossy, George. (1989). "Wiretapping Scandal Mushrooms; Congressional Hearings Scheduled," *National Law Journal*, December 25, 1989–January 1, 1990, p. 3.

Gorelick, Jamie S. (????). *Memorandum to Mary Jo White, Louis Freeh, Richard Scrugg, Jo Ann Harris; Re: Instructions on Separation of Certain Foreign Counterintelligence and Criminal Investigations.*

Gotti, John. (1992). *The Gotti Tapes*, Times Books, 1992, p. 26.

Greeley, Charles. (2002). Statement in USSH 107 *USA Patriot Act in Practice*, p. 28.

Grindler, Gary G. and Jonathan D. Schwartz. (2000). *Memorandum for the Attorney General; Subject: To Recommend that the Attorney General Authorize Certain Measures Regarding Intelligence Matters in Response to the Interim Recommendations Provided by Special Litigation Counsel Randy Bellows*, January 21, 2000.

Hager, Nicky. (1996). *Secret Power: New Zealand's Role in the International Spy Network*, Craig Potten Publishing, Nelson, New Zealand, 1996.

Halperin, Morton. (1974). In USSR 93 *Electronic Surveillance for National Security*, pp. 295–311.

Hansell, Saul. (1995). "Citibank Fraud Case Raises Computer Security Questions," *New York Times*. August 19, 1995, p. 31.

Harris, Jo Ann. (1994). Testimony in USSH 103 *Administration's Clipper Chip*.

Hart, Gary, Warren Rudman, Anne Armstrong, Norman Augustine, John Dancy, John Galvin, Leslie Gelb, Newt Gingrich, Lee Hamilton, Lionel Olner, Donald Rice, James Schlesinger, Harry Train, and Andrew Young. (2001). *Road Map for National Security: Imperative for Change*, The United States Commission for national Security/21st Century, February 15, 2001.

Headrick, Daniel R. (1991). *The Invisible Weapon, Telecommunications and International Politics, 1851–1945*, Oxford University Press.

Helsing, Cheryl. (1987). Chair, Data Security Committee, American Bankers Association, and Vice President for Corporate Preparedness and Information Security, Bank of America, Testimony in USHH *Hearings on HR 145*, pp. 112–123.

Herb, John. (1969). "250,000 War Protesters Stage Peaceful Rally in Washington; Militants Stir Clashes Later," *New York Times*, November 16, 1969, p. 1.

Hersh, Seymour. (1983). *The Price of Power, Kissinger in the Nixon White House*, Summit Books, 1983.

Hersh, Seymour. (1986). *"The Target is Destroyed": What Really Happened to Flight 007 and What America Knew About It*, Random House, 1986.

Hersh, Seymour. (1987). "Was Castro Out of Control in 1962?," in *Washington Post*, October 11, 1987, p. H1.

Hersh, Seymour. (1999). "The Intelligence Gap," *The New Yorker*, December 6, 1999, pp. 58–76.

Heymann, Philip, *Terrorism and America: A Commonsense Strategy for a Democratic Society*, MIT Press, 1998.

Hinds, Michael De Courcy. (1994). "FBI Arrests Reputed Leader of Philadelphia Mob and 23 Others," *New York Times*, March 8, 1994, p. A16.

Hoover, J. Edgar. (1941). Memo to the Attorney General, January 27, 1941, as reported in USSR 94 *Intelligence Activities: Staff Reports*, pp. 280–281.

Hoover, J. Edgar. (1950). Testimony in USHR 81 *DoJ Appropriations*, pp. 205–244.

Hoover, J. Edgar. (1956). Memo to Attorney General Brownell, December 31, 1956, as reported in USSR 94 *Intelligence Activities: Rights of Americans*, p. 63.

Hoover, J. Edgar. (1961a). Memo to Attorney General Robert Kennedy, February 14, 1961, as reported in USSR 94 *Intelligence Activities: Staff Reports*, p. 329.

Hoover, J. Edgar. (1961b). Memo to Attorney General Robert Kennedy, February 16, 1961, as reported in USSR 94 *Intelligence Activities: Staff Reports*, p. 329.

Hoover, J. Edgar. (1965). Memo to Attorney General Nicholas Katzenbach, March 30, 1965, as reported in USSR 94 *Intelligence Activities: Staff Reports*, p. 286.

Hoover, J. Edgar. (1969a). Memo to Clive Tolson, Cartha DeLoach, William Sullivan, and Mr. Bishop, May 9, 1969, 5:05 p.m., as reported in USSR 94 *Intelligence Activities: Staff Reports*, p. 324.

Hoover, J. Edgar. (1969b). Memo to the Attorney General, November 5, 1969, as reported in USSR 94 *Intelligence Activities: Staff Reports*, p. 338.

Hoover, J. Edgar. (1970). Letter to H. R. Haldeman, June 25, 1970, in USSR 94 *Intelligence Activities: Staff Reports*, p. 345.

Howell, Beryl. (2004). "Seven Weeks: The Making of the USA PATRIOT Act," *George Washington Law Review*, Vol. 72 (2004), pp. 1145–1207.

Huston, Tom. (1970). Attachment to memo, p. 2, in USSH 94 *Intelligence Activities: Huston Plan*, p. 194.

Information Security and Privacy Advisory Board. (2004). *The National Institute for Standards and Technology Computer Security Division: The Case for Adequate Funding*, June 2004, http://csrc.nist.gov/ispab/board-recommendations .html (last viewed September 1, 2006).

Information Technology Security Council, Ministry of Research and Information Technology (Denmark). (1996). "The Right to Encryption," June 11, 1996.

IDC. (2006). "Asia/Pacific (excluding Japan) Online Gaming 2005–2009 Forecast and Analysis," IDC Document Number AP654114, January 2006.

Internet Engineering Task Force, Network Working Group. (2000). *IETF Policy on Wiretapping*, IETF RFC 2804/RFC 2804, May 2000, http://www.rfc-archive.org/getrfc.php?rfc=2804 (last visited August 25, 2006).

John, Richard R. (1995). *Spreading the News: The American Postal System from Franklin to Morse*, Harvard University Press, Cambridge, Mass., 1995.

Johnson, Lyndon. (1965). "Directive to Heads of Agencies," June 30, 1965, in USSR 94 *Intelligence Activities: Staff Reports*, p. 286.

Johnson, Lyndon B. (1968). Speech, as cited in CongressionalQW 1968a, p. 1632.

Johnson, Brian. (1978). *The Secret War*, New York: Methuen, 1978.

Johnson, Clay. (2006). Executive Office of the President, Office of Management and Budget, "Memorandum for the Heads of Departments and Agencies, Subject: Protection of Sensitive Agency Information," June 23, 2006.

Jones, M. A. (1965). Memorandum to Cartha DeLoach, August 2, 1965, as reported in USSR 94 *Intelligence Activities: Staff Reports*, p. 308.

Jones, R. V. (1939). *The Wizard War—British Scientific Intelligence 1939–1945*, Coward, McCann, & Geoghegan, 1978.

Jupp, M. J. (1989). "Monolog—A Single-Line Call Logging Device," *British Telecommunications Engineering*, Vo. 8, April 1989.

Kahn, David. (1967). *The Codebreakers, The Story of Secret Writing*, New York: The Macmillan Company, 1967.

Kahn, David. (2006). "The Rise of Intelligence," Foreign Affairs, September/October 2006, pp. 125–134.

Kallstrom, James. (1997). Statement of FBI Assistant Director James Kallstrom Concerning the Second Notice of Capacity, January 14, 1997.

Kammer, Raymond. (1989). Testimony in USHH 101 *Military and Civilian Control*, pp. 219–220.

Kaplan, Robert. (2005). "How We Would Fight China," *Atlantic Monthly*, June 2005.

Katzenbach, Nicholas. (1965). Memo to J. Edgar Hoover, September 27, 1965, as reported in USSR 94 *Intelligence Activities: Staff Reports*, p. 287.

Kelling, George L. (1991). "Police," *Encyclopaedia Britannica,* 15th Edition, 1991, pp. 957–973.

Kent, Steven T. (1977). "Some Thoughts on TCP and Communication Security," MIT, Laboratory for Computer Science, Local Network Note No. 6, May 4, 1977.

Kerckhoffs, Auguste. (1883). *La Cryptographie militaire,* 1883, as cited in Kahn 1967, pp. 230.

Kerry, John and Hank Brown. (1992). "The BCCI Affair: A Report to the Committee on Foreign Relations, United States Senate," Report 102-140, One Hundred Second Congress, Second Session, 1992.

Kessler, Ronald. (1993). *The FBI,* New York: Pocket Books, 1993.

Kettle, Martin. (1998). "State-wide Informer and Secret Policing Network," *The Guardian,* March 19, 1998, p. 11.

King Jr., Neil and Ted Bridis. (2000). "FBI Wiretaps to Scan E-Mail Spark Concern," *Wall Street Journal,* July 11, 2000, p. A3.

King, Peter H. (2005a). "18 Years Waiting for a Gavel to Fall A group of Palestinians have been in legal and personal limbo for nearly two decades as the U.S. has sought to deport them. Their case foreshadowed post-9/11 policy. Series: First of two parts," *Los Angeles Times,* June 29, 2005, p. A1.

King, Peter H (2005b). "The L.A. 8, arrested in 1987 for allegedly aiding terrorists, still express bewilderment over it all. And the government still presses its case. Series: Second of two parts," *Los Angeles Times,* June 30, 2005, p. A1.

Kinzer, Stephen. (1992). "East Germans Face their Accusers," *New York Times,* April 12, 1992, p. 24–52.

Kleiman, Mark. (1993). *Against Excess: Drug Policy for Results,* Basic Books, 1993.

Kocher, Paul, Jaffe, Joshua, and Jun, Benjamin. (1999). "Differential Power Analysis," *Advances in Cryptology—Crypto 99 Proceedings,* Lecture Notes In Computer Science Vol. 1666, M. Wiener ed., Springer-Verlag, 1999.

Kollar-Kotelly, Colleen. (2002). United States Foreign Intelligence Surveillance Court, Letter to Patrick Leahy, Arlen Specter, and Charles Grassley, April 20, 2002, in USSH 107 *USA PATRIOT Act in Practice* pp. 142–143.

Kopelev, Lev. (1983). *Ease My Sorrows,* Random House, 1983.

Kravitz, David. (1991). Digital Signature Algorithm, U.S. Patent Number 5231668, applied for July 26, 1991, received July 27, 1993.

Krikorian, Greg. (1998). "D.A., Police Challenged on Wiretap Procedures," *Los Angeles Times*, April 22, 1998, p. B1.

Kris, David. (2002). Testimony in USSH 107 *USA Patriot Act in Practice* p. 28.

Kristof, Nicholas D., "Save our Spooks," New York Times, May 30, 2003.

Kuhn, Marcus. (2003). "Compromising emanations: eavesdropping risks of computer displays" University of Cambridge, Computer Laboratory, Technical Report, UCAM-CL-TR-577, December 2003.

Lacey, Mark. (1997). "Florida Couple say They Recorded Gingrich's Call," *Los Angeles Times*, January 14, 1997, p. A1.

LaMacchia, Brian and Andrew Odlyzko. (1991). "Computation of Discrete Logarithms in Prime Fields," *Design, Codes, and Cryptography*, Vol. 1, 1991, pp. 47–62.

Lambert, William. (1967). "Strange Help—Hoffa Campaign of the U.S. Senator from Missouri," *Life Magazine*, May 26, 1967.

Landau, Henry. (1937). *The Enemy Within: the Inside Story of German Sabotage in America*, G. Putnam, New York, 1937.

Landau, Susan. (1988). "Zero Knowledge and the Department of Defense," *Notices of the American Mathematical Society (Special Article Series)*, Vol. 35, No. 1 (1988), pp. 5–12.

Landau, Susan, Stephen Kent, Clinton Brooks, Scott Charney, Dorothy Denning, Whitfield Diffie, Anthony Lauck, Douglas Miller, Peter Neumann, and David Sobel. (1994). *Codes, Keys and Conflicts: Issues in U.S. Crypto Policy*, ACM Press, 1994.

Landau, Susan. (2004). "Polynomials in the Nation's Service: Using Algebra to Design the Advanced Encryption Standard," *American Mathematical Monthly*, Vol. 111, No. 2 (February 2004), pp. 89–117.

Landau, Susan. (2006). "National Security on the Line," *Journal of Telecommunications and High Technology Law*, Vol. 4, Issue 2, pp. 409–447.

Lassen, Nelson. (1937). "The History and Development of the Fourth Amendment to the United States Constitution," *The Johns Hopkins University Studies in Historical and Political Science*, Series LV, 1937.

Leahy, Patrick, Charles Grassley, and Arlen Specter. (2002). *Letter to Judge Kollar-Kotelly*, July 31, 2002.

Leahy, Patrick, Charles Grassley, and Arlen Specter. (2003). *Interim Report on*

FBI oversight in the 107th Congress by the Senate Judiciary Committee: FISA Implementation Failures, February 2003.

Lenstra, Arjen K. and Eric R. Verheul. (2000). "Selecting Cryptographic Key Sizes," PKC2000: pp. 446–465, 01/2000.

Lessig, Lawrence. (1999). *Code and Other Laws of Cyberspace*, Basic Books, New York, 1999.

Lewis, Anthony. (1961). "Robert Kennedy Vows in Georgia to Act on Rights, *New York Times*, May 7, 1961, p. 1.

Levy, Ron. (1992). Memo for Doug Steiger, May 26, in EPIC 1994.

Liberty Alliance. (2003). *Introduction to the Liberty Alliance Identity Architecture*, Version 1.0, March 2003.

Lindsey, Robert. (1979). *The Falcon and the Snowman*, Simon and Schuster, 1979.

Loftus, John and Mark Aarons. (1994). *The Secret War Against the Jews, How Western Espionage Betrayed the Jewish People*, St. Martins Press, 1994.

London School of Economics and Political Science. (2005). "The Identity Project: An Assessment of the UK Identity Cards Bill and its Implications," (interim report), London, March 2005.

Lynch, Ian. (2001). "MEPs say no to data records retention," vnunet.com, July 12, 2001, http://www.vnunet.com/vnunet/news/2115602/meps-say-records-retention?vnu_lt=vnu_art_related_articles (last viewed October 4, 2006).

Lyons, John. (1992). Testimony in USHH 102 *Threat of Economic Espionage*, pp. 163–176.

Manning, Peter K. (1977). *Police Work: The Social Organization of Policing*, MIT Press, June 1977.

Markey, Hedy Keisler and George Antheil. (1942). "Secret Communications System," U.S. Patent Number 2,292,387, August 11, 1942. (Hedy Lamarr was the stage name of Hedy Markey.)

Markoff, John. (1991). "Move on Unscrambling of Messages is Assailed," *New York Times*, April 17, 1991, p. A16.

Markoff, John. (1993). "Electronics Plan Aims to Balance Government Access with Privacy," *New York Times*, April 16, 1993, p. A1.

Markoff, John. (1996). "Cellular Industry Rejects U.S. Plan for Surveillance," *New York Times*, September 20, 1996, p. A1.

Markoff, John. (1998). "U.S. Data-Scrambling Code Cracked With Homemade Equipment," *New York Times*, July 17, 1998.

Marks, Peter. (1995). "When the Best Defense is the Prosecution's Own Tapes," *New York Times*, June 30, 1995, p. D20.

Martin, David. (1980). *Wilderness of Mirrors*, Harper and Row, New York, 1980.

Moschella, William E. (2006). U.S. Department of Justice, Office of Legislative Affairs, *Letter to J. Dennis Hastert*, April 28, 2006.

Mashberg, Tom. (1993). "N.J. Engineer Held in Blast in N.Y. Tower," *Boston Globe*, March 11, 1993, p. 1.

Massey, James L. (1969). "Shift-register synthesis and BCH decoding," IEEE Transactions on Information Theory, IT-15, 1969, pp. 122–127.

Matsui, Mitsuru. (1994). "Linear Cryptanalysis for DES Cipher," *Advances in Cryptology: EUROCRYPT '93*, T. Hellseth, ed., Springer-Verlag, Berlin, 1994, pp. 386–97.

McAuliffe, Wendy. (2001). ZDNet (UK) 14 Sep 2001 11:33 AM PT.

McConnell, John M. (1992). Letter to Willis Ware, July 23, 1992, in EPIC 1995a, p. C-14.

McConnell, John M. (1994). Testimony in USSH 103 *Administration's Clipper Chip*, pp. 95–106.

McCullagh, Declan. (2001a). "Russian Adobe Hacker Busted" *Wired News*, July 17, 2001 07:04PDT.

McCullagh, Declan. (2001b). "Senator Backs Off Backdoors" *Wired News*, October 17, 2001 02:00 PDT.

McGee, Jim. (1996a). "Tension Rises Over Digital Taps," *Washington Post*, October 27, 1996, p. H1.

McGee, Jim. (1996b). "Military Goes to War on Narcotics," *Manchester Guardian Weekly*, December 29, 1996, p. 12.

McGee, Jim. (1996c). "How Drugs Sucked in the Army," *Manchester Guardian Weekly*, December 29, 1996, p. 12.

McGrath, J. Howard. (1952). Memorandum to J. Edgar Hoover, February 26, 1952, in USSR 94 *Intelligence Activities: Staff Reports*, p. 283.

McIntosh, David and James Gattuso. (1992). Memo for Jim Jukes, May 22, 1992, in EPIC 1994.

McKinley, James, Jr. (1995a). "Suspect is Said to be Longtime Friend of Bombing Mastermind," *New York Times*, August 4, 1995, p. B5.

McKinley, James. (1995b). "The Terror Conspiracy: The Verdict," *New York Times*, October 2, 1995, p. B5.

McNulty, Lynn. (1992). Memo for the Record, August 18, 1992, in EPIC 1996, pp. C14–C19.

Meller, Paul. (2001). "European Union Set to Vote on Data Law," *New York Times*, November 13, 2001.

Menezes, Alfred, Paul van Oorschot, and Scott Vanstone. (1996). *Handbook of Applied Cryptography*, CRC Press, 1996 (fifth printing, 2001).

Meredith, Robyn. (1997). "VW Agrees to Pay G.M. $100 Million in Espionage Suit," *New York Times*, January 10, 1997, p. A1.

Merkle, Ralph C. (1978). "Secure Communications Over Insecure Channels," Communications of the ACM, Vol. 21, No. 4, April 1978, pp. 294–299.

Mikva, Abner. (1971). Testimony in USSH 92 *Federal Data Banks*, pp. 130–146.

Milonovich, Gregory M. (2003). Letter to Marlene Dortch, Re: Notice of Ex Parte Presentation, (CC Docket Nos. 02-33, 95-20, and 98-10; CS Docket No. 02-52), November 25, 2003.

Mintz, John. (1992). "Intelligence Community in Breach with Business," *Washington Post*, April 30, 1992, p. A8.

Monbiot, George. (1996). "Heavy Hand of the Law Keeps Tabs on the 'Enemy Within'," *Manchester Guardian Weekly*, September 8, 1996, p. 23.

Moore, G. C. (1972). Memo to E. S. Miller, September 27, 1972, as reported in USSR 94 *Intelligence Activities: Rights of Americans*, p. 75.

Moore, Mark. (1990). "Supply Reduction and Drug Law Enforcement," in Tonry 1990, 109–157.

Morgenthau, Henry. (1940). *The Presidential Diaries of Henry Morgenthau Jr. 1938–1945*, University Publications of America, Frederick, Maryland. (The name is mistakenly spelled 'Morganthau' on the title page.)

Moschella, William E., U.S. Department of Justice, Office of Legislative Affairs. (2006). *Letter to J. Dennis Hastert*, April 28, 2006.

Myers, F. (1979). "A Data Link Encryption System," *National Telecommunications Conference*, Washington, D.C., November 27–29, 1979, pp. 43.5.1–43.5.8.

National Research Council, Commission on Physical Sciences, Mathematics, and its Applications, Computer Science and Telecommunications Board, Committee on Maintaining Privacy and Security in Health Care Applications of the National Information Infrastructure. (1997). *For the Record: Protecting Electronic Health Information.*

National Research Council. (2006). *Engaging Privacy and Information Technology in a Digital Age,* 2006.

National Commission for the Review of Federal and State Laws relating to Wiretapping and Electronic Surveillance. (1976). *Commission Studies,* Washington, D.C.

http://www.nsa.gov/ia/industry/crypto_suite_b.cfm?MenuID=10.2.7 (last viewed October 14, 2006).

National Security Decision Directive 145. (1984). September 17, 1984, in USHH *Hearings on HR 145,* pp. 528–537.

National Telecommunications and Information Agency. (1992). "Technological, Competitiveness, and Policy Concerns.", in EPIC 1994.

National Transportation Safety Board. (2000). *Aircraft Accident Report: In-Flight Breakup Over the Atlantic Ocean of Trans World Airlines Flight 800 Boeing 747-131, N93119, Near East Moriches, New York, July 17, 1996,* National Transportation Safety Board, Washington, D.C., 2000.

Needham, Roger M. and Michael D. Schroeder. (1978). "Using Encryption for Authentication in Large Networks of Computers" (1978). *Communications of the ACM,* Vol. 21, No. 12, pp. 993–999.

Neel, Roy. (1994a). President of the U.S. Telephone Association, in USSH 103 *Digital Telephony,* pp. 53–64.

Neel, Roy. (1994b). President of the U.S. Telephone Association, in USHH 103 *Network Wiretapping Capabilities,* pp. 101–115.

Neumann, Peter. (1994). *Computer-Related Risks,* ACM Press (Addison-Wesley).

New York Times. (1879). "The Privacy of the Telegraph," (1879). March 21, p. 1, col. 3.

New York Times. (1916). "Priests Deny the Mayor's Charges; His Tapped Wire Records Falsified to Justify Himself, they Declare," May 27, 1916, p. 1.

New York Times. (1961). "Anti-Crime Bill is Passed," September 20, 1961, p. 3.

New York Times. (1971). "Vietnam Archive: Pentagon Study Traces 3 Decades of Growing U.S. Involvement," June 13, 1971, p. 1.

New York Times. (1998a). "Case in Los Angeles Raises Concerns over Secrecy of Wiretaps," August 2, 1998.

New York Times. (1998b). "Dispute Over Concealing Evidence From Wiretaps in California," November 17, 1998.

New York Times. (2005). "An Unrealistic 'Real ID'," May 4, 2005.

New York Tribune. (1876). "The Privacy of Telegrams," December 26, 1876, p. 1, col. 4.

New York Tribune. (1880). "Inviolability of Telegrams," January 11, 1880, p. 5, col. 3.

New York Tribune. (1890). "They Will Probably Answer Now," June 10, 1890, p. 3, col. 6.

North, David. (1986). "U.S. Using Disinformation Policy to Impede Technical Data Flow." *Aviation Week and Space Technology*, March 17, 1986, pp. 16–17.

O'Brien, John. (1971). Testimony in USSH 92 *Federal Data Banks*, pp. 100–130.

O'Hara, Colleen and Heather Harreld. (1997). "DOD sinks the Clipper," *Federal Computer Week*, February 17.

O'Harrow, Robert Jr. (2005). *No Place to Hide*. Free Press.

Odom, Lt. Genl. William. (1987). Testimony in USHH *Hearings on HR 145*, pp. 276–297.

OECD (Organization for Economic Cooperation and Development). (1997). "Cryptography Policy Guidelines," March 27, 1997.

Okamura, Raymond. (1981). "The Myth of Census Confidentiality," *Amerasia*, pp. 111–120.

Onion Routing, Executive Summary, http://www.onion-router.net/Summary.html last visited August 9, 2006.

Onley, Dawn and Patience Swift. (2006). "Red Storm Rising: DOD's efforts to stave off nation-state cyberattacks begin with China," *Government Computer News*, August 21, 2006, http://www.gcn.com/print/25_25/41716-1.html (last viewed on October 4, 2006).

Ostrovsky, Victor and Claire Hoy. (1990). *By Way of Deception, the Making and Unmaking of a Mossad Officer*, St. Martins Press, New York.

Parker, Donn. (1983). *Fighting Computer Crime*, Charles Scribner's, New York.

Pasztor, Andy. (1995). *When the Pentagon Was for Sale*, Charles Scribner's, New York.

Peterson, A. Padgett. (1993). "Tactical Computers Vulnerable to Malicious Software Attacks." *SIGNAL*, November, pp. 74–75.

Piller, Charles. (1993). "Privacy in Peril," *MacWorld*, July, pp. 8–14.

Pincus, Walter. (1976). "Data to be Destroyed," *Washington Post*, March 31, p. 1.

Pincus, Walter. (2006). "FBI Role in Terror Probe Questioned," *Washington Post*, September 2, 2006, p. A1.

Plum, William. (1882). *The Military Telegraph During the Civil War*, Volume 1 Chicago.

Poindexter, John. (1986). "National Policy on the Protection of Sensitive, but Unclassified Information in Federal Government Telecommunications and Automated Systems," October 29, in USHH *Hearings on HR 145*, pp. 541–545.

Polis, Richard. (1987). "European Needs and Attitudes Towards Information Security," *Fifteenth Annual Telecommunications Policy Research Conference*, September 27–30.

Poulsen, Kevin. (2003). "Slammer Worm Crashed Ohio Nuke Plant," *The Register*, August 20, 2003.

The President's Commission on Law Enforcement and the Administration of Justice. (1967). *The Challenge of Crime in a Free Society*, United States Government Printing Office, 1967.

Privacy Journal. (1992). *Compilation of State and Federal Privacy Laws*, Washington, D.C.

Pulitzer. (2006). The Pulitzer Prize Winners 2006, http://www.pulitzer.org/year/2006/national-reporting/ (last viewed October 16, 2006).

Purdom, Todd. (1995). "Clinton Assails Effort to Change His Anti-Terrorism Plan," *New York Times*, May 27, p. 24.

Quant, Kathy. (2006). "Wired in L.A.," *California Attorney's for Criminal Justice*, Vol. 99, No. 2 (1999), http://pd.co.la.ca.us/CACJ.htm (last viewed September 9, 2006).

Raber, Ellen and Michael O. Riley. (1989). "Protective Coatings for Secure Integrated Circuits, *Energy and Technology Review*, May–June, pp. 13–20.

Randall, James. (1951). *Constitutional Problems Under Lincoln*, University of Illinois Press.

Rawls, W. Lee. (1992). Assistant Attorney General, letter to Speaker of the House Thomas S. Foley, September 14.

Reed, Michael G., Paul F. Syverson, and David M. Goldschlag. (1996). "Proxies for Anonymous Routing," Twelfth Annual Computer Security Application Conference, San Diego, CA, 1996.

Reinsch, William. (1985). In USSR 105 *"PROCODE" Act.*

Reno, Janet. (1995). Office of the Attorney General, Memorandum to Assistant Attorney General, Criminal Division, Director, FBI, Counsel of Intelligence Policy, United States Attorneys Re: Procedures for Contacts Between the FBI and the Criminal Division Concerning Foreign Intelligence and Foreign Counterintelligence Investigations, July 19, 1995.

Reuter, Peter. (1985). "Eternal Hope: America's International Narcotics Efforts," *Public Interest*, Spring, pp. 79–85.

Richardson, Doug. (1991). "More Pernicious Computer Viruses Menace, Germinate." *SIGNAL*, December, p. 64.

Richardson, Margaret. (1997). "Disposition of Unauthorized Access Cases, 1994–1997" in USSH 105 *IRS and Misuse.*

Richelson, Jeffrey T. and Desmond Ball. (1985). *The Ties That Bind, Intelligence Cooperation between the UKUSA countries—the United Kingdom, the United States of America, Canada, Australia and New Zealand* Allen & Unwin.

Richelson, Jeffrey. (1987). *American Espionage and the Soviet Target*, New York: William Morrow and Company.

Richelson, Jeffrey T. (1990). *America's Secret Eyes in Space—The U.S. Keyhole Spy Satellite Program*, Harper & Row.

Risen, James and Eric Lichtblau. (2005). "Bush Lets U.S. Spy on Callers Without Courts," *New York Times*, December 16, 2005.

Risen, James. (2006). *State of War: The Secret History of the CIA and the Bush Administration.*

Rivest, R., A. Shamir, and L. Adleman. (1978). "A Method for Obtaining Digital Signatures and Public Key Cryptosystems," *Communications of the ACM*, Vol. 21 (2), pp. 120–126, February.

Rivest, Ronald. (1992). "Responses to NIST's Proposal," *Communications of the ACM*, Vol. 35 (7), July, pp. 41–47.

Robins, Natalie. (1988). "The FBI's Invasion of Libraries," *The Nation*, April 9, p. 499.

Robinson, Clarence A. Jr. (1993a). "Wireless Era Dawns—Tests extend voice and data via seamless networks to follow subscribers automatically, wherever they are located." *SIGNAL*, Volume 47 Number 7, March, pp. 19–23.

Robinson, Clarence A. Jr. (1993b). "Information Warfare Avoids Army Virus Contamination." *SIGNAL*, November, p. 75.

Rogers, Herbert L. (1987). "An Overview of the Caneware Program" paper 31, in proc. *3rd Annual Symposium on Physical / Electronic Security*, Armed Forces Communications and Electronics Association, Philadelphia Chapter, August.

Roosevelt, Franklin Delano. (1940). Memo to Attorney General Jackson, May 21, as cited in USSR 94 *Intelligence Activities: Staff Reports*, p. 279.

Rosenblum, Howard E. (1980). "Secure communication system with remote key setting—uses automatic iterative replacement of working variables of system subscribers" US Patent No. 4182933, January 8.

Rotenberg, Marc. (1996). "U.S. Lobbies OECD to Adopt Key Escrow," in *The International Privacy Bulletin*, Vol. 4, No. 2. Spring, pp. 4–7.

Ruckelshaus, William. (1974). in USSH 93 *Warrantless Wiretapping and Electronic Surveillance*, pp. 310–378.

Rydell, C. Peter and Susan Everingham. (1994). *Controlling Cocaine: Supply Versus Demand Program*, Drug Policy Research Center, RAND.

Saltzer, Jerome H., David P. Reed, and David D. Clark. (1984). "End-to-end Arguments in System Design," *ACM Transactions on Computer Systems* 2, 4 (November 1984) pages 277–288.

Sanger, David. (1987). "Rise and Fall of Data Directive," *New York Times*, March 19.

Sayers, Michael and Albert E. Kahn. (1942). *Sabotage! The Secret War Against America*, Harper and Brothers, New York, 1942.

Scheele, Carl. (1970). *Neither Snow nor Rain ...; The Story of the United States Mails*, Smithsonian Institutional Press.

Schneier, Bruce. (1996). *Applied Cryptography: Protocols, Algorithms, and Source Code in C*, Addison-Wesley. (The first edition appeared in 1993.)

Schnorr, Claus. (1989). Procedures for the Identification of Participants as well as the Generation and Verification of Electronic Signatures in a Digital Exchange System, German Patent Number 9010348.1, patent applied for February 24, patent received August 29, 1990.

Schnorr, Claus. (1990). "Efficient Identification and Signatures for Smart Cards," in *Advances in Cryptology—Crypto '89*, Springer-Verlag, New York, pp. 239–251.

Schnorr, Claus. (1991). "Method for Identifying Subscribers and for Generating and Verifying Electronic Signatures in a Data Exchange System," U.S. Patent Number 4995082, patent applied for February 23, 1990, patent received February 19, 1991.

Schrage, Michael. (1986). "U.S. Seeking to Limit Access of Soviets to Computer Data," *Washington Post*, May 27, p. A1.

Schwartau, Winn. (1994). *Information Warfare: Chaos on the Electronic Superhighway*, Thunder's Mouth Press.

Schwartz, Herman. (1972). "A Report on the Costs and Benefits of Electronic Surveillance—," American Civil Liberties Union Reports, 1972, in USSR 93 *Electronic Surveillance for National Security*, pp. 102–209.

Schwartz, Herman. (1974). Testimony in USSR 93 *Electronic Surveillance for National Security*, pp. 95–214.

Schweizer, Peter. (1993). *Friendly Spies, How America's Allies are using Economic Espionage to Steal Our Secrets*, Atlantic Monthly Press.

Scowcroft, Brent. (1992). Memorandum to Secretary of Defense Dick Cheney, Attorney General William Barr, and Director of Central Intelligence Robert Gates, January 17.

Seipp, David. (1977). *The Right to Privacy in American History*, Program in Information Resources Policy, Harvard University, Cambridge, Massachusetts, June.

Sessions, William. (1988). Testimony in USSH 100 *FBI and CISPES*, pp. 119–156.

Sessions, William. (1991). Testimony in USHH 102 *FBI Oversight*, p. 72.

Sessions, William. (1993a). Letter to Mr. George Tenet, February 9, in EPIC 1996, p. C-24.

Sessions, William. (1993b). Letter to George Tenet and Briefing Document, "Encryption: The Threat, Applications, and Potential Solutions," February 19, in EPIC 1996, pp. C44–C61.

Shackelford. (1976). as reported in USSR 94 *Intelligence Activities: Staff Reports*, p. 251, February 2.

Shane, Scott and Tom Bowman. (1995). "America's Fortress of Spies" *Baltimore Sun*, a series of 6 articles, December 3–15, 1995.

Shannon, Claude E. (1949). "Communication Theory of Secrecy Systems." *Bell System Technical Journal*, Vol. 28, No. 4 (1949), page 656–715.

Shapley, Deborah and Gina Kolata. (1977). "Cryptology: Scientists Puzzle over Threat to Open Research, Publication," *Science*, September 30, pp. 1345–1349.

Shea, Dana. (2003). "Critical Infrastructure: Control Systems and the Terrorist Threat," Congressional Research Service, Library of Congress, 21 February 2003, at CRS-7.

Shimomura, Tsutomu and John Markoff. (1996). *Takedown: The Pursuit and Capture of Kevin Mitnick, America's Most Wanted Computer Outlaw—By the Man Who Did It*, Hyperion.

Shimomura, Tsutomu. (1997). Testimony in USHR 105 *Secure Communications*.

Simmons, Gustavus J., editor. (1982). *Secure Communications and Asymmetric Cryptosystems*, AAAS Selected Symposium No. 69, Westview Press.

Simmons, Gustavus J. (1986). "Cryptology," *Encyclopaedia Britannica*, 16th Edition, pp. 913–924B.

Simpson, Jack. (1988). Testimony in USHH *Hearings on HR 145*, pp. 324–344.

Smith, John Lynn. (1971). "The Design of Lucifer, A Cryptographic Device for Data Communications," IBM White Plains, N. Y., RC 3326.

Smith, T. J. (1973). Memo to E. Miller, February 26, in USSR 94 *Intelligence Activities: Staff Reports*, p. 326.

Smith, William W. Jr. (1991). "Passive Location of Mobile Cellular Telephone Terminals" *Proceedings of the International Carnahan Conference on Security Technology*, October 1–3, Taipei, Taiwan, pp. 221–225.

Smith, Bradley F. (1993). *The Ultra-Magic Deals and the Most Secret Special Relationship, 1940–1946*, Presidio, Novato, California.

Smith, Stephen, J. Allen Crider, Henry Perritt, Jr., Mengfen Shyong, Harold Krent, Larry Reynolds, and Stephen Mencik. (2000). *Independent Review of the Carnivore System: Final Report*, IIT Research Institute, Lanham, Maryland, December 8, 2000.

Sherr, Micah, Eric Cronin and Matt Blaze. (2005). "Signaling Vulnerabilities in Wiretapping Systems," *IEEE Security and Privacy*, November/December 2005, pp. 13–25.

Socolar, Milton. (1989). Testimony of in USHH 101 *Military and Civilian Control*, pp. 36–49.

Solomon, Stephen. (1995). "American Express Applies for a New Line of Credit," *New York Times Magazine*, July 30, p. 38.

Solove, Daniel and Marc Rotenberg. (2003). *Information Privacy Law*, Aspen Publishers, New York, 2003.

Solzhenitsyn, Aleksandr I. (1968). *The First Circle* (Translated by Thomas P. Whitney.) Harper & Row.

Starr, Paul. (2003). *The Creation of the Media: Political Origins of Modern Communications*, Basic Books, New York, 2003.

Statewatch. (1997). "European Union and FBI Launch Global Surveillance System," Statewatch, P.O. Box 1516, London N16 OEW, England.

Stein, Ralph. (1971). Testimony in USSH 92 *Federal Data Banks*, pp. 244–277.

Stoll, Clifford. (1989). *The Cuckoo's Egg, Tracking a Spy Through the Maze of Computer Espionage*, New York: Doubleday.

Sullivan, William. (1964). Memo to A. Belmont, August 21, as reported in USSR 94 *Intelligence Activities: Rights of Americans*, p. 118.

Sullivan, William. (1969). Letter to J. Edgar Hoover, July 2, as reported in USSR 94 *Intelligence Activities: Staff Reports*, p. 323.

Sullivan, William. (1979). *The Bureau: My Thirty Years in Hoover's FBI*, W. W. Norton and Co.

Theoharis, Athan. (1974). Testimony in USSR 93 *Electronic Surveillance for National Security*, pp. 337–348.

Theoharis, Athan and John Cox. (1988). *The Boss. J. Edgar Hoover and the Great American Inquisition*, Temple University Press.

Thomas, Dave. (1995). "The Roswell Incident and Project Mogul, Skeptical Inquirer Vol. 19, No. 4, July–August, 1995: pp. 15–18.

Thompson, Larry D. (2001). *Memorandum. Subject: Intelligence Sharing*, August 6, 2001.

Thompson, Larry. (2002). Memo: Avoiding Collection and Investigative Use of 'Content' in the Operation of Pen Registers and Trap and Trace Devices, to Assistant Attorney General, Criminal Division, Assistant Attorney General, Antitrust Division, Assistant Attorney General, Tax Division, All United States Attorneys, Director of the Federal Bureau of Investigation, Administrator of the Drug Enforcement Agency, Commissioner of the Immigration and Naturalization Services, Director of the United States Marshals Service, May 24 2002, http://www.judiciary.house.gov/judiciary/attachd.pdf (last viewed August 27, 2004).

Thornburgh, Nathan. (2006a). "Inside the Chinese Hack Attack: How a ring of attackers, codenamed Titan Rain by investigators, probed U.S. government computers," *Time*, August 25, 2006, http://www.time.com/time/nation/article/ 0,8599,1098371,00.html (last viewed October 4, 2006).

Thornburgh, Nathan. (2006b). "The Invasion of the Chinese Cyberspies (And the Man Who Tried to Stop Them): An exclusive look at how the hackers called Titan Rain are stealing U.S. secrets," *Time*, August 29, 2006, http://www.time.com /time/magazine/article/0,9171,1098961,00.html (last viewed October 4, 2006).

Tonry, Michael and James Wilson, editors. (1990). *Drugs and Crime*, University of Chicago Press.

TRAC (Transactional Records Access Clearinghouse). (2006). "Criminal Terrorism Enforcement in the United States During the Five Years since the 9/11/01 Attacks," September 4, 2006, http://trac.syr.edu/tracreports/terrorism/169/ (last viewed September 2, 2006).

United Nations. (1985). *International Covenants on Civil and Political Rights: Human Rights Commission, Selected Decisions under the Optional Protocol, Second to Sixteenth Sessions*, United Nations Office of Public Information, New York.

USAINTC. (1969). Information Collection Plan, April 23, 1969, as reported in USSR 93 *Military Surveillance*, pp. 41–42.

United States Congress, Office of Technology Assessment. (1987). *Defending Secrets, Sharing Data: New Locks and Keys for Electronic Information*, OTA-CIT-310, Washington, D.C.: Government Printing Office, Washington, D.C.

United States Congress, Office of Technology Assessment. (1994). *Information Security and Privacy in Network Environments*, OTA-TCT-606.

United States Department of Commerce, National Bureau of Standards. (1973). "Cryptographic algorithms for protection of computer data during transmission and dormant storage: solicitation of proposals." *Federal Register* Vol. 38, No. 93, p. 12763, May 15, 1973.

United States Department of Commerce, National Bureau of Standards. (1974). "Encryption algorithms for computer data protection: reopening of solicitation." *Federal Register* Vol. 39, No. 167, p. 30961, August 27, 1974.

United States Department of Commerce, National Bureau of Standards. (1977). "Data Encryption Standard," Federal Information Processing Standard Publication No. 46.

United States Department of Commerce, National Bureau of Standards. (1980). "DES Modes of Operation," Federal Information Processing Standard 81.

United States Department of Commerce, National Bureau of Standards. (1982).

"Solicitation for Public Key Cryptographic Algorithms," *Federal Register*, Vol. 47, No. 126, p. 28445, June 30, 1982.

United States Department of Commerce, National Institute of Standards and Technology. (1989). "Comments of NSA Changes to NIST MOU of March 8," March 13, 1989 in USHH 101 *Military and Civilian Control*, p. 270.

United States Department of Commerce, National Institute of Standards and Technology. (1990a). "Memorandum for the Record, January 31, 1990," in CPSR 1993.

United States Department of Commerce, National Institute of Standards and Technology. (1990b). "Memorandum for the Record, March 26, 1990," in CPSR 1993.

United States Department of Commerce, National Institute of Standards and Technology. (1991a). "Memorandum for the Record, Twenty-Seventh Meeting of the Technical Working Group," May 28, 1991, in CPSR 1993.

United States Department of Commerce, National Institute of Standards and Technology. (1991b). *Publication XX: Announcement and Specifications for a Digital Signature Standard (DSS)*, August 19, 1991.

United States Department of Commerce, National Institute of Standards and Technology. (1991c). "Public Key Status Report," in EPIC 1996, p. C-3.

United States Department of Commerce, National Institute of Standards and Technology. (1993). "A Proposed Federal Information Processing Standard for an Escrowed Encryption Standard (EES)," *Federal Register*, Vol. 58, July 30, 1993, pp. 40791–40793.

United States Department of Commerce. (1994a). "Briefing re Escrowed Encryption Standard, Department of Commerce, February, 4, 1994."

United States Department of Commerce, National Institute of Standards and Technology. (1994b). "Approval of Federal Information Processing Standards Publication 185, Escrowed Encryption Standard," *Federal Register*, Vol. 59, No. 27, February 9, 1994.

United States Department of Commerce, National Institute of Standards and Technology. (1994c). *Federal Information Processing Standards Publication 186: Digital Signature Standard (DSS)*, May 19, 1994.

United States Department of Commerce, National Institute of Standards and Technology. (1997). "Announcing Development of a Federal Information Processing Standard for Advanced Encryption Standard," Federal Register, Vol. 62, No. 1, January 2, 1997, pp. 93–94.

United States Department of Commerce, National Institute of Standards and

Technology. (1999). Federal Information Processing Standards Publication 46-3, Data Encryption Standard (DES), Reaffirmed October 25.

United States Department of Commerce, National Institute of Standards and Technology, and United States Department of Defense, National Security Agency. (1989). "Memorandum of Understanding between the Director of the National Institute of Standards and Technology and the Director of the National Security Agency concerning the Implementation of Public Law 100–235," March 24, 1989.

United States Department of Defense, National Security Agency, Office of General Counsel. (1994). "Proposed Changes to Escrow Encryption Standard," 12 January 1994.

United States Department of Justice. (1976). *Report on Enquiry into CIA-Related Electronic Surveillance Activities*, SC-05078-76, June 30.

United States Department of Justice. (1994a). "Briefing re Escrowed Encryption Standard, Department of Commerce," February 4, 1994.

United States Department of Justice. (1994b). *Sourcebook of Criminal Justice Statistics*.

United States Department of Justice, Office of the Inspector General, Michael Bromwich. (1997). "A Review of the FBI's Performance in Uncovering the Espionage of Aldrich Hazen Ames (April 1997)," (unclassified version).

United States Department of Justice, Office of Inspector General, Audit Division. (2006a). *Implementation of the Communications Assistance for Law Enforcement Act*, Audit Report 06-13, March 2006.

United States Department of Justice, Counterterrorism Section. (2006b). *Counterterrorism White Paper*, June 22, 2006.

United States Department of State. (1997). Country Reports on Human Rights Practices for 1996.

United States Department of State. (2006). *Country Reports on Terrorism 2005*, United States Department of State Publication 11324, Office of the Coordinator for Counterterrorism, April 2006.

United States Department of War. (1943). Wartime Civil Control Administration, Western Defense Command and Fourth Army, Press Release No. 1, *Final Report: Japanese Evacuation from the West Coast 1942*, March 14, 1942, Washington 1943.

United States Foreign Intelligence Surveillance Court. (2002a). Memorandum Opinion (as Corrected and Amended), May 17, 2002, in USSH 107 *USA PATRIOT Act in Practice*, pp. 145–171.

United States Foreign Intelligence Surveillance Court. (2002b). In Re All Matters Submitted To The Foreign Intelligence Surveillance Court, Docket Numbers: Multiple, Order, as Amended, May 17, 2002, in USSH 107 *USA PATRIOT Act in Practice*, pp. 172–175.

United States Foreign Intelligence Surveillance Court. (2002c). In Re All Matters Submitted To The Foreign Intelligence Surveillance Court, Docket Numbers: Multiple, Order, May 17, 2002, USSH 107 *USA PATRIOT Act in Practice*, pp. 176–177.

United States Foreign Intelligence Surveillance Court of Review. (2002). In re: Sealed Case No. 02-001, Consolidated with 02-002, On Motions for Review of Orders of the United States Foreign Intelligence Surveillance Court (Nos. 02-662 and 02-968), November 18, 2002.

United States General Accounting Office. (1990). *Computers and Power: How the Government Obtains, Verifies, Uses and Protects Personal Data*, Briefing to the Chairman, Subcommittee of Telecommunications and Finance, Committees on Energy and Commerce, House of Representatives, August 3, 1990.

United States General Accounting Office. (1992). *Advanced Communications Technologies Pose Wiretapping Challenges*, Briefing Report to the Chairman, Subcommittee on Telecommunications and Finance, Committee on Energy and Commerce, House of Representatives, July 1992.

United States General Accounting Office. (1993a). *National Crime Information Center: Legislation Needed to Deter Misuse of Criminal Justice Information*, GAO/T-GGD-93-41, Washington, D.C., U.S. Government Printing Office, July 1993.

United States General Accounting Office. (1993b). *Communications Privacy: Federal Policy and Actions*, (Letter Report, 11/04/93, GAO/OSI-94-2).

United States General Accounting Office. (1997). *IRS Systems Security: Tax Processing and Data Still at Risk due to Serious Weaknesses*, (Letter Report, 4/8/97, GAO/AIMD-97-49).

United States General Services Administration. (1992). "Attachment to May 5, 1992 GSA memo, p. 2," in CPSR 1993.

United States House of Representatives, Committee on the Judiciary, Subcommittee on Civil and Constitutional Rights. (1991). *FBI Oversight and Authorization Request for Fiscal Year 1992*, Hearings before the Subcommittee on Civil and Constitutional Rights of the Committee on the Judiciary, One Hundred Second Congress, First Session, March 13 and 21, 1991.

United States House of Representatives, Committee on the Judiciary, Subcommittee on the Constitution. (2000). *Fourth Amendment and the Internet*, Hearing before the Subcommittee on the Constitution of the Committee on the Judiciary,

House of Representatives, One Hundred Sixth Congress, Second Session, April 6, 2000.

United States House of Representatives, Committee on Appropriations. (1950). *Department of Justice Appropriations for 1951*, Eighty-First Congress, Second Session, 1950.

United States House of Representatives, Committee on the Judiciary, Subcommittee No. 5. (1955). *Wiretapping*, Hearings on HR 762, Eighty-Fourth Congress, First Session, 1955.

United States House of Representatives, Committee on the Judiciary, Subcommittee No. 5. (1967). *Anti-Crime Program*, Hearings on HR 5037, 5038, 5384, 5385 and 5386, March 15, 16, 22, 23, April 5, 7, 10, 12, 19, 20, 26 and 27, 1967, Ninetieth Congress, First Session, 1967.

United States House of Representatives, Committee on the Judiciary. (1974). *Statement of Information, Book VII, Part 4*, Ninety-Third Congress, Second Session, 1974.

United States House of Representatives, Committee on the Judiciary. (1974). *Impeachment Inquiry*, Hearings, Ninety-Third Congress, Second Session, 1974.

United States House of Representatives, Committee on Government Operations, Subcommittee. (1987). *Computer Security Act of 1987*, Hearings on HR 145, February 25, 26, and March 17, 1987, One Hundredth Congress, First Session, 1987.

United States House of Representatives, Committee on Government Operations. (1987). House Report 100-153, Part 2, *Report on the Computer Security Act of 1987*, One Hundredth Congress, First Session, Washington, D.C.

United States House of Representatives, Committee on Government Operations, Legislative and National Security Subcommittee. (1989). *Military and Civilian Control of Computer Security Issues*, Hearings on May 4, 1989, One Hundred First Congress, First Session, 1989.

United States House of Representatives, Committee on the Judiciary, Subcommittee on Economic and Commercial Law. (1992). *The Threat of Foreign Economic Espionage to U.S. Corporations*, Hearings on April 29 and May 7, 1992, One Hundred Second Congress, Second Session.

United States House of Representatives, Committee on Foreign Affairs, Subcommittee on Economic Policy, Trade, and the Environment. (1993). *Export Controls on Mass Market Software*, Hearings on October 12, 1993, One Hundred Third Congress, First Session.

United States House of Representatives, Committee on Government Reform, Subcommittee on Government Efficiency, Financial Management and Intergovernmental Relations. (2002). *H.R. 3844, The Federal Information Security Management Act of 2002*, Serial 107-190, May 2, 2002.

United States House of Representatives, Committee on Energy and Commerce, Subcommittee on Telecommunications and Finance. (1994). *Network Wiretapping Capabilities*, Hearings on September 13, 1994, One Hundred Third Congress, Second Session.

United States House of Representatives, Committee on the Judiciary. (1994). *Report on 'Telecommunications Carrier Assistance to the Government,'* HR103-827, One Hundred Third Congress, Second Session.

United States House of Representatives, Committee on Science, Subcommittee on Technology. (1997). *Secure Communications*. One Hundred Fifth Congress, First Session. February 11, 1997.

United States House of Representatives, Committee on Appropriations, Subcommittee on Commerce, Justice, State, and the Judiciary. (1998). *FY 99 Appropriations for the Department of Justice*, One Hundred Fifth Congress, Second Session, 1998.

United States Privacy Protection Study Commission. (1977). *Personal Privacy in an Information Society*, U.S. Government Printing Office, Washington, D.C., 1977.

United States Senate, Committee on the Judiciary, Subcommittee on Constitutional Rights. (1961). *Wiretapping and Eavesdropping Legislation*, Hearings on S. 1086, S. 1221, S. 1495, and S. 1822, May 9, 10, 11 and 12, 1961, Eighty-Seventh Congress, First Session.

United States Senate, Committee on the Judiciary, Subcommittee on Administrative Practice and Procedure. (1965). *Invasions of Privacy*, Hearings Pursuant to Senate Resolution 39, Eighty-Ninth Congress, First Session, 1965.

United States Senate. (1968). *Senate Judiciary Committee Report, together with Minority, Individual, and Additional Views, on Omnibus Crime Control and Safe Streets Act of 1967*, Report No. 1097, *Senate Miscellaneous Reports on Public Bills*, Serial Set 12972-1, Ninetieth Congress, Second Session, April 29, 1968.

United States Senate, Committee on the Judiciary, Subcommittee on Constitutional Rights. (1971). *Federal Data Banks, Computers, and the Bill of Rights*, Hearings on February 23, 24, 25, March 2, 3, 4, 9, 10, 11, 15, 17, 1971, Part 1, Ninety-Second Congress, First Session.

United States Senate, Committee on the Judiciary, Staff of the Subcommittee on Constitutional Rights. (1972). *Army Surveillance of Civilians: A Documentary Analysis*, Ninety-Second Congress, Second Session, 1972.

United States Senate, Committee on the Judiciary, Subcommittee on Constitutional Rights. (1973). *Military Surveillance of Civilian Politics*, U.S. Government Printing Office, Washington, D.C., Ninety-Third Congress, First Session, 1973.

United States Senate, Select Committee on Presidential Campaign Activities. (1973). *Hearings, Phase 1: Watergate Investigation*, Ninety-Third Congress, First Session, 1973.

United States Senate, Committee on the Judiciary, Subcommittee on Constitutional Rights. (1974). *Federal Data Banks and Constitutional Rights*, Vol. I, Ninety-Third Congress, Second Session, 1974.

United States Senate, Committee on the Judiciary, Subcommittee on Administrative Practice and Procedure, and Committee on Foreign Relations, Subcommittee on Surveillance. (1974). *Warrantless Wiretapping and Electronic Surveillance*, Hearings on April 3, 8, May 8, 9, 10, and 23, 1974, Ninety-Third Congress, Second Session.

United States Senate, Committee on the Judiciary, Subcommittee on Criminal Laws and Procedures and Constitutional Rights. (1974). *Electronic Surveillance for National Security Purposes*, Hearings, October 1–3, 1974, Ninety-Third Congress, Second Session, 1974.

United States Senate. (1975). "A Resolution to Establish a Committee to Study Government Operations with Respect to Intelligence," Senate Resolution 21, Ninety-Fourth Congress, First Session, January 27, 1975.

United States Senate, Select Committee to Study Governmental Operations with Respect to Intelligence Activities. (1975). *Intelligence Activities. Senate Resolution 21. Volume 2: Huston Plan*, Hearings, September 23, 24, and 25, 1975, Ninety-Fourth Congress, First Session, 1975.

United States Senate, Select Committee to Study Governmental Operations with respect to Intelligence Activities. (1975). *Intelligence Activities. Senate Resolution 21. Volume 6: Federal Bureau of Investigation*, Hearings, November 18, 19, December 2, 3, 9, 10, 11, 1975, Ninety-Fourth Congress, First Session, 1975.

United States Senate, Select Committee to Study Governmental Operations with respect to Intelligence Activities. (1976). *Intelligence Activities and the Rights of Americans, Final Report, Book II*, Report 94-755, Ninety-Fourth Congress, Second Session, April 23, 1976.

United States Senate, Select Committee to Study Governmental Operations with respect to Intelligence Activities. (1976). *Supplementary Staff Reports on Intelligence Activities and the Rights of Americans, Final Report, Book III*, Report 94-755, Ninety-fourth Congress, Second Session, April 23, 1976.

United States Senate, Select Committee on Intelligence. (1988). *Senate Select Committee on Intelligence Inquiry into the FBI Investigation of the Commit-*

tee in Solidarity with the People of El Salvador, Hearings on February 23, April 13, September 14 and 29, 1988, One Hundredth Congress, Second Session.

United States Senate, Select Committee on Intelligence. (1989). *The FBI and CISPES*, Report 101-46, One Hundred First Congress, First Session, July, 1989.

United States Senate, Committee on the Judiciary, Subcommittee on Technology and the Law. (1994). *Administration's Clipper Chip Key Escrow Encryption Program*, Hearings, May 3, 1994, One Hundred Third Congress, Second Session.

United States Senate, Committee on the Judiciary, Subcommittee on Technology and the Law (Senate), and United States House of Representatives, Committee on the Judiciary, Subcommittee on Civil and Constitutional Rights. (1994). *Digital Telephony and Law Enforcement Access to Advanced Telecommunications Technologies and Services*, Joint Hearings on HR 4922 and S. 2375, March 18 and August 11, 1994, One Hundred Third Congress, Second Session.

United States Senate, Committee on Commerce, Science, and Transportation. (1997). *S377, Promotion of Commerce Online in the Digital Era (PROCODE) Act of 1997*, March 19, 1997, One Hundred Fifth Congress, First Session.

United States Senate, Appropriations Committee, Subcommittee on Treasury and General Government. (1997). *Overview of Internal Revenue Service Employees Misuse of Taxpayers' Files*, Hearings on April 15, 1997, One Hundred Fifth Congress, First Session.

United States Senate, Judiciary Committee. (1997). *Encryption, Key Recovery, and Privacy Protection in the Information Age*, Hearings on July 9, 1997, One Hundred Fifth Congress, First Session.

United States Senate, Judiciary Committee, Subcommittee on Technology, Terrorism, and Government Information. (1997). *The Encryption Debate: Criminals, Terrorists, and the Security Needs of Business and Industry*, Hearings on September 3, 1997, One Hundred Fifth Congress, First Session.

United States Senate, Select Committee on Intelligence. (1987). *On Nomination of William H. Webster, to be Director of Central Intelligence*, Hearings before the Senate Select Committee on Intelligence, Hearings on April 8, April 9, April 30, May 1, 1987, One Hundredth Congress, First Session.

United States Senate, Committee on the Judiciary. (2002). *The USA PATRIOT Act in Practice: Shedding Light on the FISA Process*, Hearing on September 10, 2002, S. Hrg. 107-947, One Hundred Seventh Congress, Second Session.

Van Keuren, Edgar. (1991). "Utilization of the high power microwave sources in electronics sabotage and terrorism," *Proc. International Carnahan Conference on Security Technology*, October 1–3, 1991, Taipei, Taiwan, pp. 16–20.

Van Natta Jr., Don and Desmond Butler. (2004). "How Tiny Swiss Cellphone Chips Helped Track Global Terror Web," *New York Times*, March 4, 2004, p. A1.

Van Natta Jr., Don, Elaine Sciolino, and Stephen Gray. (2006). "In Tapes, Receipts and a Diary, Details of the London Bombing Case", *New York Times*, August 28, 2006, p. A1.

Vandersypen, Lieven M. K., Matthias Steffen, Gregory Breyta, Costantino S. Yannoni, Mark H. Sherwood, and & Isaac L. Chuang. (2001). "Experimental realization of Shor's quantum factoring algorithm using nuclear magnetic resonance," *Nature*, 414 (2001), 883–887.

Wallraff, Barbara. (2000). "What Global Language?," *Atlantic Monthly*, November 2000, pp. 52–66.

Wannall, W. (1966). Memo to W. Sullivan, December 22 as reported in USSR 94 *Intelligence Activities: Staff Reports*, p. 328 and p. 346.

Ward, Daniel. (1961). State's Attorney, Cook County, Illinois, in USSH 87 *Wiretapping and Eavesdropping Legislation*, p. 400.

Ward, Philip. (2005). *The Identity Cards Bill: Bill 9 of 2005-06*, Home Affairs Section, House of Commons Library, Research Paper 05/43, June 13, 2005.

Warren, Samuel and Louis Brandeis. (1890). "The Right to Privacy," *Harvard Law Review*, pp. 193–220.

Weiner, Tim, David Johnston, and Neil Lewis. (1995). *Betrayal: The Story of Aldrich Ames, an American Spy*, Random House, New York.

Weingarten, Fred. (1997). "Cryptography: Who Holds the Key?," *SIAM News*, January/February, p. 2.

Welchman, Gordon. (1982). *The Hut Six Story*, McGraw Hill.

White, Theodore H. (1975). *Breach of Faith*, Atheneum Publishers.

White House, Office of the Press Secretary. (1993). "Statement on the Clipper Chip Initiative," in EPIC 1994 April 16, Washington, D.C.

White House, Office of the Press Secretary. (1998). "Administration Updates Encryption Policy," September 16, 1998.

White House. (2001). President's Daily Briefing Memo, August 6, 2001, http://www.cnn.com/2004/images/04/10/whitehouse.pdf (last viewed Sept. 2, 2006).

Wiener, Michael. (1993). "Efficient DES Key Search," presented at the Rump Session of CRYPTO, TR-244, School of Computer Science, Carleton University, May 1994.

Wilk, Charles K. and Donald E. Kraft. (1979). "Protected Government Telecommunications: The National Issues" National Telecommunications Conference, Washington, D.C., November 27–29, pp. 43.1.1–43.1.4.

Wilk, Charles K. and Donald E. Kraft. (1980). "The Passive Security Risk in Telecommunications" Security Management, August, pp. 52–54, 67–68.

Williams, Robert H. (1992). "Economic Spying by Foes, Friends Gains Momentum." *SIGNAL*, July, pp. 56–57.

Wines, Michael. (1990). "Security Agency Debates New Role: Economic Spying," *New York Times*, June 18, p. 1.

Wingfield, Nick, and Neil King. (2000). "EarthLink Just Says No to FBI's Carnivore," *The Wall Street Journal Online*, July 13, 2000, 5:00 PM PT, http://news .zdnet.com/2100-9595_22-522208.html (last viewed August 18, 2006).

Woolsey, James. (2000). "Why We Spy on Our Allies," *Wall Street Journal*, (March 17, 2000, p. A18).

Wright, Carroll D. and William C. Hunt. (1900). *The History and Growth of the United States Senate*, Government Printing Office, Washington.

Wright, S. Fowler. (1929). "Police and Public" Second Revised Edition, Fowler Wright Ltd., 240, High Holborn, London W.C.1.

Wright, Peter. (1987). *Spy Catcher, The Candid Autobiography of a Senior Intelligence Officer*, Viking.

Wright, Lawrence. (2006). *The Looming Tower: Al-Qaeda and the Road to 9/11*, Knopf, New York, 2006.

Yardley, Herbert O. (1931). *The American Black Chamber*, Indianapolis, Bobbs-Merrill.

Yeh, Brian T. and Charles Doyle. (2006). "USA PATRIOT Improvement and Reauthorization Act of 2005: A Legal Analysis," Congressional Research Service, March 24, 2006.

Zimmermann, Philip. (1995). *PGP: Source Code and Internals*, MIT Press.

Zimmermann, Philip. (2006). http://www.philzimmermann.com/zfoneproject/ index.html (last viewed: October 15, 2006).

Index